Lecture Notes in Statistics 90

Edited by S. Fienberg, J. Gani, K. Krickeberg, I. Olkin, and N. Wermuth

Attila Csenki

Dependability for Systems with a Partitioned State Space

Markov and Semi-Markov Theory and Computational Implementation

Springer-Verlag
New York Berlin Heidelberg London Paris
Tokyo Hong Kong Barcelona Budapest

Attila Csenki
Department of Computer Science and Applied Mathematics
Aston University
Aston Triangle, Birmingham B4 7ET
Great Britain

Library of Congress Cataloging-in-Publication Data
Csenki, Attila.
    Dependability for systems with a partitioned state space : Markov
and semi-Markov theory and computational implementation / Attila
Csenki.
        p.   cm. -- (Lecture notes in statistics ; 90)
    Includes bibliographical references and index.
    ISBN 0-387-94333-1
    1. Reliability (Engineering)--Mathematical models. 2. Markov
processes. I. Title. II. Series: Lecture notes in statistics
(Springer-Verlag) ; v. 90.
    TA169.C73 1994
    620'.00452'01176--dc20   94-25788
Printed on acid-free paper.

Camera ready copy provided by the author.
Printed and bound by Braun-Brumfield, Ann Arbor, MI.
Printed in the United States of America.

9 8 7 6 5 4 3 2 1

ISBN 0-387- 94333-1 Springer-Verlag New York Berlin Heidelberg
ISBN 3-540- 94333-1 Springer-Verlag Berlin Heidelberg New York

PREFACE

Probabilistic models of technical systems are studied here whose finite state space is partitioned into two or more subsets. The systems considered are such that each of those subsets of the state space will correspond to a certain performance level of the system. The crudest approach differentiates between 'working' and 'failed' system states only. Another, more sophisticated, approach will differentiate between the various levels of redundancy provided by the system. The dependability characteristics examined here are random variables associated with the state space's partitioned structure; some typical ones are as follows

- The sequence of the lengths of the system's working periods;
- The sequences of the times spent by the system at the various performance levels;
- The cumulative time spent by the system in the set of working states during the first m working periods;
- The total cumulative 'up' time of the system until final breakdown;
- The number of repair events during a finite time interval;
- The number of repair events until final system breakdown;
- Any combination of the above.

These dependability characteristics will be discussed within the Markov and semi-Markov frameworks.

In Chapter 1, we give a concise summary of the required theoretical background in the theory of stochastic processes and introduce those engineering systems (mainly from the power engineering field) which will serve us to demonstrate the practical potential of the theoretical results. Chapters 2 and 3 are devoted to discrete-parameter Markov chains. They are of interest for two reasons. First, they can be used to model the sequence of events in a discrete-parameter system. Second, they will be seen in Chapter 4 to be useful in a mathematical sense by allowing a key continuous-time result to be derived essentially as a corollary to the discrete-parameter formulation. Chapters 4 - 7 are devoted to continuous-time Markov chains. The main tool here is Laplace transforms. In most cases it will be possible to represent the result in the time domain by explicit Laplace transform inversion. The second technique, used in Chapter 6, is randomization (also called 'uniformization') which replaces the finite Markov process by a Poisson process and an independent discrete-time Markov chain. The former generates the holding times whereas the latter marks the sequence of state transitions in the state space of the original process. (Self-transitions for the new chain are possible.) Chapters 8 - 12 explore essentially the same kind of questions for semi-Markov processes. Closed form solutions in the time domain are not feasible here. However, numerical and symbolic Laplace transform inversion are possible given the right set of software tools.

Three software environments were used to implement the examples

- The matrix computation system MATLAB on the Apple Macintosh;
- The NAG FORTRAN 77 subroutine library on the VAX;
- The computer algebra system MAPLE on the Apple Macintosh.

Those features of the above systems are explained which were thought to be useful for their first appreciation by the reliability practitioner. Some code is also shown where it was deemed instructive. An informal discussion of future research topics is given in the Postscript.

This monograph is a coherent and self-contained account of one particular aspect of my research in reliability theory over the last couple of years. Work on the material presented here has started during my stay (1987-1990) as a research fellow with the Centre for Software Reliability, The City University, London. Most of the material was researched in my present employment at Aston University, Birmingham, where I have been a lecturer in the Department of Computer Science and Applied Mathematics since January 1990. This monograph is a Habilitation thesis for the Faculty of Mathematics and Natural Sciences of the Bergakademie Freiberg, Saxony, Federal Republic of Germany. I am most grateful to Professor Stoyan from the Bergakademie Freiberg for supporting me as an external candidate and his interest in this work.

Birmingham, April 1994 *Attila Csenki*

CONTENTS

CHAPTER 1

STOCHASTIC PROCESSES FOR DEPENDABILITY ASSESSMENT

The view taken in this monograph is that Reliability Theory is essentially an application of the theory of stochastic processes. However, the real life context in which Reliability Theory exists must also be given due consideration for three reasons. First, the theory serves a well-defined purpose: to assist in the modelling of technical systems as far as their dependability aspects are concerned. Second, there is ample scope for fertilization with ideas in the other direction. This latter point is relevant in particular for our subject since new technical systems in need of assessment from a reliability point of view are emerging with predictable regularity. Finally, the evolution of the reliability field is considerably influenced by the rapid development of the computational tools it uses (both software and hardware). Thus even though the present work is essentially a mathematical treatise, the intention was to make it of some benefit also to the reliability engineer who is perhaps more interested in how new theory might assist him in the analysis of actual systems rather than in the theory itself. In this first chapter, in addition to an informal summary of the results, the notation is established and the prerequisites from the (elementary) theory of stochastic processes are reviewed. The presentation is deliberately restricted to those aspects of the theory which will be used subsequently. The application examples are mainly power transmission reliability models.

1.1 Markov and semi-Markov processes for dependability assessment

From the large number of references on this, we would like to single out Birolini's books [BIR1] and [BIR2] for their broad coverage of the topic. The books by Bhat [BHA2, Chapter 15], Kohlas [KOH], and Iosifescu [IOS] are more mathematical, whereas those by Beasley [BEA], Frankel [FRA], Grosh [GRO], Klaassen and Peppen [KLA], and Lewis [LEW] are for the reliability engineer.

We shall be concerned in this work with models of technical systems which evolve over time. Usually, systems are considered at various levels of abstraction. At the highest level, the system is thought of as one unit which is either working (then commonly referred to as being in the 'up-state') or is idle (then also said to be 'down'). In most cases it is expedient to look at a system more closely and to identify its subsystems and their interdependence. It will then be possible to build a mathematical model of the system based on the knowledge of the behaviour of its subsystems. Failure of (sub)systems is deemed to be random, and that is why probability modelling is so important in reliability work.

Discrete-time Markov models are used to describe systems which evolve in time but for which the actually elapsed time between the events of interest is immaterial: the *sequence of events* is essential. The state space S of our models will be finite. Informally, a discrete-parameter Markov process $Z = \{ Z_n : n = 0, 1, ... \}$ with state space S can be thought of as the model of a particle moving around randomly in S, where $Z_n \in S$ stands for its position just after the nth step. This 'random' movement is governed by a probabilistic law: if the process is in $s \in S$ after the nth step, the next state to be occupied will be $r \in S$ with probability $\Pr\{ Z_{n+1}$

$= r \mid Z_n = s \} = q_{sr} \in [0, 1]$. If (as it will be assumed throughout) these probabilities are independent of n, the process Z is said to be *homogeneous*. The q_{sr} are entries of the transition probability matrix $\mathbf{Q} = \{ q_{sr} : s, r \in S \}$. Obviously, the row-sum of $\mathbf{Q}$ is unity, expressed in vector notation as $\mathbf{Q} \mathbf{1} = \mathbf{1}$, where $\mathbf{1}$ stands for a column vector of ones $(1, ..., 1)^T$ whose number of entries matches the size of $\mathbf{Q}$. The state space S will be partitioned as $S = A_1 \cup A_2$, or alternatively, as $S = A_1 \cup A_2 \cup \{\omega\}$ with pairwise disjoint sets A_1, A_2, and $\{\omega\}$. In the former case the Markov chain will be assumed *irreducible*, i.e., every state has a non-zero probability of being visited from every other state (possibly in more than one step though). Thus, the chain alternates between A_1 and A_2 indefinitely. The latter partition of S arises when Z is *absorbing*, i.e., once ω has been entered into the process stays there indefinitely. Irreducible Markov chains are used to model repairable systems and then A_1, say, is the set of working states. Absorbing chains are used to model irrepairable systems; then, ω stands for the state 'system breakdown'. In practice the initial system state Z_0 is well known but for mathematical reasons it is customary to admit the possibility of a random initial state. This is expressed in terms of the initial state vector $\boldsymbol{\alpha} = \{ \alpha_s : s \in S \}$, $\alpha_s = \Pr\{ Z_0 = s \}$, which is a column vector. If it is known that $Z_0 = s$ then, of course, the only non-zero entry of $\boldsymbol{\alpha}$ is in the *sth* position and it is unity. According to the partitioning of the state space S both $\mathbf{Q}$ and $\boldsymbol{\alpha}$ are written as block partitioned matrices. They are for irreducible and absorbing Z respectively given by

$$
\mathbf{Q} = \begin{array}{c} \\ A_1 \\ A_2 \end{array} \begin{array}{cc} A_1 & \quad A_2 \\ \begin{bmatrix} \mathbf{Q}_{A_1 A_1} & \mathbf{Q}_{A_1 A_2} \\ \hdashline \mathbf{Q}_{A_2 A_1} & \mathbf{Q}_{A_2 A_2} \end{bmatrix} \end{array}, \quad \boldsymbol{\alpha} = \begin{array}{c} A_1 \\ A_2 \end{array} \begin{bmatrix} \boldsymbol{\alpha}_{A_1} \\ \hdashline \boldsymbol{\alpha}_{A_2} \end{bmatrix}, \tag{1.1}
$$

and

$$
\mathbf{Q} = \begin{array}{c} A_1 \\ A_2 \\ \{\omega\} \end{array} \begin{array}{ccc} A_1 & A_2 & \{\omega\} \\ \begin{bmatrix} \mathbf{Q}_{A_1 A_1} & \mathbf{Q}_{A_1 A_2} & \mathbf{Q}_{A_1 \{\omega\}} \\ \hdashline \mathbf{Q}_{A_2 A_1} & \mathbf{Q}_{A_2 A_2} & \mathbf{Q}_{A_2 \{\omega\}} \\ \hdashline \mathbf{0} & \mathbf{0} & 1 \end{bmatrix} \end{array}, \quad \boldsymbol{\alpha} = \begin{array}{c} A_1 \\ A_2 \\ \{\omega\} \end{array} \begin{bmatrix} \boldsymbol{\alpha}_{A_1} \\ \hdashline \boldsymbol{\alpha}_{A_2} \\ \hdashline \alpha_\omega \end{bmatrix}, \tag{1.2}
$$

where $\mathbf{0}$ stands for the zero matrix. Notice that an irreducible Markov chain can be considered to be a limiting case of an absorbing one by assuming that $\alpha_\omega = 0$ and by letting $\max\{ q_{s\omega} : s \in S, s \neq \omega \} \to 0$. This observation will allow results for irreducible Markov chains to be established from their absorbing counterpart with relative ease.

For the quantities under consideration to be easily visualized it is expedient to attach the following meaning to the sets A_1 and A_2. $A_1 = G$ is the set of all 'good' states, i.e., the ones which correspond to the system being functional; the set $A_2 = B$ is the set of all 'bad' states, i.e., those which are associated with the system being 'down'. An irreducible process alternates between G and B indefinitely. The sequences of the lengths of sojourns in G and B will be respectively denoted by $\{ N_{G,i} : i \geq 1 \}$ and $\{ N_{B,i} : i \geq 1 \}$. If $Z_0 \in G$, we have, for example,

$N_{G,1} = n \Leftrightarrow Z_0, ..., Z_{n-1} \in G$ *and* $Z_n \in B$. The definition of these sequences is the same for absorbing Markov chains, with the proviso that $N_{G,i} = 0$ (or $N_{B,i} = 0$) if Z visits G (or B) less than i times until absorption. Initially, one of our main concerns will be the joint distribution of these sojourn times. Various quantities of interest to the reliability analyst can be expressed in term of sojourn times. For instance, for a system modelled by an absorbing Markov chain, the total amount of 'time' spent in the 'up' states, $N_{G,1} + N_{G,2} + N_{G,3} + ...$, is of interest since it is a measure of the total work done until final system breakdown. Another interesting quantity is M_B, the number of visits to B until absorption; it is a measure of the (undiscounted) total repair cost until final breakdown. It is related to the sequence of sojourn times in B by $\{ M_B = 0 \} = \{ N_{B,1} = 0 \}$ and $\{ M_B \geq m \} = \{ N_{B,m} > 0 \}$.

There are applications where it is required to consider more than one level of functionality of the system. In a parallel redundant system, for example, each level of functionality may be associated with a specific number of components being functional. An n-component system modelled by an absorbing chain will thus give rise to the partitioning $S = A_1 \cup ... \cup A_n \cup \{ \omega \}$. A quantity which will be examined is the vector $M = (M_{A_1}, ..., M_{A_n})^T$ where M_{A_i} stands for the number of visits to A_i until absorption, i.e., it is the number of periods spent by the system at level i until final breakdown. Similarly to M_B from above, M can be written in terms of the sojourn times themselves.

The behaviour of a repairable system is usually of interest for a finite length of 'time' only. Assuming again that $S = G \cup B$, then it is possible to express the number of visits of Z to B during the first n time instances, in terms of a variable $M_{\mathscr{B}}$ which counts the number of visits to some subset $\mathscr{B}$ of its state space by an auxiliary absorbing Markov chain.

Continuous-parameter Markov models are used when the time aspect of the system to be modelled is essential. The system can be still thought of as a particle moving around in a finite state space S but now the times between state transitions are exponentially distributed. To be more specific, let the Markov process $Y = \{ Y_t : t \geq 0 \}$ have the transition rate matrix Λ. The *sequence* of transitions between states is then governed by what is known as the *embedded Markov chain*, the transition probability matrix P of which is given by

$$p_{s_1 s_2} = \begin{cases} \lambda_{s_1 s_2} \left(\displaystyle\sum_{s \in S, s \neq s_1} \lambda_{s_1 s} \right)^{-1} & \text{if } s_1 \neq s_2, \\ 0 & \text{if } s_1 = s_2. \end{cases}$$

The holding time in state s is exponentially distributed with rate $- \lambda_{ss}$, which is the sum of the non-diagonal entries of the *sth* row of Λ. Sojourn times for continuous-parameter Markov chains are defined in an analogous fashion to those in discrete-time. They will be denoted by $\{ T_{G,i} : i \geq 1 \}$ and $\{ T_{B,i} : i \geq 1 \}$ if S is partitioned as $S = G \cup B$ or as $S = G \cup B \cup \{ \omega \}$. The notation used to represent the partitioned form of Λ is in close analogy to (1.1) and (1.2). We will be concerned with the joint distribution of any finite collection of these sojourn time variables. It will be shown that the continuous-time result is easily derived from its corresponding discrete-parameter version by an intuitively appealing time-discretisation argument. The sojourn time variables are of interest to the reliability analyst also because they can be used to express further dependability characteristics. One such quantity is $M_G(t)$, the number of working periods during a finite time interval for the Markov model of a repairable system. The *nth* cumulative sojourn time in G, $TS_{G,n} = T_{G,1} + ... + T_{G,n}$, and for absorbing

Markov models, the total 'up' time until failure, $TS_{G,\infty} = T_{G,1} + T_{G,2} + ...$, will also be explored. The analysis of these latter quantities will be accomplished by using Laplace transforms. As is well known, Laplace transforms are a suitable tool for the derivation of results on sums of variables whose joint distribution is known. Some ingenuity will be needed, however, to achieve closed form expressions for the cumulative distribution functions by symbolic Laplace transform inversion.

M_B, the number of periods spent in the set of repairable 'down' states until final breakdown, has already been addressed in connection with discrete-parameter absorbing Markov chains. It is also of interest in the analysis of systems modelled by *continuous-time* absorbing Markov processes; the calculation of its distribution poses no new problem, however, since M_B is dependent on the embedded Markov chain of Y, which is a discrete-time process. The dependability characteristics mentioned thus far are, with the exception of the sequences of the sojourn time variables, univariate. However, it would be desirable from the reliability analyst's point of view to be able to assess several system characteristics simultaneously. Such *compound* dependability characteristics will therefore also be considered. The one in connection with absorbing models is $\Pr\{ TS_{G,\infty} > t, M_B \leq m \}$, i.e., that the total time spent in the 'good' states is at least t *and* the number of repairable failure events until final breakdown does not exceed m. The irreducible counterpart of this measure (considered, of course, over a finite time interval $[0, t_0]$ and hence denoted by $\Pr\{ TS_{G,\infty}(t_0) > t, M_B(t_0) \leq m \}$) will also be explored.

Semi-Markov processes are used to model systems for which the holding time distributions are not necessarily memoryless. There are two equivalent ways to view such systems. The Markov renewal viewpoint starts with a discrete-parameter Markov chain on the state space. This describes the sequence of state transitions. The time aspect is then provided by an increasing sequence of transition times which is considered to be describing a renewal process in the generalized sense. The alternative viewpoint, which is the one taken in this monograph, is as follows. As before, the embedded discrete-parameter Markov chain is given by its transition probability matrix $\mathbf{R}$, say, with $r_{ss} = 0$ for all $s \in S$. The time aspect of the process is specified by its conditional holding time distribution functions $\{ F_{s,r} : s, r \in S, s \neq r \}$ where $F_{s,r}$ stands for the cumulative distribution function of the holding time in s given that the next state to be visited is r. A Markov process with transition rate matrix $\mathbf{\Lambda}$ is a special case of a semi-Markov process with $F_{s,r}(t) = 1 - \exp(\lambda_{ss}t)$, $t \geq 0$.

We start the semi-Markov theory by considering the sojourn times variables $\{ T_{A_i,j} : j = 1, 2, ... \}$, $i = 1, 2$, which will be examined in the Laplace transform domain; some of the results already obtained for Markov processes will then be rederived from this. Another focus of interest will be $M_{A_1}(t)$, the number of periods spent by Y in A_1 during $[0, t]$. Laplace transform theory will be used for examining the asymptotic behaviour of the moments of $M_G(t)$ as $t \to +\infty$. Furthermore, a numerical Laplace transform inversion scheme will be used to obtain the cumulative disrtibution function of $M_G(t)$ for finite t. A multivariate extension of this variable for semi-Markov processes whose state space is partitioned into more than two subsets will also be discussed.

The proofs of the results for semi-Markov processes crucially depend on a generalization of what is known as the 'renewal argument'. It uses the observation that semi-Markov processes are 'regenerative' in the sense that at every instance of a transition the process can be analysed without knowing its past history. The renewal argument gives rise to sets of equations in the Laplace transform domain which are solved for the transforms of the variables of interest. In addition to the variables mentioned above, we shall explore by this technique also the distribution of 'finite horizon' sojourn times. They arise by viewing the process in the time-window $[0, t_1]$. More precisely, $T_{G,n}(t_1)$, say, is defined as follows: it is equal to $T_{G,n}$ if the nth working period ends before t_1; it is $T_{G,n} - (t_2 - t_1)$ if the nth working period commences before t_1 and finishes at $t_2 > t_1$; it is zero otherwise. A reference specifically devoted to semi-Markov theory is Nollau's book [NOL]; in Chapter 3 of [NOL], the renewal technique is nicely exemplified. In its many guises, this technique will form the basis of much of the present work in both the Markov and the semi-Markov case.

From a practical viewpoint, theory should be underpinned by implementation details and application examples. Three systems were used to implement our algorithms: the matrix computation language/system MATLAB ([MAT1], [MAT2], [SCH]) and the symbolic algebra package MAPLE [CHA], both on the Apple Macintosh, and, finally, the NAG FORTRAN Subroutine Library [NUM] on the VAX mainframe. The most suitable of these seems to be MATLAB which allows reliability algorithms to be coded very concisely. A MATLAB version of an inversion scheme from the NAG FORTRAN Library is also provided.

1.2 Example systems

There are many possibilities to demonstrate the utility of Markov and semi-Markov techniques in the reliability field. The major application area in this monograph is power engineering. Books addressing this subject are [BIL1] and [SIN1].

1.2.1 Markov models

Model 1. Figure 1.1 shows the state-transition-rate diagram of the Markov model of a power transmission system from [BIL1] pp 180-203. From Figure 1.1, the system's transition-rate matrix Λ is given by

$$
\Lambda = \begin{bmatrix}
* & \nu & \lambda_2 & 0 & \lambda_1 & 0 & 0 & 0 \\
\rho & * & 0 & \lambda_2' & 0 & \lambda_1' & 0 & 0 \\
\mu_2 & 0 & * & \nu & 0 & 0 & \lambda_1 & 0 \\
0 & \mu_2' & \rho & * & 0 & 0 & 0 & \lambda_1' \\
\mu_1 & 0 & 0 & 0 & * & \nu & \lambda_2 & 0 \\
0 & \mu_1' & 0 & 0 & \rho & * & 0 & \lambda_2' \\
0 & 0 & \mu_1 & 0 & \mu_2 & 0 & * & \nu \\
0 & 0 & 0 & \mu_1' & 0 & \mu_2' & \rho & *
\end{bmatrix}.
$$

(The diagonal elements of Λ are determined by the condition that Λ is a transition-rate matrix and thus its row-sum is zero.) The system comprises two parallel power transmission lines each of which can be either in a working or a failed state. The environment is assumed to be fluctuating between normal and stormy weather conditions. The Markov assumption implies that the failures modelled here are those due to chance events (storms, earthquakes, and accidents) but not those caused by wearout (corrosion and fatigue).

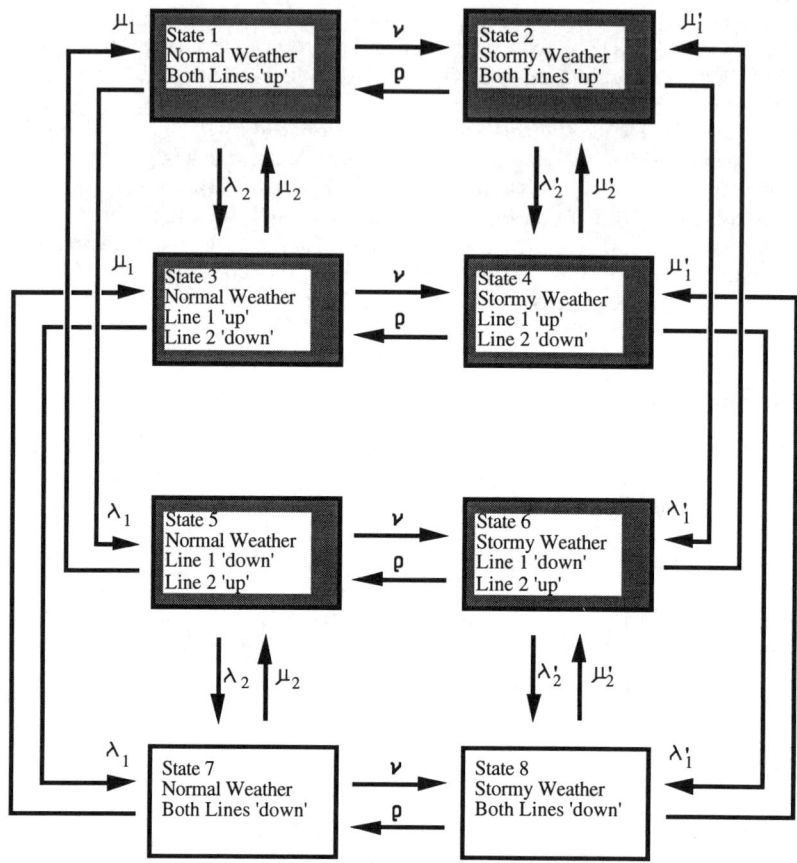

Figure 1.1 The state-transition-rate diagram of a two-component parallel power transmission system

The state space S = {1, ..., 8} is partitioned into the set of working states, G = {1, ..., 6} (shaded in Figure 1.1), and the set of failed states, B = {7, 8}. This system is irreducible since every state is accessible from any other state.

Model 2. Figure 1.2 shows the state-transition-rate diagram of a three-component power transmission system from [NAH].

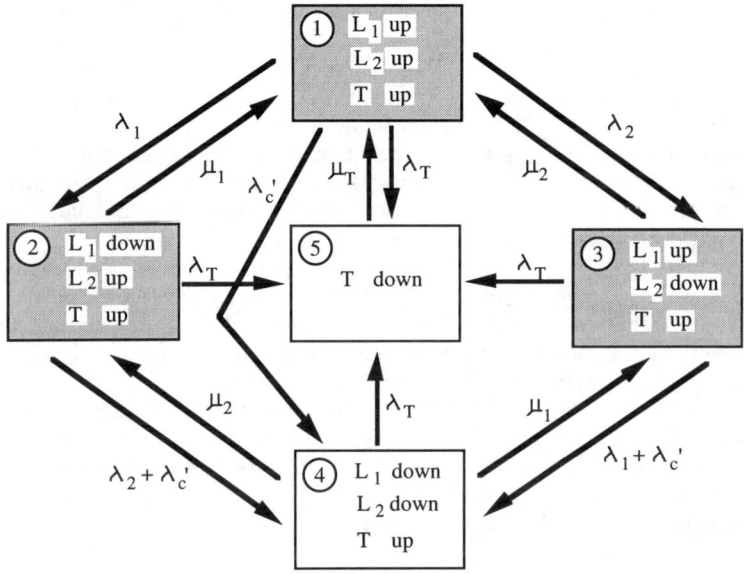

Figure 1.2 The state-transition-rate diagram of the three-component parallel power transmission system

It comprises two parallel power lines, L_1 and L_2, and a tower T. The model is Markovian with respective individual failure rates λ_1, λ_2, λ_T, and a common cause power line failure rate λ_c'. The respective repair rates are μ_1, μ_2, and μ_T. From Figure 1.2, the transition rate matrix Λ is given by

$$\Lambda = \begin{bmatrix} * & \lambda_1 & \lambda_2 & \lambda_c' & \lambda_T \\ \mu_1 & * & 0 & \lambda_2+\lambda_c' & \lambda_T \\ \mu_2 & 0 & * & \lambda_1+\lambda_c' & \lambda_T \\ 0 & \mu_2 & \mu_1 & * & \lambda_T \\ \mu_T & 0 & 0 & 0 & * \end{bmatrix}.$$

The state space is $S = \{1, ..., 5\}$ with $G = \{1, 2, 3\}$ (shaded in Figure 1.2) and $B = \{4, 5\}$. The system is irreducible.

   Model 3. The values from [NAH] for the repair rates in Model 2 are as follows: $\mu_1 = \mu_2 = 1095$/year (= 0.125/hour, i.e., mean repair time = 8 hours) and $\mu_T = 46$/year (which is equivalent to a mean tower repair time of 190.4 hours). The tower repair rate μ_T is seen to be orders of magnitude smaller than the transmission line repair rates μ_1 and μ_2. It is therefore appropriate to differentiate between 'minor' and 'major' system breakdowns. If we are interested in the system's behaviour until its first major breakdown, we convert state '5' into an absorbing state and put $\mu_T = 0$. Now, $S = G \cup B \cup \{\omega\}$ with $G = \{1, 2, 3\}$, $B = \{4\}$, and $\omega = 5$.

Model 4. This example is Billinton's Markov reliability model of a system comprising three components in a two-state fluctuating environment, [BIL1] pp 188. It is a generalisation of Model 1. The system consists of three parallel power transmission lines each of which can be either in a working or a failed state. The environment is assumed to be fluctuating between normal and stormy weather conditions. Even though it is possible to draw a state-transition-rate diagram along the lines of Figure 1.1, we want to follow a more elegant route which offers a systematic procedure of dealing with any number of components in a fluctuating environment. The model's state space is conveniently represented by the Cartesian product $E = \{0, 1\}^4$, where the entries of a state vector $(w, \ell_1, \ell_2, \ell_3) \in E$ are interpreted in the following manner: $w = 0$ is 'normal weather', $w = 1$ is 'stormy weather', $\ell_i = 1$ means 'line № i is available', and $\ell_i = 0$ means 'line № i is under repair'. The three transmission lines are assumed to be identical. Their individual normal, and stormy weather failure rates are ρ_0 and ρ_1 respectively. The component repair rate is μ for both weather conditions. For a concise representation of the state-transition-rate diagram of this continuous-time Markov model, E is written as $E = E_0 \cup E_1$ with $E_w = \{w\} \times \{0, 1\}^3$. The transition rates within E_w and between the corresponding elements of E_0 and E_1 are shown respectively in Figures 1.3 (a) and 1.3 (b). The following partitioning of E is of interest here:

$$E = U_1 \cup U_2 \cup U_3 \cup U_4,$$

where for $i =, ... 4$,

$$U_i = \{0, 1\} \times \{ (\ell_1, \ell_2, \ell_3) \in \{0, 1\}^3 : \ell_1 + \ell_2 + \ell_3 = 4 - i \}.$$

The system will be started in normal weather with all three lines in the 'up' state, i.e., in the state $(0, 1, 1, 1) \in U_1$. This is an example where the state space is partitioned into sets corresponding to specific levels of system functionality: U_i is the set of all system states where 4 - i lines are available, $i = 1, ..., 4$. Let M_{U_i}, $(i = 1, 2, 3)$ stand for the number of visits of Y to U_i until the first failure incident is observed (i.e., the first entry into U_4 occurs). M_{U_i} is clearly the number of those periods in which 4 - i lines are available during this time interval. Of interest will be the variable (M_{U_1}, M_{U_2}). The model's continuous-time nature is immaterial for the distribution of (M_{U_1}, M_{U_2}). It therefore suffices to focus attention on the absorbing Markov chain $Z = \{ Z_i : i = 0, 1, ... \}$ which is obtained from the embedded Markov chain of the initial process Y (see, e.g., Iosifescu [IOS], pp 243) by merging the states in U_4 into a single absorbing state, ω; the state space of Z is thus $S = A_1 \cup ... \cup A_4$ with $A_i = U_i$, $i = 1, 2, 3$, and $A_4 = \{\omega\}$.

(a)

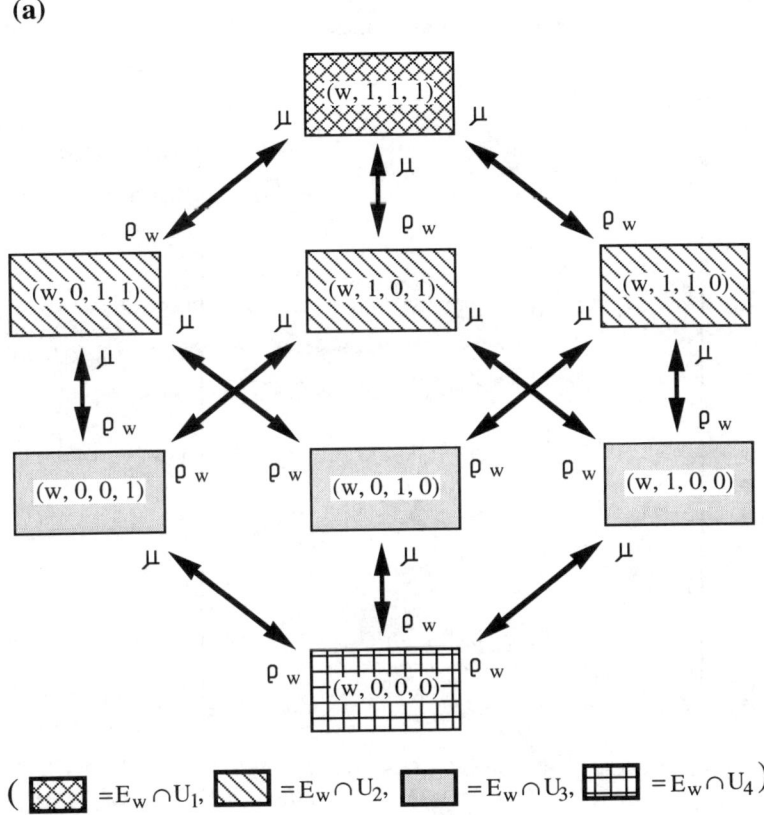

$\left(\boxed{\rlap{\diagbox}} = E_w \cap U_1, \boxed{\rlap{///}} = E_w \cap U_2, \boxed{\ } = E_w \cap U_3, \boxed{\rlap{\#}} = E_w \cap U_4 \right)$

Figure 1.3.(a) The state-transition-rate diagram of the three-component model in a fluctuating environment: transitions within E_w

(b)

$$(0, \ell_1, \ell_2, \ell_3) \xleftrightarrow[\nu_1 \quad \nu_0]{} (1, \ell_1, \ell_2, \ell_3)$$

Figure 1.3.(b) The state-transition-rate diagram of the three-component model in a fluctuating environment: transitions between elements of E_0 and E_1

It is seen from Figure 1.3 that the transition-rate matrix Λ of Y is given by

$$\Lambda = \left[\begin{array}{c|c} \Lambda_{E_0E_0} & \nu_0\mathbf{I} \\ \hline \nu_1\mathbf{I} & \Lambda_{E_1E_1} \end{array}\right],$$

(1.3)

$$\Lambda_{E_wE_w} = \left[\begin{array}{cccccccc} * & \mu & \mu & 0 & \mu & 0 & 0 & 0 \\ \rho_w & * & 0 & \mu & 0 & \mu & 0 & 0 \\ \rho_w & 0 & * & \mu & 0 & 0 & \mu & 0 \\ 0 & \rho_w & \rho_w & * & 0 & 0 & 0 & \mu \\ \rho_w & 0 & 0 & 0 & * & \mu & \mu & 0 \\ 0 & \rho_w & 0 & 0 & \rho_w & * & 0 & \mu \\ 0 & 0 & \rho_w & 0 & \rho_w & 0 & * & \mu \\ 0 & 0 & 0 & \rho_w & 0 & \rho_w & \rho_w & * \end{array}\right].$$

(1.4)

The state space enumeration assumed for (1.4) is as follows: (w, 0, 0, 0), (w, 0, 0, 1), (w, 0, 1, 0), (w, 0, 1, 1), (w, 1, 0, 0), (w, 1, 0, 1), (w, 1, 1, 0), and (w, 1, 1, 1). Also note that the diagonal elements of $\Lambda_{E_wE_w}$ are determined by the condition that (1.3) defines a transition-rate matrix.

Let us add in passing that there is a neat alternative to the above Markov-diagram-based method for the derivation of $\Lambda_{E_wE_w}$. The system of transmission lines is modelled, under the weather condition 'w', by a 3-dimensional Markov process which takes values in the product space $\{0, 1\}^3$ and the components of which are independent Markov processes each with the transition-rate matrix

$$\mathbf{J}_w = \left[\begin{array}{cc} -\mu & \mu \\ \rho_w & -\rho_w \end{array}\right].$$

For a concise representation of $\Lambda_{E_wE_w}$, the right direct product (Kronecker product) of matrices (see, e.g., Graybill [GRA]) will be used: for any $(k_1 \times n_1)$-matrix Ψ, and $(k_2 \times n_2)$-matrix Θ, their right direct product $\Psi \otimes \Theta$ is a $((k_1k_2) \times (n_1n_2))$-matrix which is defined by

$$
\Psi \otimes \Theta = \begin{bmatrix}
\theta_{11}\,\Psi & \theta_{12}\,\Psi & \cdots & \theta_{1n_2}\,\Psi \\
\theta_{21}\,\Psi & \theta_{22}\,\Psi & \cdots & \theta_{2n_2}\,\Psi \\
\vdots & \vdots & & \vdots \\
\theta_{k_21}\,\Psi & \theta_{k_22}\,\Psi & \cdots & \theta_{k_2n_2}\,\Psi
\end{bmatrix} . \tag{1.5}
$$

It is then readily seen that the transition-rate matrix of this process is given by $\mathbf{J}_w \oplus \mathbf{J}_w \oplus \mathbf{J}_w$, where

$$
\mathbf{J} \oplus \mathbf{K} = \mathbf{I} \otimes \mathbf{J} + \mathbf{K} \otimes \mathbf{I} \tag{1.6}
$$

stands for the Kronecker sum of the square matrices $\mathbf{J}$ and $\mathbf{K}$; see Keller and Qamber [KEL] and the referrences therein. (On the right-hand side of (1.6), the identity matrices are of the same size as $\mathbf{K}$ and $\mathbf{J}$ respectively. Also notice that the perhaps more usual representation of the Kronecker sum is in terms of the left direct product in which case $\mathbf{J}$ and $\mathbf{K}$ on the left-hand side of (1.6) will be interchanged.) With this notation then,

$$
\mathbf{\Lambda}_{E_w E_w} = \mathbf{J}_w \oplus \mathbf{J}_w \oplus \mathbf{J}_w - \nu_w \mathbf{I}. \tag{1.7}
$$

Since in MATLAB the Kronecker product is available, $\mathbf{\Lambda}_{E_w E_w}$ can be coded concisely if (1.7) is used instead of (1.4). Furthermore, the approach just described is applicable for systems comprising any number of parallel devices, whereas the diagram-based method quickly becomes unmanagable.

The $(S \backslash \{\omega\})$-S-submatrix of the transition-rate matrix of the absorbing Markov process Z which is obtained from Y by merging $(0,0,0,0)$ and $(1,0,0,0)$ into ω, is given by

$$
\begin{bmatrix}
\mathbf{\Lambda}_{(E_0 \backslash \{(0,0,0,0)\}) E_0} & \nu_0 \mathbf{I} \\
\mathbf{\Lambda}_{(E_1 \backslash \{(1,0,0,0)\}) \{(1,0,0,0)\}} \quad \nu_1 \mathbf{I} & \mathbf{\Lambda}_{(E_1 \backslash \{(1,0,0,0)\}) (E_1 \backslash \{(1,0,0,0)\})}
\end{bmatrix}, \tag{1.8}
$$

from which the transition probability matrix of Z can be expressed in the usual way. (Notice that the transition rates into ω are collected in the first column-vector of the matrix in (1.8).)

Model 5. This is the Markov model of a two-unit parallel system with a single repairman; see, e.g., [GRO] pp 275-281. The system's state-transition-rate diagram is shown in Figure 1.4.

12

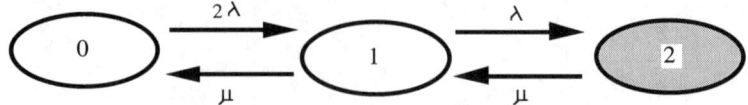

Figure 1.4 State-transition-rate diagram of the Markov model of the two-unit repairable system with a single repairman

Two identical devices run in parallel, each with constant failure rate λ. One repairman is available with repair rate μ. The system is considered 'failed' if both units are 'down'. The set of 'up' states is $G = \{0, 1\}$ and the set of 'down' states is $B = \{2\}$ (shaded in Figure 1.4). The system's transition-rate matrix Λ is given by

$$
\Lambda = \begin{array}{c} \\ 0 \\ 1 \\ 2 \end{array}
\begin{array}{ccc}
0 & 1 & 2 \\
\end{array}
\left[
\begin{array}{ccc}
-2\lambda & 2\lambda & 0 \\
\mu & -(\mu+\lambda) & \lambda \\
0 & \mu & -\mu
\end{array}
\right].
$$

It will be assumed that initially both units are operational, i.e., the blocks of the initial probability vector are $\alpha_G{}^T = (1, \ 0)$ and $\alpha_B{}^T = 0 = 0$.

1.2.2 Semi-Markov models

Model 6. The simplest semi-Markov model is given by the *alternating renewal process*, see, e.g., [BHA2] pp 286. In reliability theory, alternating renewal processes are used to model the behaviour of systems which alternate between working and repair periods indefinitely according to a two-state irreducible semi-Markov process; see, e.g., Singh and Billinton [SIN1] pp 80. The state space is $S = \{1, 2\}$ with, say, $G = \{1\}$ and $B = \{2\}$. The failure and the repair time distribution functions are respectively given by $F_{1,2}$ and $F_{2,1}$. The transition probability matrix of the embedded Markov chain is

$$
R = \left[
\begin{array}{cc}
0 & 1 \\
1 & 0
\end{array}
\right].
$$

Some renewal-theoretic aspects of this system will be considered briefly in Chapter 9. Then, in the last chapter, the system's n*th* 'up' time, $T_{G,n}(t_1)$, will be explored by assuming that the time-horizon $t_1 > 0$ is finite and that the system is started in the 'up' state at $t = 0$. Both discussions will rely on Laplace transforms for which the notation is defined as follows. The Laplace transform of the holding time in '1' (the 'good' state) is denoted by ϕ, and that in the 'bad' state '2' is γ.

Model 7. A system of two identical devices (transformers), modelled by a semi-Markov process Y, was discussed by Singh and Billinton in [SIN1], pp 178. The four possible system states are shown in Figure 1.5. Each of the devices may fail independently of the other with constant failure rate $\lambda > 0$. In state 1, the normal mode of operation, both units are working.

Upon failure of any one of them, the complete service is suspended for a random amount of time, ξ, say, during which the faulty one is identified (state 3). Subsequently (in state 2), the working unit resumes service while the faulty one undergoes repair. The rate $\mathcal{Y}(x)$ at which the system moves into the next state (state 2) is dependent on the time x already spent in state 3, i.e.,

$$\Pr\{\,\xi \leq t\,\} = 1 - \exp(-\int_0^t \mathcal{Y}(x)\,dx\,).$$

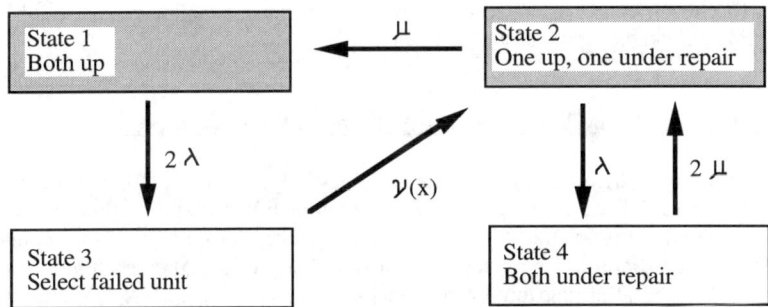

Figure 1.5 State-transition-rate diagram of the semi-Markov model of a system of two transformers

The repair rate μ is constant. State 4 stands for 'both units down' and it can be entered into and exited only via state 2 at the (constant) rates λ and 2μ respectively. The set of states in which at least one of the units is working is $G = \{1, 2\}$ (shaded), and its complement is $B = \{3, 4\}$. It will be assumed that the system is started with both units working, i.e., $\boldsymbol{\alpha}_G^T = (\,1, 0\,)$, $\boldsymbol{\alpha}_B^T = (\,0, 0\,)$. The holding time distribution functions are as follows: $F_{1,3}(t) = 1 - \exp(-2\lambda t\,)$, $F_{2,1}(t) = F_{2,4}(t) = 1 - \exp(-(\mu+\lambda)t\,)$, $F_{3,2}(t) = \Pr\{\xi \leq t\}$, and $F_{4,2}(t) = 1 - \exp(-2\mu t\,)$. $F_{s,s'}$ with $s = s'$ may be chosen arbitrarily. The transition probability matrix of the embedded Markov chain of Y is

$$\mathbf{R} = \begin{bmatrix} 0 & 0 & 1 & 0 \\ \mu/(\lambda+\mu) & 0 & 0 & \lambda/(\lambda+\mu) \\ 0 & 1 & 0 & 0 \\ 0 & 1 & 0 & 0 \end{bmatrix}.$$

For this model we shall be concerned with $M_G(t)$, the number of working periods up to some time $t > 0$.

CHAPTER 2

SOJOURN TIMES FOR DISCRETE-PARAMETER MARKOV CHAINS

It will be assumed here that $Z = \{ Z_0, Z_1, ... \}$ is a discrete-parameter Markov chain which is either irreducible or absorbing and whose transition probability matrix $\mathbf{Q}$ and initial probability vector $\boldsymbol{\alpha}$ are partitioned according to (1.1) - (1.2). Of interest will be the two sequences of sojourn times $\{ N_{A_i,j} : j \geq 1 \}$, $i = 1, 2$. In the first part of Chapter 2, the distribution theory of these variables is developed. The second part of Chapter 2 is then devoted to an application of the theory to a power transmission reliability model. The implementation details with MATLAB [MAT1], [MAT2] on the Apple Macintosh will also be discussed.

2.1 Distribution theory for sojourn times and related variables

The distribution of *individual* sojourn time variables is known from [RUB3] for irreducible chains. In Section 2.1.1, we examine the joint distribution of the first m of the variables $N_{A_1,1}$, $N_{A_1,2}$, ... This theory was first described in [CSE4]. The distribution of a large number of variables which are useful in the reliability context is deduced from this result in Section 2.1.2. In Section 2.1.3, the joint distribution of $N_{A_1,1}$, ..., $N_{A_1,m}$, $N_{A_2,1}$, ..., $N_{A_2,m}$ is obtained by a generalisation of *the renewal argument* in the probability generating function domain. Section 2.1 is concluded in 2.1.4 with a tabular summary of results.

2.1.1 Key results: sojourn times in a subset of the state space

Let Z be an absorbing Markov chain with state space $S = A_1 \cup A_2 \cup \{\omega\}$ and absorbing state ω. The following holds for the probability mass function of the first m of the variables $\{ N_{A_1,j} : j \geq 1 \}$.

THEOREM 2.1. *(a) It is* $\mathbf{Q}_{A_2A_2}{}^k \rightarrow \mathbf{0}$ *for* $k \rightarrow +\infty$ *and thus the matrix* $(\mathbf{I} - \mathbf{Q}_{A_2A_2})$ *is invertible. (b) With*

$$\mathbf{v}^T = (\mathbf{v}(A_1, A_2))^T = \boldsymbol{\alpha}_{A_2}{}^T (\mathbf{I} - \mathbf{Q}_{A_2A_2})^{-1} \mathbf{Q}_{A_2A_1} + \boldsymbol{\alpha}_{A_1}{}^T, \qquad (2.1)$$

the probability mass function of the joint distribution of $N_{A_1,1}$, ..., $N_{A_1,m}$ *is as follows. If* n_1, ..., $n_m \geq 1$, *it is*

$$\Pr\{ N_{A_1,1} = n_1, ..., N_{A_1,m} = n_m \} = \mathbf{v}^T \mathbf{Q}_{A_1A_1}{}^{n_1-1} \times$$

$$\prod_{i=2}^{m} \left\{ \mathbf{Q}_{A_1A_2} (\mathbf{I} - \mathbf{Q}_{A_2A_2})^{-1} \mathbf{Q}_{A_2A_1} \mathbf{Q}_{A_1A_1}{}^{n_i-1} \right\} (\mathbf{I} - \mathbf{Q}_{A_1A_1}) \mathbf{1}; \qquad (2.2)$$

if n_1, ..., $n_{m'} \geq 1$ *and* $n_{m'+1} = ... = n_m = 0$ *for some* $m' \in \{ 1, ..., m - 1 \}$, *it is*

$$\Pr\{\ N_{A_1,1} = n_1, ..., N_{A_1,m} = n_m\ \} =$$

$$\mathbf{v}^T \mathbf{Q}_{A_1A_1}{}^{n_1-1} \prod_{i=2}^{m'} \left\{ \mathbf{Q}_{A_1A_2} (\mathbf{I} - \mathbf{Q}_{A_2A_2})^{-1} \mathbf{Q}_{A_2A_1} \mathbf{Q}_{A_1A_1}{}^{n_i-1} \right\} \times$$

$$\left\{ \mathbf{I} - [\ \mathbf{Q}_{A_1A_1} + \mathbf{Q}_{A_1A_2} (\mathbf{I} - \mathbf{Q}_{A_2A_2})^{-1} \mathbf{Q}_{A_2A_1} \right\} \mathbf{1}; \tag{2.3}$$

if $n_1 = ... = n_m = 0$, *it is*

$$\Pr\{\ N_{A_1,1} = n_1, ..., N_{A_1,m} = n_m\ \} = 1 - \mathbf{v}^T \mathbf{1}; \tag{2.4}$$

finally, it is zero in all other cases.

PROOF OF THEOREM 2.1. *Proof of (a).* Consider the auxiliary Markov chain $X = \{\ X_0, X_1,$ ... $\}$ on $S = C \cup A_2$ (where $C = A_1 \cup \{\omega\}$) with transition probability matrix $\mathbf{P}$

$$\mathbf{P} = \begin{bmatrix} \mathbf{I} & \vdots & \mathbf{0} \\ ---- & \vdots & ---- \\ \mathbf{Q}_{A_2C} & \vdots & \mathbf{Q}_{A_2A_2} \end{bmatrix},$$

where

$$\mathbf{Q}_{A_2C} = \begin{bmatrix} \mathbf{Q}_{A_2A_1} & \vdots & \mathbf{Q}_{A_2\{\omega\}} \end{bmatrix}.$$

Let the initial probability vector of X be $\boldsymbol{\varepsilon} = (\ \mathbf{0}, \boldsymbol{\varepsilon}_{A_2}{}^T\ )^T$. Then the probability vector of X after the n*th* step, $\mathbf{p}_n = \{\ \Pr\{\ X_n = s\ \} : s \in S\ \}$, is seen to be

$$\mathbf{p}_n{}^T = \boldsymbol{\varepsilon}^T \mathbf{P}^n =$$

$$\boldsymbol{\varepsilon}^T \begin{bmatrix} \mathbf{I} & \vdots & \mathbf{0} \\ ------------------------ & \vdots & ---- \\ \{\mathbf{I} + \mathbf{Q}_{A_2A_2} + \mathbf{Q}_{A_2A_2}{}^2 + ... + \mathbf{Q}_{A_2A_2}{}^{n-1}\} \mathbf{Q}_{A_2C} & \vdots & \mathbf{Q}_{A_2A_2}{}^n \end{bmatrix} =$$

$$\left(\boldsymbol{\varepsilon}_{A_2}{}^T \left(\sum_{i=0}^{n-1} \mathbf{Q}_{A_2A_2}{}^i \right) \mathbf{Q}_{A_2C}, \boldsymbol{\varepsilon}_{A_2}{}^T \mathbf{Q}_{A_2A_2}{}^n \right). \tag{2.5}$$

Let N denote the random variable 'time to absorption into C', i.e., N is defined by $N = n \Leftrightarrow X_{n-1} \in A_2$ *and* $X_n \in C$. Then, from (2.5) it follows that for any $c \in C$ and $n = 1, 2, ...$

$$\Pr\{\ N = n, X_n = c\ \} = \Pr\{\ X_n = c\ \} - \Pr\{\ X_{n-1} = c\ \} =$$

c-entry of $\boldsymbol{\varepsilon}_{A_2}^T \mathbf{Q}_{A_2A_2}^{n-1} \mathbf{Q}_{A_2C}$. $\qquad$ (2.6)

Thus, summing in (2.5) over $c \in C$, it is seen that

$$\Pr\{\, N = n \,\} = \boldsymbol{\varepsilon}_{A_2}^T \mathbf{Q}_{A_2A_2}^{n-1} \mathbf{Q}_{A_2C}\, \mathbf{1} = \boldsymbol{\varepsilon}_{A_2}^T \mathbf{Q}_{A_2A_2}^{n-1} (\mathbf{I} - \mathbf{Q}_{A_2A_2})\, \mathbf{1},$$

from which it follows that

$$\Pr\{\, n \leq N \leq n + k \,\} = \boldsymbol{\varepsilon}_{A_2}^T \mathbf{Q}_{A_2A_2}^{n-1}\, \mathbf{1} - \boldsymbol{\varepsilon}_{A_2}^T \mathbf{Q}_{A_2A_2}^{n+k}\, \mathbf{1}. \qquad (2.7)$$

Putting in (2.7) $n = 1$ and letting $k \to +\infty$ shows that $\boldsymbol{\varepsilon}_{A_2}^T \mathbf{Q}_{A_2A_2}^{k}\, \mathbf{1} \to 0$ for *any* stochastic vector $\boldsymbol{\varepsilon}_{A_2}$, and hence $\mathbf{Q}_{A_2A_2}^{k} \to \mathbf{0}$ element-wise which implies that $(\mathbf{I} - \mathbf{Q}_{A_2A_2})$ is invertible.

Before entering the proof of (b), let us deduce from the above reasoning two results for later reference. First, define the vector $\boldsymbol{\beta}_C$ by $\beta_c = \Pr\{\, X \text{ is absorbed into } c \,\}, c \in C$. Summing in (2.6) over $n \geq 1$, we get

$$\boldsymbol{\beta}_C^T = \boldsymbol{\varepsilon}_{A_2}^T (\mathbf{I} - \mathbf{Q}_{A_2A_2})^{-1} \mathbf{Q}_{A_2C}. \qquad (2.8)$$

(2.8) can be found in Kemeny and Snell [KEM] pp 52. The second result which we want to deduce from the above reasoning is the following equation

$$\Pr\{\, N = n, X_{N-1} = a, X_N = c \,\} =$$

$$q_{ac} \times a\text{th component of } \boldsymbol{\varepsilon}_{A_2}^T \mathbf{Q}_{A_2A_2}^{n-1}, a \in A_2, c \in C. \qquad (2.9)$$

This is seen from $\Pr\{\, N = n, X_{N-1} = a, X_N = c \,\} = \Pr\{\, X_{n-1} = a, X_n = c \,\} = q_{ac}\, \Pr\{\, X_{n-1} = a \,\} = q_{ac} \times a\text{th component of } \boldsymbol{\varepsilon}_{A_2}^T \mathbf{Q}_{A_2A_2}^{n-1}$ by (2.5).

Outline proof of (b). Before proceeding more formally, let us discuss some preparatory steps and the proof outline. Starting from the original Markov chain Z, a new absorbing Markov chain $X = \{\, X_0, X_1, \dots \,\}$ will be constructed which can be imagined to be arising from the old one by letting its *i*th sojourn in A_2 (and A_1) be spent in a new (namely the *i*th) instance of A_2 (and of A_1). Having passed through this cascade of groups of states, the process X is eventually absorbed into $A_2(m+1)$, the $m+1$st instance of A_2. Absorption into $\{\omega\}$ from any of the sets $A_2(1), \dots, A_2(m)$ and $A_1(1), \dots, A_1(m)$ is possible according to $\mathbf{Q}_{A_2\{\omega\}}$ and $\mathbf{Q}_{A_1\{\omega\}}$ respectively. Thus the state space S(m) of X is the disjoint union

$$S(m) = A_2(1) \cup A_1(1) \cup \dots \cup A_2(m) \cup A_1(m) \cup A_2(m+1) \cup \{\omega\}, \qquad (2.10)$$

where $A_2(1), \dots, A_2(m+1)$ and $A_1(1), \dots, A_1(m)$ are disjoint instances of A_2 and A_1 respectively. Furthermore, transitions within each group $A_2(i)$ and $A_1(i)$, $i = 1, \dots, m$, are according to the old chain Z. The state-transition diagram of X is shown in Figure 2.1 below. The states in $A_2(m+1) \cup \{\omega\}$ are absorbing. The $A_1(1)$- and $A_2(1)$-components of the initial probability vector of X will be $\boldsymbol{\alpha}_{A_1}$ and $\boldsymbol{\alpha}_{A_2}$ respectively; all of its other components are zero. Now, the distribution sought is the joint distribution of sojourn times in $A_1(1), \dots, A_1(m)$ (shaded in Figure 2.1) until absorption.

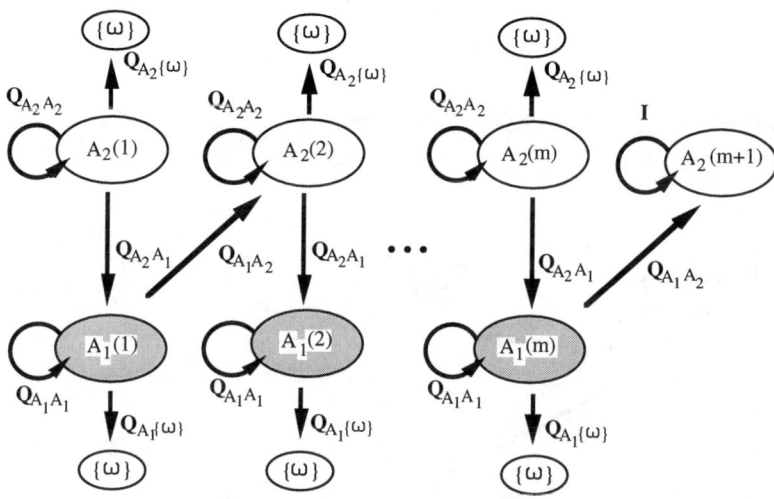

Figure 2.1 The state-transition diagram of the auxiliary Markov chain X

For the detailed proof of Theorem 2.1.(b), a lemma will be needed. It will be used to establish the initial behaviour of the chain X as it sets off from $A_1(1) \cup A_2(1)$ and eventually makes a transition into $A_2(2) \cup \{\omega\}$. Notice that upon arrival into $A_2(2)$, a new sojourn in (a new instance of) $A_1 \cup A_2$ is started and it is a probabilistic replica of the first one. This new sojourn is conditionally independent from the previous one given the knowledge of the arrival value of X in $A_2(2)$. This will give rise to a proof by induction.

LEMMA 2.2. *Let* **P** *be the transition probability matrix of an absorbing Markov chain on the partitioned state space* $A \cup B \cup \{\omega\}$ *with absorbing state* ω. *Let* C *be a second instance of* A *and let* $Y = \{Y_0, Y_1, ...\}$ *be a Markov chain on* $A \cup B \cup C \cup \{\omega\}$ *with transition probability matrix*

$$
\begin{array}{c}
\begin{array}{cccc} A & B & C & \{\omega\} \end{array} \\
\begin{array}{c} A \\ B \\ C \\ \{\omega\} \end{array}
\left[
\begin{array}{c|c|c|c}
\mathbf{P_{AA}} & \mathbf{P_{AB}} & \mathbf{0} & \mathbf{P_{A\{\omega\}}} \\
\hline
\mathbf{0} & \mathbf{P_{BB}} & \mathbf{P_{BA}} & \mathbf{P_{B\{\omega\}}} \\
\hline
\mathbf{0} & \mathbf{0} & \mathbf{I} & \mathbf{0} \\
\hline
\mathbf{0} & \mathbf{0} & \mathbf{0} & \mathbf{1}
\end{array}
\right] ;
\end{array}
$$

see Figure 2.2 below. Let $(\, \boldsymbol{\varepsilon}_A^T \, \boldsymbol{\varepsilon}_B^T \, \mathbf{0} \, \mathbf{0}\,)^T$ *be the initial probability vector of* Y *and denote by* N_B *the total time spent in B by* Y *until absorption into* $C \cup \{\omega\}$. *We define on* $\{ N_B \geq 1 \}$ *the random variable K to be the departure time from B, i.e., K is defined by* $Y_K \in B$, $Y_{K+1} \notin B$. *(K + 1 is in fact the time to absorption but in view of the context in which it will be used later the former interpretation is more helpful.) Then, for any* $b \in B$, $x \in C \cup \{\omega\}$, *and* $n \geq 1$ *we have*

Pr{ $N_B = n$, $Y_K = b$, $Y_{K+1} = x$ } =

ε_A^T { *b*th column vector of $(I - P_{AA})^{-1} P_{AB} P_{BB}^{n-1}$} p_{bx} +

ε_B^T { *b*th column vector of P_{BB}^{n-1}} p_{bx}.

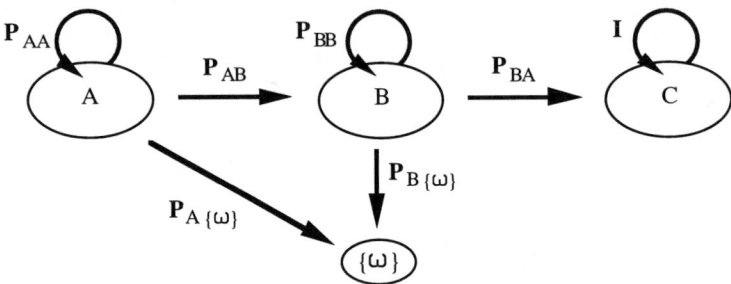

Figure 2.2 The state-transition diagram of the Markov chain Y

PROOF OF LEMMA 2.2. It is

Pr{ $N_B = n$, $Y_K = b$, $Y_{K+1} = x$ } =

$\sum_{a \in A} \varepsilon_a$ Pr{ $N_B = n$, $Y_K = b$, $Y_{K+1} = x \mid Y_0 = a$ } +

$\sum_{b' \in B} \varepsilon_{b'}$ Pr{ $N_B = n$, $Y_K = b$, $Y_{K+1} = x \mid Y_0 = b'$ }.   (2.11)

The first probability on the right hand side of (2.11) is obtained as follows. Define for $b_1 \in B$,

β_{b_1} = Pr{ Y enters B and b_1 is the point on entrance of Y into B $\mid Y_0 = a$ }.

(2.8) then shows that

$$\beta_B^T = \delta_a^T (I - P_{AA})^{-1} P_{AB}$$

where δ_a stands for the column vector whose only non-zero entry, the *a*th component, is unity. Using now (2.9), applied to the absorbing chain on $B \cup C \cup \{\omega\}$ with transition probability matrix

$$
\begin{array}{c}
 \quad\quad B \quad\quad C \quad\quad \{\omega\} \\
\begin{array}{c} B \\ C \\ \{\omega\} \end{array}
\left[
\begin{array}{c|c|c}
P_{BB} & P_{BA} & P_{B\{\omega\}} \\
\hline
0 & I & 0 \\
\hline
0 & 0 & 1
\end{array}
\right] ,
\end{array}
\qquad (2.12)
$$

we get

$$ \Pr\{ N_B = n, Y_K = b, Y_{K+1} = x \mid Y_0 = a \} = $$

$$ \sum_{b_1 \in B} \Pr\{ N_B = n, Y_K = b, Y_{K+1} = x \mid Y \text{ enters } B \text{ via } b_1, Y_0 = a \} \times $$

$$ \Pr\{ Y \text{ enters } B \text{ via } b_1 \mid Y_0 = a \} = $$

$$ \sum_{b_1 \in B} \Pr\{ N_B = n, Y_K = b, Y_{K+1} = x \mid Y \text{ enters } B \text{ via } b_1, Y_0 = a \} \, \beta_{b_1} = $$

$$ \sum_{b_1 \in B} \Pr\{ N_B = n, Y_K = b, Y_{K+1} = x \mid Y_0 = b_1 \} \, \beta_{b_1} = $$

$$ \sum_{b_1 \in B} p_{bx} \times b\textit{th} \text{ component of } \delta_{b'}^T P_{BB}^{\,n-1} \, \beta_{b_1} = $$

$$ p_{bx} \times b\textit{th} \text{ component of } \beta_B^T P_{BB}^{\,n-1} = $$

$$ p_{bx} \times b\textit{th} \text{ component of } \delta_a^T (I - P_{AA})^{-1} P_{AB} P_{BB}^{\,n-1} = $$

$$ p_{bx} \times (a, b) \text{ entry of } (I - P_{AA})^{-1} P_{AB} P_{BB}^{\,n-1} . $$

Thus,

$$ \sum_{a \in A} \varepsilon_a \Pr\{ N_B = n, Y_K = b, Y_{K+1} = x \mid Y_0 = a \} = $$

$$ \varepsilon_A^T \{ b\textit{th} \text{ column vector of } (I - P_{AA})^{-1} P_{AB} P_{BB}^{\,n-1} \} \, p_{bx} . $$

The second probability on the right hand side of (2.11) follows from (2.9) applied to the absorbing Markov chain with transition probability matrix as in (2.12) and initial probability vector $\delta_{b'}$:

$\Pr\{ N_B = n, Y_K = b, Y_{K+1} = x \mid Y_0 = b' \} =$

$p_{bx} \times b$th component of $\boldsymbol{\delta}_b^T \, \mathbf{P}_{BB}^{n-1} = p_{bx} \times (b', b)$ -entry of $\mathbf{P}_{BB}^{n-1}$.

Thus,

$$\sum_{b' \in B} \varepsilon_{b'} \Pr\{ N_B = n, Y_K = b, Y_{K+1} = x \mid Y_0 = b' \} =$$

$\boldsymbol{\varepsilon}_B^T \{ b$th column vector of $\mathbf{P}_{BB}^{n-1} \} \, p_{bx}$. ∎

Proof of Theorem 2.1.(b). In order to be able to represent the transition probability matrix of the auxiliary Markov chain X concisely, let us introduce a block partitioned matrix $\boldsymbol{\Delta}_n$ which is defined in terms of its blocks $\boldsymbol{\Phi}_1, \boldsymbol{\Phi}_2, \ldots$ and $\boldsymbol{\Psi}_1, \boldsymbol{\Psi}_2, \ldots$ by

$$\boldsymbol{\Delta}_n(\boldsymbol{\Phi}_1, \ldots, \boldsymbol{\Phi}_n; \boldsymbol{\Psi}_1, \ldots, \boldsymbol{\Psi}_{n-1}) =$$

$$\begin{bmatrix} \boldsymbol{\Phi}_1 & \boldsymbol{\Psi}_1 & 0 & \ldots & 0 & 0 \\ 0 & \boldsymbol{\Phi}_2 & \boldsymbol{\Psi}_2 & \ldots & 0 & 0 \\ \vdots & \vdots & \vdots & & & \vdots \\ 0 & 0 & 0 & \ldots & \boldsymbol{\Phi}_{n-1} & \boldsymbol{\Psi}_{n-1} \\ 0 & 0 & 0 & \ldots & 0 & \boldsymbol{\Phi}_n \end{bmatrix}. \tag{2.13}$$

Matrices of the type $\boldsymbol{\Delta}_n$ in (2.13) will be encountered repeatedly not only here but in subsequent sections too. It is easily seen from Figure 2.1 that the transition probability matrix of X can be written as

$$\begin{bmatrix} \boldsymbol{\Delta} & \mathbf{u} \\ 0 & 1 \end{bmatrix}, \tag{2.14}$$

with $\boldsymbol{\Delta}$ defined as

$$\boldsymbol{\Delta} = \boldsymbol{\Delta}_{2m+1}(\mathbf{Q}_{A_2A_2}, \mathbf{Q}_{A_1A_1}, \ldots, \mathbf{Q}_{A_2A_2}, \mathbf{Q}_{A_1A_1}, \mathbf{I}; \mathbf{Q}_{A_2A_1}, \mathbf{Q}_{A_1A_2}, \ldots, \mathbf{Q}_{A_2A_1}, \mathbf{Q}_{A_1A_2}),$$

and

$$\mathbf{u}^T = \left[\mathbf{Q}_{A_2\{\omega\}}^T \mid \mathbf{Q}_{A_1\{\omega\}}^T \mid \ldots \mid \mathbf{Q}_{A_2\{\omega\}}^T \mid \mathbf{Q}_{A_1\{\omega\}}^T \mid 0 \right].$$

(The representation of the stochastic matrix in (2.14) corresponds to the partitioning of S(m) in

(2.10).)

To prove (2.2), take $n_1, ..., n_m \geq 1$; the case $m \geq 2$ will be examined first. The event

$$\{ N_{A_1,1} = n_1, ..., N_{A_1,m} = n_m \}$$

can be described in terms of the new chain X as the path $(A_2(1) \rightarrow) \; A_1(1) \rightarrow A_2(2) \rightarrow A_1(2) \rightarrow ... \rightarrow A_1(m) \rightarrow A_2(m+1) \cup \{\omega\}$ such that the time spent in $A_1(i)$ is n_i, $i = 1, ...,m$. *On this event*, the departure time K_i from $A_1(i)$ is defined by $K_i = k$ if $X_k \in A_1(i)$ *and* $X_{k+1} \in A_2(i+1)$; X_k and X_{k+1} are the departure value from $A_1(i)$ and arrival value in $A_2(i+1)$ respectively. Denote by N_A the time spent until absorption by X in $A \in \{ A_1(1), ..., A_1(m) \}$. The event of interest is now partitioned according to the departure and arrival values of X as follows

$$\{ N_{A_1,1} = n_1, ..., N_{A_1,m} = n_m \} =$$

$$\left\{ \bigcup_{b_1 \in A_1(1)} \bigcup_{a_2 \in A_2(2)} ... \bigcup_{b_m \in A_1(m)} \bigcup_{a_{m+1} \in A_2(m+1)} \right.$$

$$\left. \left\{ \bigcap_{i=1}^{m} \{ N_{A_1(i)} = n_i, X_{K_i} = b_i, X_{K_i+1} = a_{i+1} \} \right\} \right\} \cup$$

$$\left\{ \bigcup_{b_1 \in A_1(1)} \bigcup_{a_2 \in A_2(2)} ... \bigcup_{a_m \in A_2(m)} \bigcup_{b_m \in A_1(m)} \right.$$

$$\left\{ \bigcap_{i=1}^{m-1} \{ N_{A_1(i)} = n_i, X_{K_i} = b_i, X_{K_i+1} = a_{i+1} \} \cap \right.$$

$$\left. \left. \{ N_{A_1(m)} = n_m, X_{K_m} = b_m, X_{K_m+1} = \omega \} \right\} \right\} . \qquad (2.15)$$

The above is a disjoint union of events. The initial probability vector of X is ($\boldsymbol{\alpha}_{A_2}^{\mathrm{T}}, \boldsymbol{\alpha}_{A_1}^{\mathrm{T}}, \mathbf{0}, ..., \mathbf{0}, 0$). For the first bracketed event on the right hand side of (2.15), we have

$$\Pr\left\{ \bigcap_{i=1}^{m} \{ N_{A_1(i)} = n_i, X_{K_i} = b_i, X_{K_i+1} = a_{i+1} \} \right\} = \qquad (2.16)$$

$$\Pr\left\{ \bigcap_{i=1}^{m} \{ N_{A_1(i)} = n_i, X_{K_i} = b_i, X_{K_i+1} = a_{i+1} \} \mid X_{K_1+1} = a_2 \right\} \times$$

$$\Pr\{ X_{K_1+1} = a_2 \} =$$

$$\Pr\{ \bigcap_{i=2}^{m} \{ N_{A_1(i)} = n_i, X_{K_i} = b_i, X_{K_i+1} = a_{i+1} \} \mid X_{K_1+1} = a_2 \} \times$$

$$\Pr\{ N_{A_1(1)} = n_1, X_{K_1} = b_1 \mid X_{K_1+1} = a_2 \} \; \Pr\{ X_{K_1+1} = a_2 \} = \qquad (2.17)$$

$$\Pr\{ \bigcap_{i=2}^{m} \{ N_{A_1(i)} = n_i, X_{K_i} = b_i, X_{K_i+1} = a_{i+1} \} \mid X_{K_1+1} = a_2 \} \times$$

$$\Pr\{ N_{A_1(1)} = n_1, X_{K_1} = b_1, X_{K_1+1} = a_2 \}. \qquad (2.18)$$

Transition to (2.17) is achieved by noting that due to the Markov property the vector comprising the sojourn times and departure and arrival values associated with $A_1(2)$, ..., $A_1(m)$, $A_2(3)$, ..., $A_2(m+1)$ and the pair consisting of the sojourn time and departure value associated with $A_1(1)$ are conditionally independent given the arrival value in $A_2(2)$. By the Markov property, for the first probability in (2.18) we have

$$\Pr\{ \bigcap_{i=2}^{m} \{ N_{A_1(i)} = n_i, X_{K_i} = b_i, X_{K_i+1} = a_{i+1} \} \mid X_{K_1+1} = a_2 \} =$$

$$\Pr\{ \bigcap_{i=1}^{m-1} \{ N_{A_1(i)} = n_{i+1}, X_{K_i} = b_{i+1}, X_{K_i+1} = a_{i+2} \} \mid X_0 = a_2 \}, \qquad (2.19)$$

which then by (2.16) - (2.18) can be written as

$$\Pr\{ \bigcap_{i=1}^{m-1} \{ N_{A_1(i)} = n_{i+1}, X_{K_i} = b_{i+1}, X_{K_i+1} = a_{i+2} \} \mid X_0 = a_2 \} =$$

$$\Pr\{ \bigcap_{i=2}^{m-1} \{ N_{A_1(i)} = n_{i+1}, X_{K_i} = b_{i+1}, X_{K_i+1} = a_{i+2} \} \mid X_{K_1+1} = a_3 \} \times$$

$$\Pr\{ N_{A_1(1)} = n_2, X_{K_1} = b_2, X_{K_1+1} = a_3 \mid X_0 = a_2 \}. \qquad (2.20)$$

(2.20) is now rewritten by applying (2.19) (with m replaced by m-1) to the first term on its right hand side

$$\Pr\{ \bigcap_{i=1}^{m-1} \{ N_{A_1(i)} = n_{i+1}, X_{K_i} = b_{i+1}, X_{K_i+1} = a_{i+2} \} \mid X_0 = a_2 \} =$$

$$\Pr\{ \bigcap_{i=1}^{m-2} \{ N_{A_1(i)} = n_{i+2}, X_{K_i} = b_{i+2}, X_{K_i+1} = a_{i+3} \} \mid X_0 = a_3 \} \times$$

$$\Pr\{\ N_{A_1(1)} = n_2,\ X_{K_1} = b_2,\ X_{K_1+1} = a_3\ |\ X_0 = a_2\ \}. \tag{2.21}$$

From (2.21) it follows by induction that

$$\Pr\{\ \bigcap_{i=1}^{m-1} \{\ N_{A_1(i)} = n_{i+1},\ X_{K_i} = b_{i+1},\ X_{K_i+1} = a_{i+2}\ \}\ |\ X_0 = a_2\} =$$

$$\prod_{j=2}^{m} \Pr\{\ N_{A_1(1)} = n_j,\ X_{K_1} = b_j,\ X_{K_1+1} = a_{j+1}\ |\ X_0 = a_j\ \}. \tag{2.22}$$

Combining (2.16) - (2.18), (2.19), and (2.22), it is seen that

$$\Pr\{\ \bigcap_{i=1}^{m} \{\ N_{A_1(i)} = n_i,\ X_{K_i} = b_i,\ X_{K_i+1} = a_{i+1}\ \}\} =$$

$$\left\{ \prod_{j=2}^{m} \Pr\{\ N_{A_1(1)} = n_j,\ X_{K_1} = b_j,\ X_{K_1+1} = a_{j+1}\ |\ X_0 = a_j\ \} \right\} \times$$

$$\Pr\{\ N_{A_1(1)} = n_1,\ X_{K_1} = b_1,\ X_{K_1+1} = a_2\ \}. \tag{2.23}$$

The probabilities on the right hand side of (2.23) are identically structured. To obtain an expression for the right hand (unconditional) term, we apply Lemma 2.2 to the Markov chain with $A = A_2(1)$, $B = A_1(1)$, $C = A_2(2)$, $\mathbf{P} = \mathbf{Q}$, and $\boldsymbol{\varepsilon} = \boldsymbol{\alpha}$, to get

$$\Pr\{\ N_{A_1(1)} = n_1,\ X_{K_1} = b_1,\ X_{K_1+1} = a_2\ \} =$$

$$\boldsymbol{\alpha}_{A_2}^{T} \{\ b_1 th \text{ column vector of } (\mathbf{I} - \mathbf{Q}_{A_2A_2})^{-1}\ \mathbf{Q}_{A_2A_1}\ \mathbf{Q}_{A_1A_1}{}^{n_1-1}\}\ q_{b_1 a_2} +$$

$$\boldsymbol{\alpha}_{A_1}^{T} \{\ b_1 th \text{ column vector of } \mathbf{Q}_{A_1A_1}{}^{n_1-1}\}\ q_{b_1 a_2}.$$

By the same token, we have for the probabilities in the bracketed term in (2.23)

$$\Pr\{\ N_{A_1(1)} = n_j,\ X_{K_1} = b_j,\ X_{K_1+1} = a_{j+1}\ |\ X_0 = a_j\ \} =$$

$$\boldsymbol{\delta}_{a_j}^{T} \{\ b_j th \text{ column vector of } (\mathbf{I} - \mathbf{Q}_{A_2A_2})^{-1}\ \mathbf{Q}_{A_2A_1}\ \mathbf{Q}_{A_1A_1}{}^{n_j-1}\}\ q_{b_j a_{j+1}} =$$

$$\{\ (a_j, b_j)\text{-entry of } (\mathbf{I} - \mathbf{Q}_{A_2A_2})^{-1}\ \mathbf{Q}_{A_2A_1}\ \mathbf{Q}_{A_1A_1}{}^{n_j-1}\}\ q_{b_j a_{j+1}}.$$

Thus,

$$\Pr\left\{ \bigcap_{i=1}^{m} \{ N_{A_1(i)} = n_i, X_{K_i} = b_i, X_{K_i+1} = a_{i+1} \} \right\} =$$

$$\left\{ \prod_{j=2}^{m} \{ (a_j, b_j)\text{-entry of } (\mathbf{I} - \mathbf{Q}_{A_2A_2})^{-1} \mathbf{Q}_{A_2A_1} \mathbf{Q}_{A_1A_1}{}^{n_j-1} \} q_{b_j a_{j+1}} \right\} \times$$

$$\left\{ \boldsymbol{\alpha}_{A_2}{}^T \{ b_1 th \text{ column vector of } (\mathbf{I} - \mathbf{Q}_{A_2A_2})^{-1} \mathbf{Q}_{A_2A_1} \mathbf{Q}_{A_1A_1}{}^{n_1-1} \} q_{b_1 a_2} + \right.$$

$$\left. \boldsymbol{\alpha}_{A_1}{}^T \{ b_1 th \text{ column vector of } \mathbf{Q}_{A_1A_1}{}^{n_1-1} \} q_{b_1 a_2} \right\},$$

which shows that the first union of events in (2.15) has the probability

$$\Pr\left\{ \bigcup_{b_1 \in A_1(1)} \bigcup_{a_2 \in A_2(2)} \cdots \bigcup_{b_m \in A_1(m)} \bigcup_{a_{m+1} \in A_2(m+1)} \right.$$

$$\left\{ \bigcap_{i=1}^{m} \{ N_{A_1(i)} = n_i, X_{K_i} = b_i, X_{K_i+1} = a_{i+1} \} \right\} \Bigg\} =$$

$$\boldsymbol{\alpha}_{A_2}{}^T \left\{ \prod_{i=1}^{m} \{ (\mathbf{I} - \mathbf{Q}_{A_2A_2})^{-1} \mathbf{Q}_{A_2A_1} \mathbf{Q}_{A_1A_1}{}^{n_i-1} \mathbf{Q}_{A_1A_2} \} \right\} \mathbf{1} +$$

$$\boldsymbol{\alpha}_{A_1}{}^T \mathbf{Q}_{A_1A_1}{}^{n_1-1} \mathbf{Q}_{A_1A_2} \left\{ \prod_{i=2}^{m} \{ (\mathbf{I} - \mathbf{Q}_{A_2A_2})^{-1} \mathbf{Q}_{A_2A_1} \mathbf{Q}_{A_1A_1}{}^{n_i-1} \mathbf{Q}_{A_1A_2} \} \right\} \mathbf{1} =$$

$$\mathbf{v}^T \mathbf{Q}_{A_1A_1}{}^{n_1-1} \mathbf{Q}_{A_1A_2} \left\{ \prod_{i=2}^{m} \{ (\mathbf{I} - \mathbf{Q}_{A_2A_2})^{-1} \mathbf{Q}_{A_2A_1} \mathbf{Q}_{A_1A_1}{}^{n_i-1} \mathbf{Q}_{A_1A_2} \} \right\} \mathbf{1}. \quad (2.24)$$

We now turn to the second union of events in (2.15). By a similar reasoning it follows again by Lemma 2.2 that its probability is

$$\mathbf{v}^T \mathbf{Q}_{A_1A_1}{}^{n_1-1} \mathbf{Q}_{A_1A_2} \left\{ \prod_{i=2}^{m-1} \{ (\mathbf{I} - \mathbf{Q}_{A_2A_2})^{-1} \mathbf{Q}_{A_2A_1} \mathbf{Q}_{A_1A_1}{}^{n_i-1} \mathbf{Q}_{A_1A_2} \} \right\} \times$$

$$(\mathbf{I} - \mathbf{Q}_{A_2A_2})^{-1} \mathbf{Q}_{A_2A_1} \mathbf{Q}_{A_1A_1}{}^{n_m-1} \mathbf{Q}_{A_1}\{\omega\}. \quad (2.25)$$

(2.24) and (2.25) sum to

$$\mathbf{v}^T \mathbf{Q}_{A_1A_1}^{n_1-1} \mathbf{Q}_{A_1A_2} \left\{ \prod_{i=2}^{m-1} \left\{ (\mathbf{I} - \mathbf{Q}_{A_2A_2})^{-1} \mathbf{Q}_{A_2A_1} \mathbf{Q}_{A_1A_1}^{n_i-1} \mathbf{Q}_{A_1A_2} \right\} \right\} \times$$

$$(\mathbf{I} - \mathbf{Q}_{A_2A_2})^{-1} \mathbf{Q}_{A_2A_1} \mathbf{Q}_{A_1A_1}^{n_m-1} (\mathbf{Q}_{A_1A_2}\mathbf{1} + \mathbf{Q}_{A_1\{\omega\}}),$$

which is easily seen to be (2.2) by

$$\mathbf{Q}_{A_1A_2}\mathbf{1} + \mathbf{Q}_{A_1\{\omega\}} = (\mathbf{I} - \mathbf{Q}_{A_1A_1})\mathbf{1}. \tag{2.26}$$

To show that (2.2) also holds for m = 1, notice that in this case (2.15) takes the form

$$\{ N_{A_1,1} = n \} = \left\{ \bigcup_{b \in A_1(1)} \bigcup_{a \in A_2(2)} \{ N_{A_1(1)} = n, X_{K_1} = b, X_{K_1+1} = a \} \right\} \cup$$

$$\left\{ \bigcup_{b \in A_1(1)} \{ N_{A_1(1)} = n, X_{K_1} = b, X_{K_1+1} = \omega \} \right\},$$

from which by Lemma 2.2 and (2.26) we have

$$\Pr\{ N_{A_1,1} = n \} = \sum_{b \in A_1(1)} \sum_{a \in A_2(2)} \Pr\{ N_{A_1(1)} = n, X_{K_1} = b, X_{K_1+1} = a \} +$$

$$\sum_{b \in A_1(1)} \Pr\{ N_{A_1(1)} = n, X_{K_1} = b, X_{K_1+1} = \omega \} =$$

$$\sum_{\substack{b \in A_1(1) \\ a \in A_2(2)}} \left\{ \boldsymbol{\alpha}_{A_2}^T \{ b th \text{ column vector of } (\mathbf{I} - \mathbf{Q}_{A_2A_2})^{-1} \mathbf{Q}_{A_2A_1} \mathbf{Q}_{A_1A_1}^{n-1} \} q_{ba} + \right.$$

$$\left. \boldsymbol{\alpha}_{A_1}^T \{ b th \text{ column vector of } \mathbf{Q}_{A_1A_1}^{n-1} \} q_{ba} \right\} +$$

$$\sum_{b \in A_1(1)} \left\{ \boldsymbol{\alpha}_{A_2}^T \{ b th \text{ column vector of } (\mathbf{I} - \mathbf{Q}_{A_2A_2})^{-1} \mathbf{Q}_{A_2A_1} \mathbf{Q}_{A_1A_1}^{n-1} \} q_{b\omega} + \right.$$

$$\left. \boldsymbol{\alpha}_{A_1}^T \{ b th \text{ column vector of } \mathbf{Q}_{A_1A_1}^{n-1} \} q_{b\omega} \right\} =$$

$$\boldsymbol{\alpha}_{A_2}^T (\mathbf{I} - \mathbf{Q}_{A_2A_2})^{-1} \mathbf{Q}_{A_2A_1} \mathbf{Q}_{A_1A_1}^{n-1} \mathbf{Q}_{A_1A_2}\mathbf{1} + \boldsymbol{\alpha}_{A_1}^T \mathbf{Q}_{A_1A_1}^{n-1} \mathbf{Q}_{A_1\{\omega\}} =$$

$$\mathbf{v}^T \mathbf{Q}_{A_1A_1}^{n-1} (\mathbf{Q}_{A_1A_2}\mathbf{1} + \mathbf{Q}_{A_1\{\omega\}}) = \mathbf{v}^T \mathbf{Q}_{A_1A_1}^{n-1} (\mathbf{I} - \mathbf{Q}_{A_1A_1})\mathbf{1}, \tag{2.27}$$

which is (2.2) for m = 1. Summing over n ≥ 1 in (2.27) also implies (2.4) since $N_{A_1,1} = 0 \Leftrightarrow$

$N_{A_1,1} = ... = N_{A_1,m} = 0$. To prove (2.3), we may assume that $m' = m - 1 \geq 1$ since $N_{A_1,m'+1} = 0 \Leftrightarrow N_{A_1,m'+1} = ... = N_{A_1,m} = 0$. It is thus $n_1, ..., n_{m-1} \geq 1$ and $n_m = 0$. We have by (2.2) the following

$$\Pr\{ N_{A_1,1} = n_1, ..., N_{A_1,m-1} = n_{m-1}, N_{A_1,m} = 0 \} =$$

$$\Pr\{ N_{A_1,1} = n_1, ..., N_{A_1,m-1} = n_{m-1} \} -$$

$$\sum_{n_m=1}^{\infty} \Pr\{ N_{A_1,1} = n_1, ..., N_{A_1,m-1} = n_{m-1}, N_{A_1,m} = n_m \} =$$

$$\mathbf{v}^T \mathbf{Q}_{A_1A_1}^{n_1-1} \prod_{i=2}^{m-1} \left\{ \mathbf{Q}_{A_1A_2} (\mathbf{I} - \mathbf{Q}_{A_2A_2})^{-1} \mathbf{Q}_{A_2A_1} \mathbf{Q}_{A_1A_1}^{n_i-1} \right\} \times$$

$$\left\{ (\mathbf{I} - \mathbf{Q}_{A_1A_1}) - \mathbf{Q}_{A_1A_2} (\mathbf{I} - \mathbf{Q}_{A_2A_2})^{-1} \mathbf{Q}_{A_2A_1} \right\} \mathbf{1}.$$

To conclude the proof, it suffices to note that all the other events hitherto not considered carry zero probability due to the following two equivalences: $N_{A_1,i} = 0$ for all $i \geq m + 1 \Leftrightarrow N_{A_1,m+1} = 0$, and $N_{A_1,1}, ..., N_{A_1,m+1} \geq 1 \Leftrightarrow N_{A_1,m+1} \geq 1$. ∎

Upon close inspection of the proof of Theorem 2.1, it is seen that the requirement that Z should be an absorbing chain is not essential. The crucial property of Z is that neither in A_1 nor in A_2 should there be contained a closed set of states. (Roughly speaking, a subset of the state space is said to be closed if it cannot be left once it has been entered; see also [IOS], Ch. 2.4.) This certainly holds for *irreducible* chains. Bearing this in mind, Theorem 2.1 implies the following corollary.

COROLLARY 2.3. *Let Z be an irreducible Markov chain on* $S = A_1 \cup A_2$. *Then, (a) The matrix* $(\mathbf{I} - \mathbf{Q}_{A_2A_2})$ *is invertible; (b) With* $\mathbf{v}$ *defined by (2.1), the probability mass function of the joint distribution of* $N_{A_1,1}, ..., N_{A_1,m}$ *is given for* $n_1, ..., n_m \geq 1$ *by*

$$\Pr\{ N_{A_1,1} = n_1, ..., N_{A_1,m} = n_m \} =$$

$$\mathbf{v}^T \mathbf{Q}_{A_1A_1}^{n_1-1} \mathbf{Q}_{A_1A_2} \prod_{i=2}^{m} \left\{ (\mathbf{I} - \mathbf{Q}_{A_2A_2})^{-1} \mathbf{Q}_{A_2A_1} \mathbf{Q}_{A_1A_1}^{n_i-1} \mathbf{Q}_{A_1A_2} \right\} \mathbf{1}. \tag{2.28}$$

PROOF OF COROLLARY 2.3. Apply Theorem 2.1 to the Markov chain on $A_1 \cup A_2 \cup \{\omega\}$ which is identical to Z on $A_1 \cup A_2$ and all transition probabilities of which to ω are equal to zero. (2.28) is then seen to be (2.2) by $(\mathbf{I} - \mathbf{Q}_{A_1A_1}) \mathbf{1} = \mathbf{Q}_{A_1A_2} \mathbf{1}$. Similarly, (2.3) and (2.4) are seen to vanish. ∎

In the next two corollaries, the second of which is due to Rubino and Sericola [RUB3], the distribution of individual sojourn times will be established.

COROLLARY 2.4. *Under the assumptions and with the notation of Theorem 2.1, the probability mass function of the mth sojourn in* A_1, $N_{A_1,m}$, *is given by*

$$\Pr\{ N_{A_1,m} = n \} = \begin{cases} 1 - \mathbf{v}^T \mathbf{H}^{m-1} \mathbf{1} & \text{for } n = 0, \\ \mathbf{v}^T \mathbf{H}^{m-1} \mathbf{Q}_{A_1A_1}^{\ n-1} (\mathbf{I} - \mathbf{Q}_{A_1A_1}) \mathbf{1} & \text{for } n \geq 1, \end{cases} \quad (2.29)$$

with

$$\mathbf{H} = \mathbf{H}(A_1, A_2) = (\mathbf{I} - \mathbf{Q}_{A_1A_1})^{-1} \mathbf{Q}_{A_1A_2} (\mathbf{I} - \mathbf{Q}_{A_2A_2})^{-1} \mathbf{Q}_{A_2A_1}. \quad (2.30)$$

COROLLARY 2.5. *Under the assumptions and with the notation of Corollary 2.3, the probability mass function of the mth sojourn in* A_1, $N_{A_1,m}$, *is given for* $n \geq 1$ *by*

$$\Pr\{ N_{A_1,m} = n \} = \mathbf{v}^T \mathbf{H}^{m-1} \mathbf{Q}_{A_1A_1}^{\ n-1} (\mathbf{I} - \mathbf{Q}_{A_1A_1}) \mathbf{1}, \quad (2.31)$$

where $\mathbf{H}$ *is defined by (2.30).*

PROOF OF COROLLARIES 2.4 AND 2.5. (2.31) follows from (2.29) by noting that for an irreducible chain $\mathbf{H}$ is a stochastic matrix and thus the first expression on the right hand side of (2.29) evaluates to zero. ($\mathbf{H}$ is seen to have only non-negative entries by a power series expansion of the matrices $(\mathbf{I} - \mathbf{Q}_{A_1A_1})^{-1}$ and $(\mathbf{I} - \mathbf{Q}_{A_2A_2})^{-1}$). Using $\mathbf{Q}_{A_1A_1}^{\ k} \to \mathbf{0}$ for $k \to +\infty$, see Theorem 2.1 (a), (2.29) for $n \geq 1$ is seen from

$$\Pr\{ N_{A_1,m} = n \} = \Pr\{ N_{A_1,1} \geq 1, ..., N_{A_1,m-1} \geq 1, N_{A_1,m} = n \} =$$

$$\sum_{n_1,...,n_{m-1}=1}^{\infty} \Pr\{ N_{A_1,1} = n_1, ..., N_{A_1,m-1} = n_{m-1}, N_{A_1,m} = n \} =$$

$$\mathbf{v}^T (\mathbf{I} - \mathbf{Q}_{A_1A_1})^{-1} \left\{ \mathbf{Q}_{A_1A_2} (\mathbf{I} - \mathbf{Q}_{A_2A_2})^{-1} \mathbf{Q}_{A_2A_1} (\mathbf{I} - \mathbf{Q}_{A_1A_1})^{-1} \right\}^{m-2} \times$$

$$\mathbf{Q}_{A_1A_2} (\mathbf{I} - \mathbf{Q}_{A_2A_2})^{-1} \mathbf{Q}_{A_2A_1} \mathbf{Q}_{A_1A_1}^{\ n-1} (\mathbf{I} - \mathbf{Q}_{A_1A_1}) \mathbf{1} =$$

$$\mathbf{v}^T \mathbf{H}^{m-1} \mathbf{Q}_{A_1A_1}^{\ n-1} (\mathbf{I} - \mathbf{Q}_{A_1A_1}) \mathbf{1}. \quad (2.32)$$

(2.29) for $n = 0$ is obtained by summing (2.32) over $n \geq 1$. ∎

2.1.2 Distribution theory for variables related to the sojourn time vector

Corollary 2.4 allows us to obtain the probability mass function of the number of visits to A_1 until absorption, a result which was first established by Rubino and Sericola [RUB4].

COROLLARY 2.6. *Under the assumptions and with the notation of Theorem 2.1, the probability mass function of* M_{A_1}, *the number of visits in* A_1 *until absorption is given by*

$$\Pr\{ M_{A_1} = m \} = \begin{cases} 1 - \mathbf{v}^T \mathbf{1} & \text{for } m = 0, \\ \mathbf{v}^T \mathbf{H}^{m-1} (\mathbf{I} - \mathbf{H}) \mathbf{1} & \text{for } m \geq 1, \end{cases}$$

where $\mathbf{H}$ *is defined by (2.30). Notice in particular that*

$$\Pr\{ M_{A_1} \geq m \} = \mathbf{v}^T \mathbf{H}^{m-1} \mathbf{1} \text{ for } m \geq 1. \tag{2.33}$$

PROOF OF COROLLARY 2.6. By virtue of the equivalence $M_{A_1} = 0 \Leftrightarrow N_{A_1,1} = 0$, we have

$$\Pr\{ M_{A_1} = 0 \} = \Pr\{ N_{A_1,1} = 0 \} = 1 - \mathbf{v}^T \mathbf{1}$$

by (2.29). For $m \geq 1$, the equivalence $M_{A_1} \geq m \Leftrightarrow N_{A_1,m} > 0$, again in conjunction with (2.29), gives

$$\Pr\{ M_{A_1} = m \} = \Pr\{ M_{A_1} \geq m \} - \Pr\{ M_{A_1} \geq m + 1 \} =$$

$$\Pr\{ N_{A_1,m} > 0 \} - \Pr\{ N_{A_1,m+1} > 0 \} = \mathbf{v}^T \mathbf{H}^{m-1} \mathbf{1} - \mathbf{v}^T \mathbf{H}^m \mathbf{1} = \mathbf{v}^T \mathbf{H}^{m-1} (\mathbf{I} - \mathbf{H}) \mathbf{1}.$$

∎

For a system modelled by an absorbing Markov chain, a measure of the total amount of work delivered by the system until final breakdown is the total 'time' spent in the set of good states A_1, say, until absorption. This is the limit of the sum of the first m sojourns of Z in A_1, i.e.,

$$NS_{A_1,m} = N_{A_1,1} + \ldots + N_{A_1,m} \to NS_{A_1,\infty} = \sum_{k=1}^{\infty} N_{A_1,k}, \text{ for } m \to \infty. \tag{2.34}$$

COROLLARY 2.7. *The probability generating function of the total 'time'* $NS_{A_1,\infty} = N_{A_1,1} + N_{A_1,2} + \ldots$ *spent in the set* A_1 *by an absorbing Markov chain Z with state space* $S = A_1 \cup A_2 \cup \{\omega\}$ *is given by*

$$E\left[z^{NS_{A_1,\infty}}\right] = 1 - \mathbf{v}^T \mathbf{1} + \mathbf{v}^T (\mathbf{I} - z \mathbf{M})^{-1} (\mathbf{I} - \mathbf{M}) \mathbf{1} z, \tag{2.35}$$

where $\mathbf{v}$ *is defined by (2.1) and*

$$\mathbf{M} = \mathbf{M}(A_1, A_2) = \mathbf{Q}_{A_1A_1} + \mathbf{Q}_{A_1A_2} (\mathbf{I} - \mathbf{Q}_{A_2A_2})^{-1} \mathbf{Q}_{A_2A_1}. \tag{2.36}$$

(2.35) holds for all points in the closed unit disc of the complex plane, i.e., $z \in \mathbb{C}, |z| \le 1$.

PROOF OF COROLLARY 2.7. From Theorem 2.1, the probability generating function of $NS_{A_1,m}$ is given by

$$E\left[z^{NS_{A_1,m}}\right] = \sum_{n_1,\ldots,n_m=0}^{\infty} z^{n_1 + \ldots + n_m} \Pr\{ N_{A_1,1} = n_1, \ldots, N_{A_1,m} = n_m \} =$$

$$\mathbf{v}^T (\mathbf{I} - z\,\mathbf{Q}_{A_1A_1})^{-1} \left\{ \mathbf{Q}_{A_1A_2} (\mathbf{I} - \mathbf{Q}_{A_2A_2})^{-1} \mathbf{Q}_{A_2A_1} (\mathbf{I} - z\,\mathbf{Q}_{A_1A_1})^{-1} \right\}^{m-1} (\mathbf{I} - \mathbf{Q}_{A_1A_1})\,\mathbf{1}\, z^m +$$

$$\sum_{m'=1}^{m-1} \mathbf{v}^T (\mathbf{I} - z\,\mathbf{Q}_{A_1A_1})^{-1} \left\{ \mathbf{Q}_{A_1A_2} (\mathbf{I} - \mathbf{Q}_{A_2A_2})^{-1} \mathbf{Q}_{A_2A_1} (\mathbf{I} - z\,\mathbf{Q}_{A_1A_1})^{-1} \right\}^{m'-1} \times$$

$$\left\{ \mathbf{I} - [\, \mathbf{Q}_{A_1A_1} + \mathbf{Q}_{A_1A_2} (\mathbf{I} - \mathbf{Q}_{A_2A_2})^{-1} \mathbf{Q}_{A_2A_1} \right\}\,\mathbf{1}\, z^{m'} + 1 - \mathbf{v}^T \mathbf{1} =$$

$$1 - \mathbf{v}^T \mathbf{1} + \mathbf{v}^T (z\,\mathbf{H}(z))^{m-1} (\mathbf{I} - z\,\mathbf{Q}_{A_1A_1})^{-1} (\mathbf{I} - \mathbf{Q}_{A_1A_1})\,\mathbf{1}\, z +$$

$$\sum_{m'=1}^{m-1} \mathbf{v}^T (z\,\mathbf{H}(z))^{m'-1} (\mathbf{I} - z\,\mathbf{Q}_{A_1A_1})^{-1} \left\{ \mathbf{I} - [\, \mathbf{Q}_{A_1A_1} + \mathbf{Q}_{A_1A_2} (\mathbf{I} - \mathbf{Q}_{A_2A_2})^{-1} \mathbf{Q}_{A_2A_1} \right\}\,\mathbf{1}\, z,$$

$$\tag{2.37}$$

with

$$\mathbf{H}(z) = (\mathbf{I} - z\,\mathbf{Q}_{A_1A_1})^{-1} \mathbf{Q}_{A_1A_2} (\mathbf{I} - \mathbf{Q}_{A_2A_2})^{-1} \mathbf{Q}_{A_2A_1}.$$

(Notice that for $|z| \le 1$ we have $(z\,\mathbf{Q}_{A_1A_1})^k \to \mathbf{0}$ as $k \to +\infty$.) If it can be shown that

$$(z\,\mathbf{H}(z))^m \to \mathbf{0} \text{ as } m \to +\infty, \tag{2.38}$$

then (2.37) tends for $m \to +\infty$ to

$$1 - \mathbf{v}^T \mathbf{1} + \mathbf{v}^T (\mathbf{I} - z\,\mathbf{H}(z))^{-1} (\mathbf{I} - z\,\mathbf{Q}_{A_1A_1})^{-1} (\mathbf{I} - \mathbf{Q}_{A_1A_1}) (\mathbf{I} - \mathbf{H}(1))\,\mathbf{1}\, z =$$

$$1 - \mathbf{v}^T \mathbf{1} + \mathbf{v}^T \left\{ (\mathbf{I} - z\,\mathbf{Q}_{A_1A_1}) (\mathbf{I} - z\,\mathbf{H}(z)) \right\}^{-1} (\mathbf{I} - \mathbf{Q}_{A_1A_1}) (\mathbf{I} - \mathbf{H}(1))\,\mathbf{1}\, z =$$

$$1 - \mathbf{v}^T \mathbf{1} + \mathbf{v}^T (\mathbf{I} - z\,\mathbf{M})^{-1} (\mathbf{I} - \mathbf{M})\,\mathbf{1}\, z.$$

To see finally (2.38), first we note that by a power series expansion of $\mathbf{H}(z)$ we have element-wise $| z\, \mathbf{H}(z) | \leq |z|\, \mathbf{H}(|z|)$, and therefore, again element-wise, $|(z\, \mathbf{H}(z))^m| \leq |z|^m\, \mathbf{H}(|z|)^m \leq (\mathbf{H}(1))^m$. It is $\mathbf{H}(1) = \mathbf{H}$ by (2.30), and $\mathbf{v}^T\, \mathbf{H}^{m-1}\, \mathbf{1} \to 0$ as $m \to +\infty$ by (2.33) for *any* stochastic vector $\mathbf{v}$, since $\mathbf{v} = \boldsymbol{\alpha}_{A_1}$ whenever $\boldsymbol{\alpha}_{A_2} = \mathbf{0}$. Thus (2.38) holds. ∎

The probability mass function of $NS_{A_1,\infty}$ immediately follows from Corollary 2.7 by a power series expansion of (2.35) about $z = 0$.

COROLLARY 2.8. *The probability mass function of* $NS_{A_1,\infty}$ *is given by*

$$\Pr\{ NS_{A_1,\infty} = n \} = \begin{cases} 1 - \mathbf{v}^T\, \mathbf{1} & \text{if } n = 0, \\[2mm] \mathbf{v}^T\, \mathbf{M}^{(n-1)}\, (\mathbf{I} - \mathbf{M})\, \mathbf{1} & \text{if } n \geq 1, \end{cases}$$

where $\mathbf{v}$ *and* $\mathbf{M}$ *are defined by (2.1) and (2.36) respectively.* ∎

It is easily seen from (2.35) by induction that for $k \geq 1$

$$\frac{d^k}{dz^k} E[z^{NS_{A_1,\infty}}] = k!\, \mathbf{v}^T\, (\mathbf{I} - z\, \mathbf{M})^{-(k+1)}\, \mathbf{M}^{(k-1)}\, (\mathbf{I} - \mathbf{M})\, \mathbf{1}. \tag{2.39}$$

From (2.39) we have the following for the factorial moments of $NS_{A_1,\infty}$.

COROLLARY 2.9. *The kth factorial moment of* $NS_{A_1,\infty}$ *is given for* $k = 1, 2, \ldots$ *by*

$$FM_k(NS_{A_1,\infty}) = E[NS_{A_1,\infty}\, (NS_{A_1,\infty} - 1) \ldots (NS_{A_1,\infty} - k + 1)] =$$

$$k!\, \mathbf{v}^T\, (\mathbf{I} - \mathbf{M})^{-(k+1)}\, \mathbf{M}^{(k-1)}\, (\mathbf{I} - \mathbf{M})\, \mathbf{1},$$

where $\mathbf{v}$ *and* $\mathbf{M}$ *are defined by (2.1) and (2.36) respectively.* ∎

Having obtained the distribution of $NS_{A_1,\infty}$ in Corollary 2.8 by a power series expansion of its probability generating function, it would seem reasonable to suppose that the same method now applied to (2.37) would yield the probability mass function of $NS_{A_1,m}$. Due to its relatively complicated structure, however, (2.37) does not lend itself to such an approach. Instead, the method employed is the 'equivalent auxiliary Markov chain' technique already encountered in the proof of Theorem 2.1. Let us define for $i, n \geq 1$ the square matrix $\mathbf{E}(n, i)$, of size n, by its entries

$$e(n, i)_{jk} = \begin{cases} 1 & \text{for } k - j + 1 = i, \\[2mm] 0 & \text{otherwise,} \end{cases} \tag{2.40}$$

i.e., $\mathbf{E}(n, 1) = \mathbf{I}$ (identity matrix),

$$\mathbf{E}(n, 2) = \begin{bmatrix} 0 & 1 & 0 & 0 & 0 & 0 & 0 & 0 & 0 \\ 0 & 0 & 1 & 0 & 0 & 0 & 0 & 0 & 0 \\ 0 & 0 & 0 & 1 & 0 & 0 & 0 & 0 & 0 \\ 0 & 0 & 0 & 0 & 1 & 0 & 0 & 0 & 0 \\ \cdot & \cdot & \cdot & \cdot & \cdot & \cdot & \cdot & \cdot \\ \cdot & \cdot & \cdot & \cdot & \cdots & \cdot & \cdot & \cdot \\ \cdot & \cdot & \cdot & \cdot & \cdot & \cdot & \cdot & \cdot \\ 0 & 0 & 0 & 0 & 0 & 0 & 1 & 0 & 0 \\ 0 & 0 & 0 & 0 & 0 & 0 & 0 & 1 & 0 \\ 0 & 0 & 0 & 0 & 0 & 0 & 0 & 0 & 1 \\ 0 & 0 & 0 & 0 & 0 & 0 & 0 & 0 & 0 \end{bmatrix}, \quad \mathbf{E}(n, 3) = \begin{bmatrix} 0 & 0 & 1 & 0 & 0 & 0 & 0 & 0 \\ 0 & 0 & 0 & 1 & 0 & 0 & 0 & 0 \\ 0 & 0 & 0 & 0 & 1 & 0 & 0 & 0 \\ 0 & 0 & 0 & 0 & 0 & 0 & 0 & 0 \\ \cdot & \cdot & \cdot & \cdot & \cdot & \cdot & \cdot \\ \cdot & \cdot & \cdot & \cdot & \cdots & \cdot & \cdot \\ \cdot & \cdot & \cdot & \cdot & \cdot & \cdot & \cdot \\ 0 & 0 & 0 & 0 & 0 & 0 & 1 & 0 \\ 0 & 0 & 0 & 0 & 0 & 0 & 0 & 1 \\ 0 & 0 & 0 & 0 & 0 & 0 & 0 & 0 \\ 0 & 0 & 0 & 0 & 0 & 0 & 0 & 0 \end{bmatrix}, \quad \ldots$$

(Note that $\mathbf{E}(n, i) = \mathbf{0}$ for $i \geq n + 1$.) We have now the following for $NS_{A_1,m}$.

COROLLARY 2.10. *Let Z be an irreducible (or absorbing) Markov chain on $A_1 \cup A_2$ (or on $A_1 \cup A_2 \cup \{\omega\}$ in which case ω is the absorbing state of Z). Let Z be started according to the stochastic vector $(\boldsymbol{\alpha}_{A_1}{}^T, \boldsymbol{\alpha}_{A_2}{}^T)^T$ on $A_1 \cup A_2$. Then, the probability mass function of the total 'time' spent by Z in A_1 during its first m sojourns in A_1 is given by*

$$\Pr\{ NS_{A_1,m} = n \} = \begin{cases} 1 - \tilde{\mathbf{v}}_m{}^T & \text{if } n = 0, \\[2ex] \tilde{\mathbf{v}}_m{}^T \, \tilde{\mathbf{M}}_m{}^{(n-1)} \, (\mathbf{I} - \tilde{\mathbf{M}}_m) \, \mathbf{1} & \text{if } n \geq 1, \end{cases} \tag{2.41}$$

where

$$\tilde{\mathbf{v}}_m{}^T = (\boldsymbol{\alpha}_{A_2}{}^T, \mathbf{0}, \ldots, \mathbf{0}) \, (\mathbf{I} - \mathbf{Q}_{A_2 A_2} \otimes \mathbf{E}(m, 1))^{-1} \, \mathbf{Q}_{A_2 A_1} \otimes \mathbf{E}(m, 1) + (\boldsymbol{\alpha}_{A_1}{}^T, \mathbf{0}, \ldots, \mathbf{0}),$$

$$\tag{2.42}$$

$$\tilde{\mathbf{M}}_m = \mathbf{Q}_{A_1 A_1} \otimes \mathbf{E}(m, 1) + \mathbf{Q}_{A_1 A_2} \otimes \mathbf{E}(m, 2) \, (\mathbf{I} - \mathbf{Q}_{A_2 A_2} \otimes \mathbf{E}(m, 1))^{-1} \, \mathbf{Q}_{A_2 A_1} \otimes \mathbf{E}(m, 1);$$

$$\tag{2.43}$$

$\otimes$ *stands for the right direct product of matrices, see (1.5). The inverse matrices on the right hand side of (2.42) and (2.43) exist. Furthermore, the row vectors on the right hand side of (2.42) contain m - 1 zero vectors of size $| A_2 |$ and $| A_1 |$ respectively.*

PROOF OF COROLLARY 2.10. Consider the auxiliary absorbing Markov chain X defined by its state-transition diagram in Figure 2.3 below.

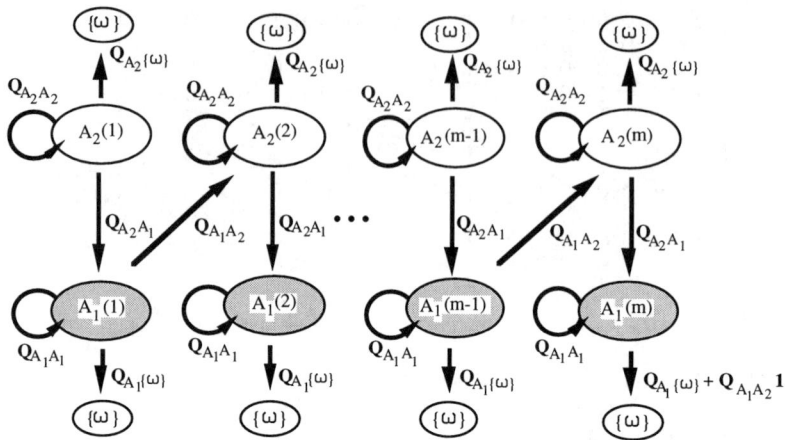

Figure 2.3 The state-transition diagram of the auxiliary Markov chain X

($\mathscr{A}_2$ = unshaded sets of states, $\mathscr{A}_1$ = shaded sets of states, ω = absorbing state.)

Its state space is $\mathscr{A}_1 \cup \mathscr{A}_2 \cup \{\omega\}$ with $\mathscr{A}_1 = A_1(1) \cup ... \cup A_1(m)$, and $\mathscr{A}_2 = A_2(1) \cup ... \cup A_2(m)$, where $A_i(j)$, $j = 1, ..., m$, are disjoint instances of A_i, $i = 1, 2$. Put $\mathbf{Q}_{A_1\{\omega\}} = \mathbf{0}$ and $\mathbf{Q}_{A_2\{\omega\}} = \mathbf{0}$ if Z is not absorbing. X is started in $A_1(1) \cup A_2(1)$ according to $(\boldsymbol{\alpha}_{A_1}{}^T, \boldsymbol{\alpha}_{A_2}{}^T)^T$. It is easily seen that the transition probability matrix $\mathbf{P}$ of X is given by

$$
\mathbf{P} = \begin{array}{c} \\ \mathscr{A}_1 \\ \mathscr{A}_2 \\ \{\omega\} \end{array}
\begin{array}{ccc} \mathscr{A}_1 & \mathscr{A}_2 & \{\omega\} \end{array}
\left[\begin{array}{c|c|c}
\mathbf{Q}_{A_1A_1} \otimes \mathbf{E}(m, 1) & \mathbf{Q}_{A_1A_2} \otimes \mathbf{E}(m, 2) & \mathbf{P}_{\mathscr{A}_1\{\omega\}} \\ \hline
\mathbf{Q}_{A_2A_1} \otimes \mathbf{E}(m, 1) & \mathbf{Q}_{A_2A_2} \otimes \mathbf{E}(m, 1) & \mathbf{P}_{\mathscr{A}_2\{\omega\}} \\ \hline
\mathbf{0} & \mathbf{0} & 1
\end{array} \right], \qquad (2.44)
$$

where $\mathbf{P}_{\mathscr{A}_1\{\omega\}}$ and $\mathbf{P}_{\mathscr{A}_2\{\omega\}}$ are defined by the condition that $\mathbf{P}$ is a stochastic matrix. (2.41) follows from Corollary 2.8 by noting that the following equality of *distributions* holds: $\mathscr{L}(NS_{A_1,m}) = \mathscr{L}(NS_{\mathscr{A}_1,\infty})$. (2.42) and (2.43) follow from (2.1) and (2.36) respectively by (2.44) with $\tilde{\mathbf{v}}_m = \mathbf{v}(\mathscr{A}_1, \mathscr{A}_2)$ and $\mathbf{M}_m = \mathbf{M}(\mathscr{A}_1, \mathscr{A}_2)$. ∎

NOTE. For m = 1, the probability mass function of $NS_{A_1,m} = N_{A_1,1}$ is available from Corollary 2.4, too. In this case (2.29) and (2.41) coincide since $\tilde{\mathbf{v}}_1 = \mathbf{v}$, and $\mathbf{M}_1 = \mathbf{Q}_{A_1A_1} \otimes \mathbf{E}(1, 1) = \mathbf{Q}_{A_1A_1}$ because of $\mathbf{E}(1, 2) = \mathbf{0}$.

An alternative view to that taken in (2.34) is that the total amount of time spent by Z in A_1 until absorption is the limit as $n \to \infty$ of the cumulative time spent by Z in A_1 during the first n 'time instances', i.e.,

$$
NS_{A_1,\infty} = L_{A_1,\infty} = \lim_{n\to\infty} L_{A_1,n}, \qquad (2.45)
$$

where

$$L_{A_1,n} = \sum_{i=0}^{n-1} I_{\{ Z_i \in A_1 \}},$$ (2.46)

with $I_{\{...\}}$ standing for the indicator function of the subscript set $\{...\}$. The variable $L_{A_1,n}$ in (2.46) will be termed the *finite-horizon cumulative sojourn time* of Z in A_1 during $\{ 0, 1, ..., n - 1 \}$. This variable is finite irrespective of whether or not Z is absorbing. We now turn our attention to the probability generating and probability mass functions and factorial moments of $L_{A_1,n}$.

In what follows the Markov chain Z may be absorbing, irreducible, or possibly neither of these two.

COROLLARY 2.11. *For* $n \geq 2$, *the probability generating function of* $L_{A_1,n}$ *from (2.46) is given, for* $z \in \mathbb{C}, |z| \leq 1$, *by*

$$E\left[z^{L_{A_1,n}}\right] = 1 - v_n^T 1 + v_n^T (I - z M_n)^{-1} (I - M_n) 1 z,$$ (2.47)

where v_n *and* M_n *are respectively defined by*

$$v_n^T = (v_n(A_1, A_2))^T = (\alpha_{A_2}^T, 0, ..., 0) (I - Q_{A_2A_2} \otimes E(n, 2))^{-1} Q_{A_2A_1} \otimes E(n, 2) +$$

$$+ (\alpha_{A_1}^T, 0, ..., 0),$$ (2.48)

$$M_n = M_n(A_1, A_2) =$$

$$Q_{A_1A_1} \otimes E(n, 2) + Q_{A_1A_2} \otimes E(n, 2) (I - Q_{A_2A_2} \otimes E(n, 2))^{-1} Q_{A_2A_1} \otimes E(n, 2);$$ (2.49)

$\otimes$ *stands for the right direct product of matrices, see (1.5), and* $E(n, 2)$ *is defined by (2.40). The inverse matrices on the right hand side of (2.48) and (2.49) exist. Furthermore, the row vectors on the right hand side of (2.48) contain* $n - 1$ *zero vectors of size* $| A_2 |$ *and* $| A_1 |$ *respectively.*

PROOF OF COROLLARY 2.11. Corollary 2.7 will be applied to an equivalent auxiliary Markov chain $X = \{ X_0, X_1, ... \}$ which is such that at the most after $n - 1$ transitions it is absorbed into ω. Figure 2.4 (overleaf) shows the state-transition diagram of X. Its state space is $\mathcal{S} = \mathcal{A}_1 \cup \mathcal{A}_2 \cup \{\omega\}$, where $\mathcal{A}_1$ and $\mathcal{A}_2$ are defined respectively by $\mathcal{A}_1 = A_1(0) \cup A_1(1) \cup ... \cup A_1(n-1)$, $\mathcal{A}_2 = A_2(0) \cup A_2(0) \ ... \cup A_2(n-1)$, such that the sets of states $A_1(0), ..., A_1(n-1)$ and $A_2(0), ..., A_2(n-1)$ are disjoint instances of A_1 and A_2 respectively. X is started in $A_1(0) \cup A_2(0)$ according to $(\alpha_{A_1}^T, \alpha_{A_2}^T)^T$, i.e., the $\mathcal{A}_i$ - component of the initial probability vector of X is $(\alpha_{A_i}^T, 0, ..., 0)^T$, $i = 1, 2$. It is easily verified that a concise representation of the transition probability matrix P of X is given by

34

$$
P = \begin{array}{c} \\ \mathscr{A}_1 \\ \mathscr{A}_2 \\ \{\omega\} \end{array}
\begin{array}{ccc}
\mathscr{A}_1 \qquad\qquad \mathscr{A}_2 \qquad\qquad \{\omega\}
\end{array}
\left[
\begin{array}{c:c:c}
\mathbf{Q}_{A_1A_1} \otimes \mathbf{E}(n,2) & \mathbf{Q}_{A_1A_2} \otimes \mathbf{E}(n,2) & \mathbf{P}_{\mathscr{A}_1\{\omega\}} \\
\hdashline
\mathbf{Q}_{A_2A_1} \otimes \mathbf{E}(n,2) & \mathbf{Q}_{A_2A_2} \otimes \mathbf{E}(n,2) & \mathbf{P}_{\mathscr{A}_2\{\omega\}} \\
\hdashline
\mathbf{0} & \mathbf{0} & 1
\end{array}
\right], \qquad (2.50)
$$

where $\mathbf{P}_{\mathscr{A}_1\{\omega\}}$ and $\mathbf{P}_{\mathscr{A}_2\{\omega\}}$ are defined by the condition that $\mathbf{P}$ is a stochastic matrix. (2.47) now immediately follows by Corollary 2.7 via the equality of *distributions* $\mathscr{L}(L_{A_1,n}) = \mathscr{L}(NS_{\mathscr{A}_1,\infty})$. (2.48) and (2.49) follow from (2.1) and (2.36) respectively by (2.50). ∎

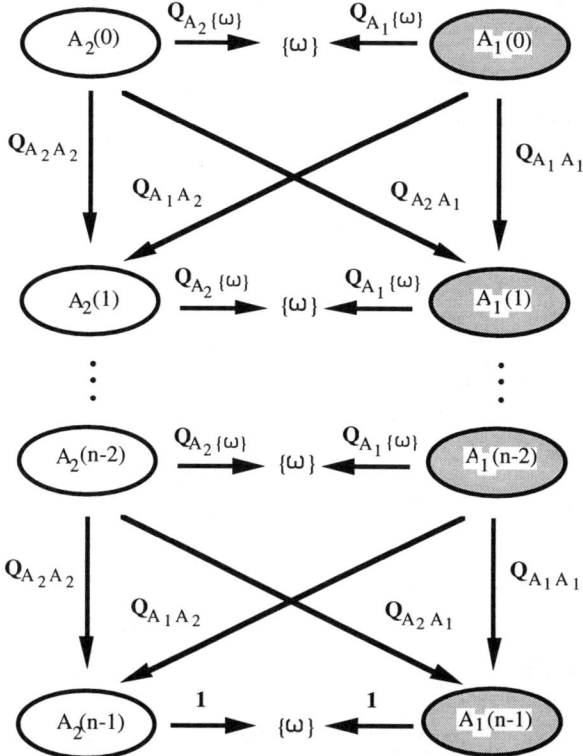

Figure 2.4 The state-transition diagram of the auxiliary Markov chain X
($\mathscr{A}_2$ = unshaded sets of states, $\mathscr{A}_1$ = shaded sets of states, ω = absorbing state.)

The following two corollaries readily follow from Corollary 2.11.

COROLLARY 2.12. *The probability mass function of* $L_{A_1,1}$ *is given by*

$$\Pr\{\, L_{A_1,1} = \ell \,\} = \begin{cases} \boldsymbol{\alpha}_{A_2}{}^T \mathbf{1} & \text{if } \ell = 0, \\ \boldsymbol{\alpha}_{A_1}{}^T \mathbf{1} & \text{if } \ell = 1, \\ 0 & \text{if } \ell \geq 2. \end{cases} \tag{2.51}$$

For $n \geq 2$*, the probability mass function of* $L_{A_1,n}$ *is given by*

$$\Pr\{\, L_{A_1,n} = \ell \,\} = \begin{cases} 1 - \mathbf{v}_n{}^T \mathbf{1} & \text{if } \ell = 0, \\ \mathbf{v}_n{}^T \mathbf{M}_n{}^{(\ell-1)} (\mathbf{I} - \mathbf{M}_n) \mathbf{1} & \text{if } \ell \in \{1, \ldots, n\}, \\ 0 & \text{if } \ell \geq n+1. \end{cases} \tag{2.52}$$

COROLLARY 2.13. *The kth factorial moment of* $L_{A_1,1}$ *is given by*

$$FM_k(L_{A_1,1}) = E[L_{A_1,1} (L_{A_1,1} - 1)\ldots(L_{A_1,1} - k + 1)] = \begin{cases} \boldsymbol{\alpha}_{A_1}{}^T \mathbf{1} & \text{if } k = 1, \\ 0 & \text{if } k \geq 2. \end{cases} \tag{2.53}$$

whereas for $n \geq 2$ *we have*

$$FM_k(L_{A_1,n}) = E[L_{A_1,n} (L_{A_1,n} - 1) \ldots (L_{A_1,n} - k + 1)] =$$

$$k! \, \mathbf{v}_n{}^T (\mathbf{I} - \mathbf{M}_n)^{-(k+1)} \mathbf{M}_n{}^{(k-1)} (\mathbf{I} - \mathbf{M}_n) \mathbf{1}. \tag{2.54}$$

PROOF OF COROLLARIES 2.12 AND 2.13. (2.51) and (2.53) obviously hold. For $n \geq 2$, it is seen from (2.47) by induction that for $k \geq 1$

$$\frac{d^k}{dz^k} E[z^{L_{A_1,n}}] = k! \, \mathbf{v}_n{}^T (\mathbf{I} - z \mathbf{M}_n)^{-(k+1)} \mathbf{M}_n{}^{(k-1)} (\mathbf{I} - \mathbf{M}_n) \mathbf{1}. \tag{2.55}$$

Putting $z = 0$ in (2.55) shows that (2.51) holds. (2.53) follows from (2.55) with $z = 1$. ∎

NOTE. Corollaries 2.11 - 2.13 first appeared in [CSE5]. The distribution of the continuous-time counterpart of $L_{A_1,n}$ was considered by Sericola [SER]. Closed form expressions for the distribution of the variable $L_{A_1,n}$ seem, however, not to have been considered in the literature prior to [CSE5]. The closest result to ours is in [IOS] pp 162, where a formula is given for the characteristic function of the vector $\{\, \nu_n(s) : s \in S \,\}$ with $\nu_n(s)$ standing for the 'occupation time in state s during the first n steps'. From there, of course, the characteristic function of

$$L_{A_1,1} = \sum_{s \in A_1} \nu_n(s)$$

is available.

The variable M_{A_1} in Corollary 2.6 is the number of visits by Z to A_1 until absorption. An alternative interpretation of M_{A_1} is given by

$$M_{A_1} = K_{A_1,\infty} = \lim_{n \to \infty} K_{A_1,n}, \tag{2.56}$$

where

$$K_{A_1,n} = I_{\{ Z_0 \in A_1 \}} + \sum_{i=0}^{n-1} I_{\{ Z_{i-1} \in A_2, Z_i \in A_1 \}}, \ n \geq 1, \tag{2.57}$$

is the number of visits to A_1 by $\{ Z_0, ..., Z_{n-1} \}$, i.e., by Z during the first n 'time' instances. Notice the respective correspondence between (2.45), (2.56) and (2.46), (2.57): $K_{A_1,n}$ is the the finite-horizon counterpart of M_{A_1} just as $L_{A_1,n}$ was the finite-horizon version of $NS_{A_1,\infty}$. Similarly to $L_{A_1,n}$, $K_{A_1,n}$ is also finite even if Z is not absorbing. The next corollary deals with the probability mass function of $K_{A_1,n}$. Z may be any Markov chain, i.e., it need not be absorbing or irreducible.

COROLLARY 2.14. *The probability mass function of* $K_{A_1,1}$ *is given by*

$$\Pr\{ K_{A_1,n} = k \} = \begin{cases} \boldsymbol{\alpha}_{A_2}^T \mathbf{1} & \text{if } k = 0, \\ \boldsymbol{\alpha}_{A_1}^T \mathbf{1} & \text{if } k = 1, \\ 0 & \text{if } k \geq 2. \end{cases}$$

For $n \geq 2$, *the probability mass function of* $K_{A_1,n}$ *is given by*

$$\Pr\{ K_{A_1,n} = k \} = \begin{cases} 1 - \mathbf{v}_n^T \mathbf{1} & \text{if } k = 0, \\ \mathbf{v}_n^T \mathbf{H}_n^{(k-1)} (\mathbf{I} - \mathbf{H}_n) \mathbf{1} & \text{if } k \in \{ 1, ..., \lfloor (n+1)/2 \rfloor \}, \\ 0 & \text{if } k \geq \lfloor (n+1)/2 \rfloor, \end{cases}$$

where $\lfloor (n+1)/2 \rfloor$ *stands for the integer part of* $(n+1)/2$, *i.e., it is the largest integer not exceeding* $(n+1)/2$. $\mathbf{v}_n$ *and* $\mathbf{H}_n$ *are respectively defined by (2.48) and*

$$\mathbf{H}_n = (\mathbf{I} - \mathbf{Q}_{A_1A_1} \otimes \mathbf{E}(n, 2))^{-1} \mathbf{Q}_{A_1A_2} \otimes \mathbf{E}(n, 2) \times$$

$$(\mathbf{I} - \mathbf{Q}_{A_2A_2} \otimes \mathbf{E}(n, 2))^{-1} \mathbf{Q}_{A_2A_1} \otimes \mathbf{E}(n, 2). \tag{2.58}$$

PROOF OF COROLLARY 2.14. Corollary 2.14 follows from Corollary 2.6 along the lines of the proof of Corollary 2.11 with the same auxiliary Markov chain X; see Figure 2.4. ∎

In both (2.49) and (2.58), matrices of the form $\mathbf{I} - \mathbf{R} \otimes \mathbf{E}(n, 2)$ need to be inverted. These matrices may be very large and hence it is useful from the computational point of view to have a simple expression for their inverse. This is accomplished in the following proposition.

PROPOSITION 2.15. *The matrix* $\mathbf{I} - \mathbf{R} \otimes \mathbf{E}(n, 2)$ *is invertible for any square matrix* $\mathbf{R}$, *and the inverse is*

$$(\mathbf{I} - \mathbf{R} \otimes \mathbf{E}(n, 2))^{-1} = \sum_{i=0}^{n-1} \mathbf{R}^i \otimes \mathbf{E}(n, i{+}1). \tag{2.59}$$

PROOF OF PROPOSITION 2.15. For $i, j \geq 1$, we have $\mathbf{E}(n, i)\,\mathbf{E}(n, j) = \mathbf{E}(n, i{+}j{-}1)$; thus, for any two compatible matrices $\mathbf{R}_1$ and $\mathbf{R}_2$,

$$(\mathbf{R}_1 \otimes \mathbf{E}(n, i))\,(\mathbf{R}_2 \otimes \mathbf{E}(n, j)) = (\mathbf{R}_1\,\mathbf{R}_2) \otimes \mathbf{E}(n, i{+}j{-}1). \tag{2.60}$$

From (2.60), for any square matrix $\mathbf{R}$, it is seen that

$$(\mathbf{R} \otimes \mathbf{E}(n, 2))^k = \mathbf{R}^k \otimes \mathbf{E}(n, k{+}1), \ k = 0, 1, \dots . \tag{2.61}$$

Also, for any matrix $\mathbf{R}$ and $i \geq n + 1$, we have

$$\mathbf{R} \otimes \mathbf{E}(n, i) = \mathbf{0}. \tag{2.62}$$

Now (2.61) and (2.62) imply (2.59). ∎

2.1.3 The joint distribution of sojourn times in A_1 and A_2 by the generalised renewal argument

The reasoning thus far has been mainly of a probabilistic nature. Probability generating functions were by and large used only as a tool for deducing the distribution of the sum of the components of a random vector from their joint distribution (see the proof of Corollary 2.7). In this section, the joint distribution of the variables $N_{A_1,1}, \dots, N_{A_1,m}, N_{A_2,1}, \dots, N_{A_2,m}$ will be examined by means of probability generating functions and by using a generalisation of what is known as the 'renewal argument'. Assume that the Markov chain Z is irreducible (or absorbing) and it takes values in $S = A_1 \cup A_2$ (or in $S = A_1 \cup A_2 \cup \{\omega\}$ in which case ω is the absorbing state). If Z is irreducible, we supplement its state space by ω and augment its transition probability matrix $\mathbf{Q}$ and initial probability vector $\boldsymbol{\alpha}$ according to (1.2) with $\mathbf{Q}_{A_1\{\omega\}} = \mathbf{0}$, $\mathbf{Q}_{A_2\{\omega\}} = \mathbf{0}$, and $\alpha_\omega = 0$. The A_i - component of the initial probability vector of Z is $\boldsymbol{\alpha}_{A_i}$ with $\boldsymbol{\alpha}_{A_1}^T \mathbf{1} + \boldsymbol{\alpha}_{A_2}^T \mathbf{1} = 1$. For $z \in D = \{\, z \in \mathbb{C} : |z| \leq 1\,\}$, and $S', S'' \in \{\, A_1, A_2, \{\omega\}\,\}$, $S' \neq \{\omega\}$, $S' \neq S''$, define

$$\boldsymbol{\Phi}_{S'S''}(z) = z\,(\mathbf{I} - z\,\mathbf{Q}_{S'S'})^{-1}\,\mathbf{Q}_{S'S''}. \tag{2.63}$$

In the next theorem, the probability generating function of any finite collection of sojourn times in A_1 *and* A_2 is stated. The following notation will be used: for $i \in \{\,1, 2\,\}$, $c(i) \in \{\,1, 2\,\}$ is defined by $\{\,c(i)\,\} = \{1, 2\} \setminus \{i\}$, i.e., $c(i) = 3 - i$.

THEOREM 2.16. *For* $m \in \{1, 2, \ldots\}$, *the probability generating function of the first* $2m$ *sojourns times in* A_1 *and* A_2 *is given for* $z_1, z_2 \in D^m$ *by*

$$E\left[\prod_{i=1}^{2} \prod_{\ell=1}^{m} z_{i\ell}^{N_{A_i,\ell}}\right] = \sum_{i=1}^{2} \boldsymbol{\alpha}_{A_i}^{T} \boldsymbol{\Psi}_{A_i;m}(z_1 ; z_2),$$

where

$$\boldsymbol{\Psi}_{A_i;m}(z_1 ; z_2) = \left\{ E\left[\prod_{i=1}^{2} \prod_{\ell=1}^{m} z_{i\ell}^{N_{A_i,\ell}} \,\bigg|\, Z_0 = s\right] : s \in A_i \right\}, \tag{2.64}$$

with

$$\boldsymbol{\Psi}_{A_i;m}(z_1 ; z_2) = \sum_{\ell=0}^{m-1} \prod_{k=1}^{\ell} \left(\boldsymbol{\Phi}_{A_iA_{c(i)}}(z_{ik})\boldsymbol{\Phi}_{A_{c(i)}A_i}(z_{c(i)k})\right) \boldsymbol{\Phi}_{A_i\{\omega\}}(z_{i\,\ell+1}) +$$

$$\sum_{\ell=0}^{m-1} \prod_{k=1}^{\ell} \left(\boldsymbol{\Phi}_{A_iA_{c(i)}}(z_{ik})\boldsymbol{\Phi}_{A_{c(i)}A_i}(z_{c(i)k})\right) \boldsymbol{\Phi}_{A_iA_{c(i)}}(z_{i\,\ell+1})\boldsymbol{\Phi}_{A_{c(i)}\{\omega\}}(z_{c(i)\,\ell+1}) +$$

$$\prod_{k=1}^{m} \left(\boldsymbol{\Phi}_{A_iA_{c(i)}}(z_{ik})\boldsymbol{\Phi}_{A_{c(i)}A_i}(z_{c(i)k})\right) \mathbf{1}. \tag{2.65}$$

For irreducible Z, *all but the last term on the right hand side of (2.65) are zero.*

PROOF OF THEOREM 2.16. The proof is by induction on m. As a preliminary step, we state an alternative representation of the matrices in (2.63). Then, a recurrence relation for the entries of $\boldsymbol{\Psi}$ in (2.64) will be derived from which (2.65) readily follows.

We first note that if Z starts in some $a \in A_i$, the first sojourn of Z in A_i is of length K_i ($= N_{A_i,1}$) which is defined by $K_i = k \Leftrightarrow Z_0 \in A_i, \ldots, Z_{k-1} \in A_i, Z_k \notin A_i$. The first 'arrival value' of Z into $S \setminus A_i$ is then Z_{K_i}. From Lemma 2.2 it is easily deduced that for $S'' \in \{A_1, A_2, \{\omega\}\}\setminus\{A_i\}$ the expression in (2.63) is the following array of generating functions

$$\boldsymbol{\Phi}_{A_iS''}(z) = \left\{ \sum_{n=0}^{\infty} \Pr\{ N_{A_i,1} = n, Z_{K_i} = s \mid Z_0 = a \} z^n : a \in A_i, s \in S'' \right\}. \tag{2.66}$$

To find a recurrence relation for the entries of $\boldsymbol{\Psi}$, we note that for $m \geq 1$ and $a \in A_1$, the a*th* entry of $\boldsymbol{\Psi}_{A_1;m}$ is given by

$$\Psi_{a;m}(\mathbf{z}_1 ; \mathbf{z}_2) = \sum_{\substack{n_{11},\dots,n_{1m}=0 \\ n_{21},\dots,n_{2m}=0}}^{\infty} \Pr\left\{ \bigcap_{i=1}^{2} \bigcap_{\ell=1}^{m} \{ N_{A_i,\ell} = n_{i\ell} \} \mid Z_0 = a \right\} \prod_{i=1}^{2} \prod_{\ell=1}^{m} z_{i\ell}^{n_{i\ell}}, \quad (2.67)$$

with

$$\Pr\left\{ \bigcap_{i=1}^{2} \bigcap_{\ell=1}^{m} \{ N_{A_i,\ell} = n_{i\ell} \} \mid Z_0 = a \right\} =$$

$$\sum_{s \in S\backslash A_1} \Pr\left\{ \bigcap_{i=1}^{2} \bigcap_{\ell=1}^{m} \{ N_{A_i,\ell} = n_{i\ell} \} \cap \{ Z_{K_1} = s \} \mid Z_0 = a \right\} =$$

$$\sum_{s \in S\backslash A_1} \Pr\left\{ \bigcap_{\ell=1}^{m-1} \{ N_{A_1,\ell} = n_{1\,\ell+1} \} \cap \bigcap_{\ell=1}^{m} \{ N_{A_2,\ell} = n_{2\ell} \} \mid Z_0 = s \right\} \times$$

$$\Pr\{ N_{A_1,1} = n_{11}, Z_{K_1} = s \mid Z_0 = a \}. \qquad (2.68)$$

(The last step in (2.68) is a simple instance of what can be termed a 'generalized renewal argument'.) From (2.66) - (2.68) it follows for $a \in A_1$ that

$$\Psi_{a;m}(\mathbf{z}_1 ; \mathbf{z}_2) = \sum_{s \in S\backslash A_1} \Psi_{s;m}(z_{12}, z_{13}, \dots, z_{1m}, 1; z_{21}, \dots, z_{2m})\, \phi_{as}(z_{11}),$$

or, since $\Psi_{\omega;m} \equiv 1$,

$$\Psi_{a;m}(\mathbf{z}_1 ; \mathbf{z}_2) =$$

$$\phi_{a\omega}(z_{11}) + \sum_{s \in A_2} \Psi_{s;m}(z_{12}, z_{13}, \dots, z_{1m}, 1; z_{21}, \dots, z_{2m})\, \phi_{as}(z_{11}). \qquad (2.69)$$

A corresponding equation holds of course for $\Psi_{s;m}(\mathbf{z}_1 ; \mathbf{z}_2)$, $s \in A_2$. Applying now this equation to $\Psi_{s;m}$ on the right-hand side of (2.69), we get

$$\Psi_{a;m}(\mathbf{z}_1 ; \mathbf{z}_2) = \phi_{a\omega}(z_{11}) + \sum_{s \in A_2} \phi_{as}(z_{11})\, \phi_{s\omega}(z_{21}) +$$

$$\sum_{\substack{s_1 \in A_1 \\ s_2 \in A_2}} \phi_{as_2}(z_{11})\, \phi_{s_2 s_1}(z_{21})\, \Psi_{s_1;m}(z_{12}, \dots, z_{1m}, 1; z_{22}, \dots, z_{2m}, 1),$$

which for $m = 1$ and $m \geq 2$ reads respectively as

$$\Psi_{a;1}(z_{11} ; z_{21}) =$$

$$\Phi_{a\omega}(z_{11}) + \sum_{s \in A_2} \Phi_{as}(z_{11}) \Phi_{s\omega}(z_{21}) + \sum_{\substack{s_1 \in A_1 \\ s_2 \in A_2}} \Phi_{as_2}(z_{11}) \Phi_{s_2 s_1}(z_{21}), \qquad (2.70)$$

and

$$\Psi_{a;m}(\mathbf{z_1} ; \mathbf{z_2}) = \Phi_{a\omega}(z_{11}) + \sum_{s \in A_2} \Phi_{as}(z_{11}) \Phi_{s\omega}(z_{21}) +$$

$$\sum_{\substack{s_1 \in A_1 \\ s_2 \in A_2}} \Phi_{as_2}(z_{11}) \Phi_{s_2 s_1}(z_{21}) \Psi_{s_1;m-1}(z_{12}, ..., z_{1m}; z_{22}, ..., z_{2m}). \qquad (2.71)$$

The matrix form of (2.70) is (2.65) for $m = 1$. The required recurrence relation for the induction step is (2.71); the induction step itself is straightforward. ∎

Theorem 2.16 can be used to obtain the probability mass function of the joint distribution of sojourn times in A_1 and A_2. For the sake of notational simplicity we shall confine attention to the irreducible case; then, for $i \in \{1, 2\}$ we have

$$E\left[\prod_{i=1}^{2} \prod_{\ell=1}^{m} z_{i\ell}^{N_{A_i,\ell}} \right] = \sum_{i=1}^{2} \boldsymbol{\alpha}_{A_i}^{T} \prod_{k=1}^{m} \left(\boldsymbol{\Phi}_{A_i A_{c(i)}}(z_{ik}) \boldsymbol{\Phi}_{A_{c(i)} A_i}(z_{c(i)k}) \right) \mathbf{1} =$$

$$\sum_{i=1}^{2} \boldsymbol{\alpha}_{A_i}^{T} \prod_{k=1}^{m} \left\{ z_{ik} (\mathbf{I} - z_{ik} \mathbf{Q}_{A_i A_i})^{-1} \mathbf{Q}_{A_i A_{c(i)}} z_{c(i)k} (\mathbf{I} - z_{c(i)k} \mathbf{Q}_{A_{c(i)} A_{c(i)}})^{-1} \mathbf{Q}_{A_{c(i)} A_i} \right\} \mathbf{1} =$$

$$\sum_{\substack{j_{11},...,j_{1m}=0 \\ j_{21},...,j_{2m}=0}}^{\infty} \sum_{i=1}^{2} \boldsymbol{\alpha}_{A_i}^{T} \prod_{k=1}^{m} \left\{ \mathbf{Q}_{A_i A_i}^{j_{ik}} \mathbf{Q}_{A_i A_{c(i)}} \mathbf{Q}_{A_{c(i)} A_{c(i)}}^{j_{c(i)k}} \mathbf{Q}_{A_{c(i)} A_i} \right\} \mathbf{1} \times$$

$$\prod_{k=1}^{m} \left\{ z_{ik}^{(j_{ik}+1)} z_{c(i)k}^{(j_{c(i)k}+1)} \right\}. \qquad (2.72)$$

The following corollary follows from (2.72) immediately.

COROLLARY 2.17. *For irreducible Z, the probability mass function of the vector of sojourn times* $(N_{A_1,1}, ..., N_{A_1,m}, N_{A_2,1}, ..., N_{A_2,m})$ *is given by*

$$\text{Pr}\{ N_{A_1,1} = n_{11}, \ldots, N_{A_1,m} = n_{1m}, N_{A_2,1} = n_{21}, \ldots, N_{A_2,m} = n_{2m} \} =$$

$$\sum_{i=1}^{2} \boldsymbol{\alpha}_{A_i}^{T} \prod_{k=1}^{m} \left\{ \mathbf{Q}_{A_iA_i}^{(n_{ik}-1)} \mathbf{Q}_{A_iA_{c(i)}} \mathbf{Q}_{A_{c(i)}A_{c(i)}}^{(n_{c(i)k}-1)} \mathbf{Q}_{A_{c(i)}A_i} \right\} \mathbf{1}, \qquad (2.73)$$

where $n_{11}, \ldots, n_{1m}, n_{21}, \ldots, n_{2m} \geq 1$. ∎

Notice that by summing over $n_{21}, \ldots, n_{2m} \geq 1$ in (2.73), Corollary 2.3 easily follows once again.

2.1.4 Tabular summary of results about sojourn times and related variables

In Table 2.1 a guide is shown to the results on sojourn times obtained in Section 2.1.

Variable	Characteristic	Reference
$(N_{A_1,1}, \ldots, N_{A_1,m})$	pmf	Theorem 2.1 and Corollary 2.3
$N_{A_1,m}$	pmf	Corollaries 2.4 and 2.5
$M_{A_1} = K_{A_1,\infty}$	pmf	Corollary 2.6
$NS_{A_1,\infty}$	pgf	Corollary 2.7
$NS_{A_1,\infty}$	pmf	Corollary 2.8
$NS_{A_1,\infty} = L_{A_1,\infty}$	fm	Corollary 2.9
$NS_{A_1,m}$	pmf	Corollary 2.10
$L_{A_1,n}$	pgf	Corollary 2.11
$L_{A_1,n}$	pmf	Corollary 2.12
$L_{A_1,n}$	fm	Corollary 2.13
$K_{A_1,n}$	pmf	Corollary 2.14
$(N_{A_1,1}, \ldots, N_{A_1,m}, N_{A_2,1}, \ldots, N_{A_2,m})$	pgf	Theorem 2.16
$(N_{A_1,1}, \ldots, N_{A_1,m}, N_{A_2,1}, \ldots, N_{A_2,m})$	pmf	Corollary 2.17

Table 2.1 A guide to the results on sojourn times in Section 2.1

2.2 An application: the sequence of repair events for a three-unit power transmission model

In this section a certain aspect of the power transmission reliability Model 2 from Section 1.2.1 will be examined. It is an elaboration of an example from [CSE5]. The system whose state-transition-rate diagram is shown in Figure 1.2 is started at time $t = 0$ with all its components in the 'up' state, i.e., in state 1. We will be interested in *completed repair events*, i.e., in the occurrence of the transitions $2 \to 1$, $3 \to 1$, $5 \to 1$, $4 \to 2$, and $4 \to 3$. The first two of these transitions may be termed *minor* since the system's performance is not immediately affected by them. The other three are *major* repair events since they are a precondition for the system being functional. The sequence of completed repair events is the system characteristic which will be considered here. More precisely, the variable under consideration will be the number of major repair events in a finite sequence of repair events of a fixed length $n \in \{1, 2, ...\}$, say. It will be denoted by R_n, and it takes values in $\{0, ..., n\}$.

<u>2.2.1 The number of major repair events R_n in a repair sequence of length n</u>

The continuous parameter nature of the initial model is immaterial for the analysis of R_n. We may thus consider the embedded Markov chain [IOS] pp 243 instead; its transition probability matrix U, say, is

$$U = \begin{bmatrix} 0 & \lambda_1/c_1 & \lambda_2/c_1 & \lambda_c'/c_1 & \lambda_T/c_1 \\ \mu_1/c_2 & 0 & 0 & (\lambda_2+\lambda_c')/c_2 & \lambda_T/c_2 \\ \mu_2/c_3 & 0 & 0 & (\lambda_1+\lambda_c')/c_3 & \lambda_T/c_3 \\ 0 & \mu_2/c_4 & \mu_1/c_4 & 0 & \lambda_T/c_4 \\ 1 & 0 & 0 & 0 & 0 \end{bmatrix},$$

with

$$c_1 = \lambda_1 + \lambda_2 + \lambda_c' + \lambda_T, \quad c_2 = \lambda_2 + \lambda_c' + \mu_1 + \lambda_T,$$

$$c_3 = \lambda_1 + \lambda_c' + \mu_2 + \lambda_T, \quad c_4 = \mu_1 + \mu_2 + \lambda_T.$$

The random variable R_n is associated with transitions between, rather that with visits to, particular states. R_n is, however, expressible as an appropriately defined sojourn time of the expanded Markov chain [IOS] pp 173-175. Its state space is the Cartesian product $\{1, ..., 5\}^2$, and its transition probability matrix P is given by

$$P_{(i,j),(k,l)} = \begin{cases} u_{kl} & \text{for } j = k, \\ 0 & \text{for } j \neq k. \end{cases} \tag{2.74}$$

The system's starting condition now implies the following initial probability vector β, say, for the expanded Markov chain

$$\beta_{(i,j)} = \begin{cases} u_{1j} & \text{if } i = 1 \text{ and } j = 2, ..., 5, \\ 0 & \text{otherwise.} \end{cases} \tag{2.75}$$

Repair events are seen to correspond to visits by the expanded chain to the elements of the set $S = \{(2, 1), (3, 1), (4, 2), (4, 3), (5, 1)\}$. Put $S^c = \{1, ..., 5\}^2 \setminus S$, and denote by $W = \{W_0, W_1, ...\}$ the expanded Markov chain which is defined by (2.74) and (2.75). Then, the Markov chain of interest, Z, is the sequence of consecutive arrival values of W into S. More formally, Z is defined recursively via the (random) sequence of indices $N_0 < N_1 < ...$, where

$$Z_0 = W_{N_0} \text{ if } W_0, ..., W_{N_0-1} \in S^c, W_{N_0} \in S,$$

and for $k \geq 1$,

$$Z_k = W_{N_k} \text{ if } W_{N_{k-1}+1}, ..., W_{N_k-1} \in S^c, W_{N_k} \in S.$$

The Markov property for Z is easily verified. The transition probability matrix $\mathbf{Q}$ and the initial probability vector $\boldsymbol{\alpha}$ of Z is given in the following lemma.

LEMMA 2.18. *The transition probability matrix* $\mathbf{Q}$ *and the initial probability vector* $\boldsymbol{\alpha}$ *of* Z *are given respectively by*

$$\mathbf{Q} = \mathbf{P}_{SS} + \mathbf{P}_{SS^c} (\mathbf{I} - \mathbf{P}_{S^cS^c})^{-1} \mathbf{P}_{S^cS}, \tag{2.76}$$

$$\boldsymbol{\alpha}^T = \boldsymbol{\beta}_S^T + \boldsymbol{\beta}_{S^c}^T (\mathbf{I} - \mathbf{P}_{S^cS^c})^{-1} \mathbf{P}_{S^cS}. \tag{2.77}$$

PROOF OF LEMMA 2.18. Without loss of generality we may assume that $N_0 = 0$ and thus $Z_0 = W_0$. Therefore,

$$\Pr\{ Z_1 = s_1 \mid Z_0 = s_0 \} = \Pr\{ W_1 = s_1 \mid W_0 = s_0 \} +$$

$$\Pr\{ W_1 \notin S, W_2 = s_1 \mid W_0 = s_0 \} +$$

$$\Pr\{ W_1, W_2 \notin S, W_3 = s_1 \mid W_0 = s_0 \} + ... =$$

$$P_{s_0 s_1} + \sum_{r_1 \in S^c} P_{s_0 r_1} P_{r_1 s_1} + \sum_{r_1 \in S^c} \sum_{r_2 \in S^c} P_{s_0 r_1} P_{r_1 r_2} P_{r_2 s_1} + ... =$$

(s_0, s_1) - entry of $\left\{ \mathbf{P}_{SS} + \mathbf{P}_{SS^c} \mathbf{P}_{S^cS} + \mathbf{P}_{SS^c} \mathbf{P}_{S^cS^c} \mathbf{P}_{S^cS} + ... \right\} =$

(s_0, s_1) - entry of $\left\{ \mathbf{P}_{SS} + \mathbf{P}_{SS^c} (\mathbf{I} - \mathbf{P}_{S^cS^c})^{-1} \mathbf{P}_{S^cS} \right\}$,

which is (2.76). The proof of (2.77) is along similar lines. ∎

By construction, Z is such that each of its transitions corresponds to a completed repair event of the original system. Visits of Z to the states in $A_2 = \{ (2, 1), (3, 1) \}$ and $A_1 = \{ (4, 2), (4, 3), (5, 1) \}$ stand for minor and major repair events respectively. The theory in Section 2.1 is applicable to the chain Z thus defined. By construction, $R_n = L_{A_1,n}$.

2.2.2 Numerical results

(2.51) - (2.54) have been implemented in MATLAB [MAT1] on the Apple Macintosh Plus. Implementation issues will be discussed and an assessment will be given of MATLAB in Section 2.2.3. The code is shown in Section 2.2.4.

The three sets of values assumed for the rate parameters (in year^{-1}) are shown in Table 2.2 overleaf.

set №	λ_1	λ_2	λ_T	λ_c'	μ_1	μ_2	μ_T
1	0.25	0.25	0.01	0.02	1095	1095	46
2	0.25	0.25	0.025	0.05	1095	1095	46
3	0.25	0.25	0.025	0.05	547.5	547.5	46

Table 2.2 Parameter combinations for Model 2

The first two parameter combinations were quoted by Nahman and Mijuskovic [NAH] as based on a real failure data set whereas our set № 3 above is considered here in order to assess the model sensitivity in the repair rate parameters μ_1 and μ_2. Tables 2.3 - 2.6 (starting overleaf) show the results of the full analyses for the sets № 1 and № 2 for the variables R_1, ..., R_{10}. The computing time on the Apple Macintosh Plus for each pair of the tables was 21 min. As an example, Figure 2.5 (overleaf) shows the probability mass function of R_{10}. It is seen that the probability mass of R_{10} moves towards larger values as the failure rates λ_T and λ_c' are increased. This is what we would expect intuitively since these two parameters are those associated solely with system failure and thus with a subsequent major repair event. (Notice, however, that a combination of individual failures of line 1 and line 2 can also lead to system failure.) The factorial moments are of course also obtainable from the probability mass functions; the Tables 2.4 and 2.6 are, however, based on the MATLAB implementation of (2.53) and (2.54). This affords a convenient tool for the validation of the (implementation of the) method.

For $n \to \infty$, the sequence of 'interval occupancies' $\{ E[R_n] / n : n = 1, 2, ... \}$ (i.e., the mean proportion of 'time' spent in $\{ 0, 1, ... n-1 \}$) converges to $\pi_{A_1}^T \mathbf{1}$ where π stands for the equilibrium probability vector of Z. This follows from

$$\frac{E[R_n]}{n} = \sum_{s \in A_1} \frac{1}{n} \sum_{i=0}^{n-1} E[I_{\{ Z_i = s \}}] = \sum_{s \in A_1} \frac{1}{n} \sum_{i=0}^{n-1} Pr\{ Z_i = s \} \to \pi_{A_1}^T \mathbf{1},$$

as $n \to \infty$.

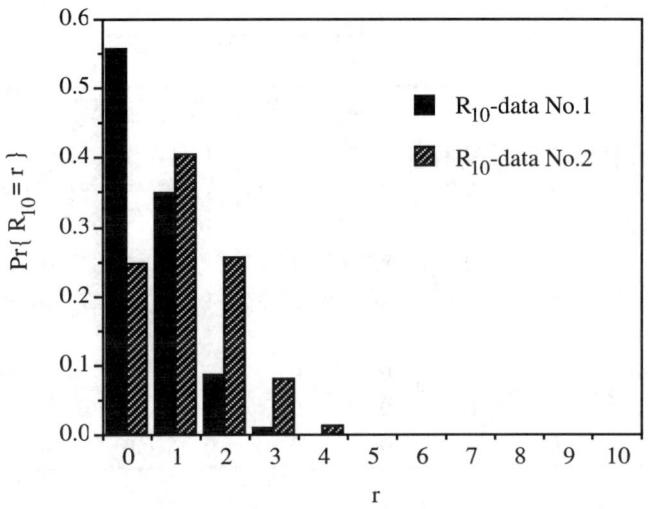

Figure 2.5 The probability mass function of R_{10} for parameter sets 1 and 2

r	$i = 1$	$i = 2$	$i = 3$	$i = 4$	$i = 5$	$i = 6$	$i = 7$	$i = 8$	$i = 9$	$i = 10$
0	9.432_1	8.895_1	8.390_1	7.913_1	7.463_1	7.039_1	6.639_1	6.261_1	5.905_1	5.570_1
1	5.684_2	1.094_1	1.558_1	1.965_1	2.321_1	2.630_1	2.897_1	3.125_1	3.317_1	3.478_1
2	0	1.083_3	5.253_3	1.206_2	2.100_2	3.166_2	4.364_2	5.662_2	7.030_2	8.444_2
3	0	0	2.046_5	1.601_4	5.791_4	1.402_3	2.719_3	4.592_3	7.059_3	1.014_2
4	0	0	0	3.886_7	4.176_6	2.100_5	6.837_5	1.692_4	3.498_4	6.379_4
5	0	0	0	0	7.305_9	1.006_7	6.516_7	2.708_6	8.358_6	2.094_5
6	0	0	0	0	0	1.380_{10}	2.313_9	1.835_8	9.339_8	3.494_7
7	0	0	0	0	0	0	2.608_{12}	5.146_{11}	4.839_{10}	2.923_9
8	0	0	0	0	0	0	0	4.928_{14}	1.119_{12}	1.217_{11}
9	0	0	0	0	0	0	0	0	9.312_{16}	2.391_{14}
10	0	0	0	0	0	0	0	0	0	1.760_{17}

Table 2.3 Values of $\Pr\{\ R_i = r\ \}$ for $i = 1, ..., 10$; parameter combination № 1
(Negative powers of 10 are indicated by a subscript, e.g., 9.432_1 stands for 9.432×10^{-1}.)

k	i = 1	i = 2	i = 3	i = 4	i = 5	i = 6	i = 7	i = 8	i = 9	i = 10
1	5.684_2	1.115_1	1.663_1	2.211_1	2.759_1	3.306_1	3.854_1	4.402_1	4.950_1	5.497_1
2	0	2.166_3	1.063_2	2.508_2	4.553_2	7.197_2	1.044_1	1.429_1	1.873_1	2.378_1
3	0	0	1.228_4	9.701_4	3.575_3	8.920_3	1.799_2	3.178_2	5.126_2	7.743_2
4	0	0	0	9.279_6	1.011_4	5.162_4	1.720_3	4.393_3	9.432_3	1.795_2
5	0	0	0	0	8.766_7	1.218_5	7.987_5	3.383_4	1.071_3	2.771_3
6	0	0	0	0	0	9.938_8	1.678_6	1.347_5	6.970_5	2.665_4
7	0	0	0	0	0	0	1.315_8	2.613_7	2.484_6	1.523_5
8	0	0	0	0	0	0	0	1.897_9	4.545_8	4.994_7
9	0	0	0	0	0	0	0	0	3.379_{10}	8.742_9
10	0	0	0	0	0	0	0	0	0	6.385_{11}

Table 2.4 Values of $FM_k(R_i)$ for $i = 1, \ldots, 10$; parameter combination № 1
(Negative powers of 10 are indicated by a subscript, e.g., 9.432_1 stands for 9.432×10^{-1}.)

r	i = 1	i = 2	i = 3	i = 4	i = 5	i = 6	i = 7	i = 8	i = 9	i = 10
0	8.693_1	7.557_1	6.569_1	5.711_1	4.964_1	4.316_1	3.752_1	3.261_1	2.835_1	2.464_1
1	1.307_1	2.386_1	3.161_1	3.692_1	4.031_1	4.218_1	4.287_1	4.266_1	4.177_1	4.039_1
2	0	5.711_3	2.676_2	5.783_2	9.390_2	1.314_1	1.677_1	2.012_1	2.308_1	2.560_1
3	0	0	2.487_4	1.882_3	6.490_3	1.478_2	2.675_2	4.201_2	5.990_2	7.966_2
4	0	0	0	1.083_5	1.131_4	5.455_4	1.648_3	3.918_3	7.569_3	1.285_2
5	0	0	0	0	4.716_7	6.285_6	3.911_5	1.550_4	4.524_4	1.065_3
6	0	0	0	0	0	2.053_8	3.328_7	2.542_6	1.238_5	4.400_5
7	0	0	0	0	0	0	8.941_{10}	1.707_8	1.547_7	8.957_7
8	0	0	0	0	0	0	0	3.894_{11}	8.554_{10}	8.972_9
9	0	0	0	0	0	0	0	0	1.695_{12}	4.213_{11}
10	0	0	0	0	0	0	0	0	0	7.383_{14}

Table 2.5 Values of $Pr\{ R_i = r \}$ for $i = 1, \ldots, 10$; parameter combination № 2
(Negative powers of 10 are indicated by a subscript, e.g., 9.432_1 stands for 9.432×10^{-1}.)

k	i = 1	i = 2	i = 3	i = 4	i = 5	i = 6	i = 7	i = 8	i = 9	i = 10
1	1.307_1	2.500_1	3.703_1	4.906_1	6.108_1	7.310_1	8.513_1	9.715_1	1.092	1.212
2	0	1.142_2	5.501_2	1.271_1	2.281_1	3.581_1	5.169_1	7.047_1	9.213_1	1.167
3	0	0	1.492_3	1.155_2	4.169_2	1.021_1	2.033_1	3.557_1	5.697_1	8.558_1
4	0	0	0	2.599_4	2.772_3	1.385_2	4.524_2	1.136_1	2.405_1	4.528_1
5	0	0	0	0	5.659_5	7.690_4	4.935_3	2.047_2	6.360_2	1.618_1
6	0	0	0	0	0	1.478_5	2.441_4	1.917_3	9.709_3	3.638_2
7	0	0	0	0	0	0	4.506_6	8.760_5	8.143_4	4.884_3
8	0	0	0	0	0	0	0	1.570_6	3.511_5	3.772_4
9	0	0	0	0	0	0	0	0	6.152_7	1.556_5
10	0	0	0	0	0	0	0	0	0	2.679_7

Table 2.6 Values of $FM_k(R_i)$ for i = 1, ..., 10; parameter combination № 2
(Negative powers of 10 are indicated by a subscript, e.g., 9.432_1 stands for 9.432×10^{-1}.)

For the parameter set № 2, for example, **Q** is (by MATLAB)

$$\begin{bmatrix} 4.347\times10^{-1} & 4.347\times10^{-1} & 4.360\times10^{-2} & 4.360\times10^{-2} & 4.360\times10^{-2} \\ 4.347\times10^{-1} & 4.347\times10^{-1} & 4.360\times10^{-2} & 4.360\times10^{-2} & 4.360\times10^{-2} \\ 9.997\times10^{-1} & 0 & 1.369\times10^{-4} & 1.369\times10^{-4} & 2.283\times10^{-5} \\ 0 & 9.997\times10^{-1} & 1.369\times10^{-4} & 1.369\times10^{-4} & 2.283\times10^{-5} \\ 4.347\times10^{-1} & 4.347\times10^{-1} & 4.360\times10^{-2} & 4.360\times10^{-2} & 4.360\times10^{-2} \end{bmatrix}.$$

From this, the corresponding equilibrium probability vector is

$$\boldsymbol{\pi}^T = (\ 4.400\times10^{-1},\ 4.400\times10^{-1},\ 4.011\times10^{-2},\ 4.011\times10^{-2},\ 4.001\times10^{-2}\ ).$$

Thus, as $n \to \infty$,

$$\frac{E[R_n]}{n} \to \boldsymbol{\pi}_{A_1}^T \mathbf{1} = 1.202\times10^{-1}.$$

The first line in Table 2.6 shows the values of $FM_1(R_n) = E[R_n]$ for n = 1, ..., 10. Table 2.7 (overleaf) shows the values for R_{10} for the parameter set № 3. This set arises from № 2 by halving the individual line repair rates μ_1 and μ_2. It is seen that the probability mass function and the factorial moments of R_{10} are hardly affected by the change. This is because those two parameters are primarily associated with the duration of a repair but not with its nature (minor/major). There is a secondary effect, however, which is also seen from Table 2.5. By decreasing the repair rates μ_1 and μ_2, the sojourns in the states '2' and '3' will tend to be longer. These repair periods are critical in the sense that whenever the only remaining power line fails, system failure ensues. For this reason, as a secondary effect, we observe a slight stochastic increase in the distribution of R_{10}.

r	$\Pr\{ R_{10} = r \}$	$FM_r(R_{10})$
0	2.457×10^{-1}	-
1	4.036×10^{-1}	1.214
2	2.565×10^{-1}	1.172
3	8.004×10^{-2}	8.614×10^{-1}
4	1.296×10^{-2}	4.571×10^{-1}
5	1.077×10^{-3}	1.638×10^{-1}
6	4.469×10^{-5}	3.696×10^{-2}
7	9.126×10^{-7}	4.976×10^{-3}
8	9.157×10^{-9}	3.849×10^{-4}
9	4.301×10^{-11}	1.588×10^{-5}
10	7.528×10^{-14}	2.732×10^{-7}

Table 2.7 Data for R_{10}; parameter combination № 3

Note that even though Proposition 2.15 has not been used in the present MATLAB implementation, it should be useful for larger problems where computing time is at a premium.

2.2.3 Implementation with MATLAB

This section is intended as a first account of our system of implementation, MATLAB. The purpose here (and in later chapters) will be to document those features of MATLAB which are specifically important in reliability work. However, no exhaustive description of it was envisaged since several books are available on this topic (see, e.g., [MAT1], [MAT2], and [SCH]).

MATLAB is a system and language with many built-in matrix calculation routines which are very useful in Markov modelling. The built-in functions used in the present example are shown in Table 2.8 overleaf. The MATLAB control structures used in Section 2.2.4, i.e., the `for` loop and selection by the `if` statement, correspond to those from other languages. It is seen from the code in Section 2.2.4 that code written in MATLAB is essentially self-documenting and very similar to the usual mathematical notation. The only step possibly deserving some attention is the transition from the transition probability matrix $\mathbf{U}$ to the transition probability matrix $\mathbf{P}$ of the expanded chain. The expanded chain has the two-dimensional state space $\{1, ..., 5\}^2$ which is enumerated as shown in Figure 2.6 (overleaf).

MATLAB function/rule/construct	Example
matrix multiplication	`A*B`
matrix inversion	`inv(A)`
any integer power of a square matrix	`A^(-5)`
Kronecker product of two matrices	
A and **B**, i.e., $\mathbf{A} \otimes \mathbf{B}$ (see (1.5))	`kron(B,A)`
transpose of a matrix **A**	`A'`
identity matrix of a certain size	`eye(20)`
matrix of a given size with all entries unity	`ones(4,6)`
submatrix of a matrix	`P(22:25,7:10)`
ith row vector of matrix **A**	`A(i,:)`
ith column vector of matrix **A**	`A(:,i)`
definition of a matrix in terms of submatrices	`Q=[A B; C D]`
CPU time	`fix(clock)`
shows variable/text on the VDU	`disp('Computing Q');`
shows text on VDU and then assignes value	
to variable by user input via keyboard	`a=input('Enter a');`
comment line is preceeded by "%"	`% comment...`
supression of display is by ";"	`Q=[A B; C D];`
deletion of variable **v** from the workspace	`clear v`

Table 2.8 MATLAB constructs used to implement (2.51) - (2.54)

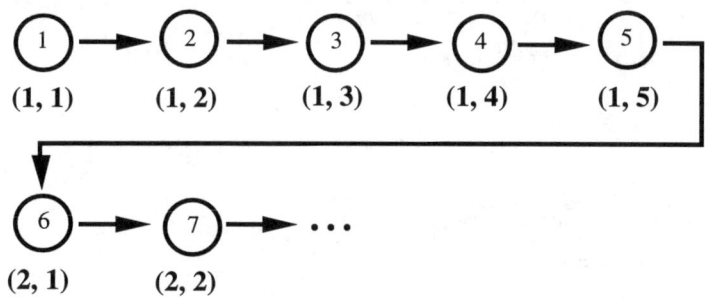

Figure 2.6 Enumeration of $\{1, ..., 5\}^2$

Then, by (2.74), **P** is seen to be a 25 by 25 block partitioned matrix with

$$
\mathbf{P} = \begin{bmatrix} \boxed{\mathbf{P}_{11}} & \boxed{\mathbf{P}_{12}} & \boxed{\mathbf{P}_{13}} & \boxed{\mathbf{P}_{14}} & \boxed{\mathbf{P}_{15}} \\ \boxed{\mathbf{P}_{11}} & \boxed{\mathbf{P}_{12}} & \boxed{\mathbf{P}_{13}} & \boxed{\mathbf{P}_{14}} & \boxed{\mathbf{P}_{15}} \\ \boxed{\mathbf{P}_{11}} & \boxed{\mathbf{P}_{12}} & \boxed{\mathbf{P}_{13}} & \boxed{\mathbf{P}_{14}} & \boxed{\mathbf{P}_{15}} \\ \boxed{\mathbf{P}_{11}} & \boxed{\mathbf{P}_{12}} & \boxed{\mathbf{P}_{13}} & \boxed{\mathbf{P}_{14}} & \boxed{\mathbf{P}_{15}} \\ \boxed{\mathbf{P}_{11}} & \boxed{\mathbf{P}_{12}} & \boxed{\mathbf{P}_{13}} & \boxed{\mathbf{P}_{14}} & \boxed{\mathbf{P}_{15}} \end{bmatrix} , \tag{2.78}
$$

where

$$
\mathbf{P}_{11} = \begin{bmatrix} u_{11} & u_{12} & u_{13} & u_{14} & u_{15} \\ 0 & 0 & 0 & 0 & 0 \\ 0 & 0 & 0 & 0 & 0 \\ 0 & 0 & 0 & 0 & 0 \\ 0 & 0 & 0 & 0 & 0 \end{bmatrix}, \mathbf{P}_{12} = \begin{bmatrix} 0 & 0 & 0 & 0 & 0 \\ u_{21} & u_{22} & u_{23} & u_{24} & u_{25} \\ 0 & 0 & 0 & 0 & 0 \\ 0 & 0 & 0 & 0 & 0 \\ 0 & 0 & 0 & 0 & 0 \end{bmatrix}, \dots .
$$

The MATLAB implementation in Section 2.2.4 of (2.78) is then accomplished by assembling $\mathbf{P}$ from the row vectors of $\mathbf{U}$ using the appropriate MATLAB matrix constructs. Similarly, $\boldsymbol{\beta}$ in (2.75) is a column vector of size 25 with $\boldsymbol{\beta}^T = (u_{11}, u_{12}, u_{13}, u_{14}, u_{15}, 0, \dots, 0)$.

2.2.4 MATLAB code

```
% MATLAB IMPLEMENTATION OF (2.51)-(2.54)
lamt = input('Enter lambdat in 1/year... ');
lamc = input('Enter lambdacdash in 1/year ...');
fix(clock)
disp('Other parameter values assumed ...');
lam1 = 0.25
lam2 = 0.25
mu1 = 1095
mu2 = 1095
mut = 46

disp('Now calculating Q and alpha ...');
c1 = lam1 + lam2 + lamc + lamt;
c2 = lam2 + lamc + mu1 + lamt;
c3 = lam1 + lamc + mu2 + lamt;
c4 = mu1 + mu2 + lamt;
U = [ 0         lam1/c1   lam2/c1   lamc/c1          lamt/c1
      mu1/c2    0         0         (lam2 + lamc)/c2 lamt/c2
      mu2/c3    0         0         (lam1 + lamc)/c3 lamt/c3
      0         mu2/c4    mu1/c4    0                lamt/c4
      1         0         0         0                0      ];

unit = eye(5);
e1 = unit(:,1);
e2 = unit(:,2);
e3 = unit(:,3);
e4 = unit(:,4);
e5 = unit(:,5);
% P IS A SQUARE MATRIX OF SIZE 25.
% ENUMERATION: 1 <-> (1,1), 2 <-> (1,2), 3 <-> (1,3), 4 <-> (1,4),
% 5 <-> (1,5), 6 <-> (2,1), 7 <-> (2,2), ..., 25 <-> (5,5).
```

```
P = [ e1*e1'*U e2*e2'*U e3*e3'*U e4*e4'*U e5*e5'*U
      e1*e1'*U e2*e2'*U e3*e3'*U e4*e4'*U e5*e5'*U
      e1*e1'*U e2*e2'*U e3*e3'*U e4*e4'*U e5*e5'*U
      e1*e1'*U e2*e2'*U e3*e3'*U e4*e4'*U e5*e5'*U
      e1*e1'*U e2*e2'*U e3*e3'*U e4*e4'*U e5*e5'*U ]; % Use (2.78)

beta = [ U(1,:) zeros(1,20) ]';                        % Use (2.75)

% S = {(2,1),(3,1),(4,2),(4,3),(5,1)}, I.E., S = {6,11,17,18,21}
% A2 = {(2,1),(3,1)} AND A1 = {(4,2),(4,3),(5,1)},
% I.E., A2 = {6,11} AND A1 = {17,18,21}

PSS = [ P(6,6)   P(6,11)   P(6,17)   P(6,18)   P(6,21)
        P(11,6)  P(11,11)  P(11,17)  P(11,18)  P(11,21)
        P(17,6)  P(17,11)  P(17,17)  P(17,18)  P(17,21)
        P(18,6)  P(18,11)  P(18,17)  P(18,18)  P(18,21)
        P(21,6)  P(21,11)  P(21,17)  P(21,18)  P(21,21)];

PSCSC=[P(1:5,1:5)   P(1:5,7:10)   P(1:5,12:16)   P(1:5,19:20)   P(1:5,22:25)
 P(7:10,1:5)   P(7:10,7:10)   P(7:10,12:16)   P(7:10,19:20)   P(7:10,22:25)
 P(12:16,1:5)  P(12:16,7:10)  P(12:16,12:16)  P(12:16,19:20)  P(12:16,22:25)
 P(19:20,1:5)  P(19:20,7:10)  P(19:20,12:16)  P(19:20,19:20)  P(19:20,22:25)
 P(22:25,1:5)  P(22:25,7:10)  P(22:25,12:16)  P(22:25,19:20)  P(22:25,22:25)];

PSSC=[P(6:6,1:5)    P(6:6,7:10)    P(6:6,12:16)    P(6:6,19:20)    P(6:6,22:25)
 P(11:11,1:5)  P(11:11,7:10)  P(11:11,12:16)  P(11:11,19:20)  P(11:11,22:25)
 P(17:18,1:5)  P(17:18,7:10)  P(17:18,12:16)  P(17:18,19:20)  P(17:18,22:25)
 P(21:21,1:5)  P(21:21,7:10)  P(21:21,12:16)  P(21:21,19:20)  P(21:21,22:25)];

PSCS = [P(1:5,6:6)    P(1:5,11:11)    P(1:5,17:18)    P(1:5,21:21)
        P(7:10,6:6)   P(7:10,11:11)   P(7:10,17:18)   P(7:10,21:21)
        P(12:16,6:6)  P(12:16,11:11)  P(12:16,17:18)  P(12:16,21:21)
        P(19:20,6:6)  P(19:20,11:11)  P(19:20,17:18)  P(19:20,21:21)
        P(22:25,6:6)  P(22:25,11:11)  P(22:25,17:18)  P(22:25,21:21)];

Q = PSS + PSSC*inv(eye(20) - PSCSC)*PSCS; % Use (2.76)

betas = [ beta(6:6,1:1)
          beta(11:11,1:1)
          beta(17:18,1:1)
          beta(21:21,1:1) ];

betasc = [ beta(1:5,1:1)
           beta(7:10,1:1)
           beta(12:16,1:1)
           beta(19:20,1:1)
           beta(22:25,1:1) ];

alpha = betas + PSCS'*(inv(eye(20) - PSCSC))'*betasc;

alphaa = alpha(1:2,1:1)
alphab = alpha(3:5,1:1)

QA2A2 = Q(1:2,1:2)
QA1A1 = Q(3:5,3:5)
QA2A1 = Q(1:2,3:5)
QA1A2 = Q(3:5,1:2)

nmax = input('Enter maximum length of repair event sequence ...');

for n=1:nmax
n
   if n>1
      En = zeros(n,n);
      for j=1:n
```

```
            for k=1:n
                if k-j==1
                    En(j,k) = 1;
                end
            end
        end

        TEMP = inv(eye(2*n) - kron(En,QA2A2))*kron(En,QA2A1);
        Mn = kron(En,QA1A1) + kron(En,QA1A2)*TEMP;
        vnt = [alphaa' zeros(1,2*(n-1))]*TEMP + [alphab' zeros(1,3*(n-1))];
        clear TEMP;
    end

% CALCULATE AND DISPLAY THE PMF OF Ln ...
    disp('Pr{ Ln = 0 }')
    if n==1                          % Pr{ Ln = 0 }
        alphaa'*ones(2,1)
    else
        1 - vnt*ones(3*n,1)
    end
    if n==1                          % Pr{ Ln = el } FOR el = 1,2,...,n
        disp('Pr{ Ln = 1 }')
        alphab'*ones(3,1)
    else
        for el=1:n
            disp('Pr{ Ln = el } for '); el
            vnt*Mn^(el-1)*(eye(3*n) - Mn)*ones(3*n,1)
        end
    end

% CALCULATE AND DISPLAY THE FACTORIAL MOMENTS OF Ln ...
    if n==1
        disp('The first factorial moment of L1 is')
        alphab'*ones(3,1)
    else
        fact = 1;
        for k=1:n
         fact = fact*k;
         disp('The k-th factorial moment of Ln is for'); k
         fact*vnt*((eye(3*n)-Mn)^(-k-1))*(Mn^(k-1))*(eye(3*n)-Mn)*ones(3*n,1)
        end
    end
end
fix(clock)
```

CHAPTER 3

THE NUMBER OF VISITS UNTIL ABSORPTION TO SUBSETS OF THE STATE SPACE BY A DISCRETE-PARAMETER MARKOV CHAIN: THE MULTIVARIATE CASE

In Corollary 2.6, we have considered the variable M_{A_1}, the number of visits to A_1 until absorption by a chain $Z = \{ Z_0, Z_1, ... \}$ whose state space is partitioned as $A_1 \cup A_2 \cup \{\omega\}$ with ω being the absorbing state. There, the probability mass function of M_{A_1} was deduced from our main results on sojourn times. On the other hand, the power of the generalised renewal argument was demonstrated in Section 2.1.3. In this chapter, these two threads will be combined to obtain results on the *joint* distribution of the number of visits to a subset of the state space whose set of transient states is partitioned into possibly more than two subsets.

Thus assume that Z is a discrete-parameter Markov chain, and that its state space S is partitioned as $S = A_1 \cup ... \cup A_n \cup A_{n+1}$, with n (≥ 2) (non-empty) transient sets $A_1, ..., A_n$, and the set A_{n+1} comprising one single absorbing state, ω. The variable of interest here is the random vector $M = (M_{A_1}, ..., M_{A_n})^T$, where M_{A_i} is the number of visits to A_i by Z until absorption. In Section 3.1, a closed form expression is obtained for the vector of probability generating functions of M, $g(z_1, ..., z_n)$, which is defined component-wise for $s \in S$ and $z \in D^n$, where D is the closed unit disc in the complex plane, $D = \{ z \in \mathbb{C} : |z| \leq 1 \}$, by

$$g_s(z) = \sum_{m_1,...,m_n=0}^{\infty} \Pr\{ M = (m_1, ..., m_n)^T \mid Z_0 = s \} z_1^{m_1} ... z_n^{m_n} .$$

As a corollary, the probability mass function of $L = M^T 1 = M_{A_1} + ... + M_{A_n}$ will also be deduced. For certain systems in reliability applications, L, 'the number of class transitions until final breakdown', can be used to express the number of completed repair events until final breakdown; it is a quantity which may well be used to supplement the usual system dependability characteristics like reliability, (point or interval) availability, etc. In Section 3.2 then, attention is restricted to the case $n \in \{ 2, 3 \}$. For $n = 2$, the probability generating function of $M = (M_{A_1}, M_{A_2})^T$ is restated in an alternative form, from which then the probability mass function of the first marginal distribution is deduced. For $n = 3$, an alternative form for the probability generating function of $M = (M_{A_1}, M_{A_2}, M_{A_3})^T$ is derived, and in turn, the probability mass function is established for the bivariate distribution of the first two components of M. (For some problems in reliability modelling it suffices to consider (M_{A_1}, M_{A_2}) in lieu of M since due to the model's specific transition structure, the third component, M_{A_3}, may be expressible in terms of the first two. This is the case in particular for a process with a state-transition diagram such as in Figure 3.1 (overleaf).

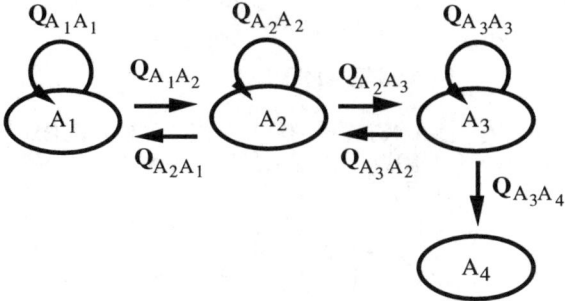

Figure 3.1 A special state-transition structure for $n = 3$

Then,

$$M_{A_3} = M_{A_2} - M_{A_1} + 1, \tag{3.1}$$

if $Z_0 \in A_1$. Notice, however, that (3.1) does not hold if the initial state of Z is not in A_1: if, for example, $Z_0 \in A_2$, $Z_1 \in A_1$, $Z_2 \in A_2$, $Z_3 \in A_3$, and $Z_4 \in A_4$, then $M_{A_3} = 1$ but $M_{A_2} - M_{A_1} + 1 = 2$.) Section 3.3 is a tabular summary of results from 3.1 and 3.2. In Section 3.4, the theory is applied to analyse the transient behaviour of Model 4 from Section 1.2.1.

This chapter is based on the original research reported in [CSE8]. An early related work is that by Bhat [BHA1]. Bhat [BHA1] derived the probability generating function of the joint distribution of transition frequencies $\{ f(s, s') : s, s' \in S; s, s' \neq \omega \}$, where $f(s, s')$ is the number of transitions $s \to s'$ by Z until absorption. Even though it is true that, in principle, the knowledge of the joint distribution of the variables $f(s, s')$ suffices for the derivation of the joint distribution of class frequencies, the formulae in Bhat [BHA1] are rather unwieldy and they do not lend themselves readily for such work in the general setting.

3.1 The probability generating function of M and the probability mass function of L

To start with, some remarks on the notation: **g** and the transition probability matrix of Z, **Q**, are partitioned as follows

$$\mathbf{g} = (\mathbf{g}_{A_1}{}^T, ..., \mathbf{g}_{A_n}{}^T, \mathbf{g}_{A_{n+1}}{}^T),$$

$$\mathbf{Q} = \begin{bmatrix} \mathbf{Q}_{A_1A_1} & \cdots & \mathbf{Q}_{A_1A_n} & \mathbf{Q}_{A_1A_{n+1}} \\ \cdot & \cdot & \cdot & \cdot \\ \cdot & \cdot & \cdot & \cdot \\ \cdot & \cdot & \cdot & \cdot \\ \mathbf{Q}_{A_nA_1} & \cdots & \mathbf{Q}_{A_nA_n} & \mathbf{Q}_{A_nA_{n+1}} \\ \mathbf{0} & \cdots & \mathbf{0} & 1 \end{bmatrix}.$$

The matrices $\Phi_{A_iA_j}$ are defined by

$$\Phi_{A_iA_j} = (\mathbf{I} - \mathbf{Q}_{A_iA_i})^{-1}\,\mathbf{Q}_{A_iA_j}, \quad i = 1, ..., n; \ j = 1, ..., n+1;$$

they exist because of Theorem 2.1 (a).

The following holds for the probability generating function of $\mathbf{M}$.

THEOREM 3.1. *The probability generating function of* $\mathbf{M}$ *is given by*

$$g_{S\backslash A_{n+1}}(\mathbf{z}) =$$

$$(\mathbf{I} - \mathbf{V}_{(S\backslash A_{n+1})(S\backslash A_{n+1})}(\mathbf{z}))^{-1}\,(\ \mathbf{V}_{A_1A_{n+1}}(\mathbf{z})^{\mathrm{T}}, ..., \mathbf{V}_{A_nA_{n+1}}(\mathbf{z})^{\mathrm{T}})^{\mathrm{T}}, \tag{3.2}$$

$$g_{A_{n+1}}(\mathbf{z}) \equiv 1, \tag{3.3}$$

where $\mathbf{V}(\mathbf{z})$ *is defined by*

$$\mathbf{V}_{A_iA_j}(\mathbf{z}) = \begin{cases} z_i\,\Phi_{A_iA_j}\,for\ i,\,j \in \{1, ..., n+1\},\,i \neq j,\,i \neq n+1, \\[2mm] 0 \qquad\quad otherwise. \end{cases}$$

(3.2) *holds for all* $\mathbf{z} \in \mathbb{C}^n$ *whenever its right hand side exists; a neighbourhood of* $\mathbf{0} \in \mathbb{C}^n$ *is contained in that set.*

PROOF OF THEOREM 3.1. $g_\omega(z_1, ..., z_n) \equiv 1$ since $\Pr\{\ \mathbf{M} = \mathbf{0} \mid Z_0 = \omega\ \} = 1$; (3.3) thus holds. For (3.2) the generalised renewal argument (already encountered in the proof of Theorem 2.16) will be applied. Define the measures $\kappa_{i;a,s}$ (for $i = 1, ..., n$, $a \in A_i$, and $s \in S \backslash A_i$) on the set of positive integers by

$$\kappa_{i;a,s}(\{k\}) = \Pr\{\ K_i = k,\ Z \text{ visits 's' immediately after leaving } A_i \mid Z_0 = a\ \},$$

where the variable K_i is defined as the 'length of the first sojourn in A_i', provided Z starts in some $a \in A_i$:

$$K_i = k \iff Z_0 \in A_i,\ Z_1 \in A_i, ...,\ Z_{k-1} \in A_i,\ Z_k \notin A_i\ .$$

Let now $i \in \{1, ..., n\}$ be fixed. Then, for $m_0, ..., m_n \geq 0$, $m_i \geq 1$, and $a \in A_i$ we have

$$\Pr\{\ M_{A_1} = m_1, ..., M_{A_n} = m_n \mid Z_0 = a\ \}$$

$$\sum_{k=1}^{\infty}\ \sum_{s\in S\backslash A_i}\ \Pr\{\{M_{A_1} = m_1, ..., M_{A_n} = m_n\} \cap \{\ K_i = k,\ Z_{K_i} = s\ \} \mid Z_0 = a\ \} =$$

$$\sum_{k=1}^{\infty} \sum_{s \in S \backslash A_i} \Pr\{ \{ M_{A_i} = m_i - 1 \} \cap \bigcap_{\substack{j=1 \\ j \neq i}}^{n} \{ M_{A_j} = m_j \} \mid Z_0 = s \} \, \kappa_{i;a,s}(\{k\}) =$$

$$\sum_{s \in S \backslash A_i} \Pr\{ \{ M_{A_i} = m_i - 1 \} \cap \bigcap_{\substack{j=1 \\ j \neq i}}^{n} \{ M_{A_j} = m_j \} \mid Z_0 = s \} \, P\{ Z_{K_i} = s \mid Z_0 = a \}. \qquad (3.4)$$

(Notice that (3.4) also holds for $m_i = 0$ in which case both sides are zero.) From (3.4) it follows that

$$g_a(z_1, ..., z_n) = \sum_{s \in S \backslash A_i} \Pr\{ Z_{K_i} = s \mid Z_0 = a \} \, g_s(z_1, ..., z_n) \, z_i \,. \qquad (3.5)$$

It is easily seen from (2.8) that

$$\{ \Pr\{ Z_{K_i} = s \mid Z_0 = a \} : a \in A_i, s \in S \backslash A_i \} = (\mathbf{I} - \mathbf{Q}_{A_i A_i})^{-1} \, \mathbf{Q}_{A_i (S \backslash A_i)}. \qquad (3.6)$$

(3.5) and (3.6) (in conjunction with (3.3)) imply that

$$g_{A_i}(\mathbf{z}) = z_i \, (\mathbf{I} - \mathbf{Q}_{A_i A_i})^{-1} \left\{ \sum_{\substack{j=1 \\ j \neq i}}^{n} \mathbf{Q}_{A_i A_j} \, g_{A_j}(\mathbf{z}) + \mathbf{Q}_{A_i \{\omega\}} \right\} =$$

$$\sum_{j=1}^{n} \mathbf{V}_{A_i A_j}(\mathbf{z}) \, g_{A_j}(\mathbf{z}) + \mathbf{V}_{A_i A_{n+1}}(\mathbf{z}). \qquad (3.7)$$

(3.2) now follows from (3.7) immediately. ∎

COROLLARY 3.2. *The probability mass function of* $L = M_{A_1} + ... + M_{A_n}$, *the number of class transitions until absorption into* ω, *is given by*

$$\{ \Pr\{ L = \ell \mid Z_0 = s \} : s \in S \backslash \{\omega\} \} =$$

$$\begin{cases} \mathbf{0} & \textit{for } \ell = 0, \\[2mm] (\mathbf{V}_{(S \backslash A_{n+1})(S \backslash A_{n+1})}(\mathbf{1}))^{\ell-1} \, (\boldsymbol{\Phi}_{A_1 A_{n+1}}{}^{\mathrm{T}}, ..., \boldsymbol{\Phi}_{A_n A_{n+1}}{}^{\mathrm{T}})^{\mathrm{T}} & \textit{for } \ell \geq 1. \end{cases} \qquad (3.8)$$

PROOF OF COROLLARY 3.2. The probability generating function of L is obtained from (3.2) as

$$\{ E[\, z^L \,|\, Z_0 = s\,] : s \in S \setminus \{\omega\} \} =$$

$$z\, (\mathbf{I} - z\, \mathbf{V}_{(S \setminus A_{n+1})(S \setminus A_{n+1})}(\mathbf{1}))^{-1}\, (\Phi_{A_1 A_{n+1}}{}^T, \ldots, \Phi_{A_n A_{n+1}}{}^T)^T,$$

from which (3.8) follows by a power series expansion about $z = 0$. ∎

3.2 Further results for $n \in \{2, 3\}$

The case $n = 2$ will be examined first. (3.2) is equivalently restated as follows.

COROLLARY 3.3. *For $n = 2$, the A_i-entries ($i = 1, 2$) of the probability generating function of* **M** *are given by*

$$g_{A_i}(z_1, z_2) = z_i\, (\mathbf{I} - z_1\, z_2\, \Phi_{A_i A_{c(i)}}\, \Phi_{A_{c(i)} A_i})^{-1}\, \Phi_{A_i A_3} +$$

$$z_1\, z_2\, \Phi_{A_i A_{c(i)}}\, (\mathbf{I} - z_1\, z_2\, \Phi_{A_{c(i)} A_i}\, \Phi_{A_i A_{c(i)}})^{-1}\, \Phi_{A_{c(i)} A_3}, \tag{3.9}$$

where $c(i)$ is defined for $i \in \{1, 2\}$ by $\{c(i)\} = \{1, 2\} \setminus \{i\}$.

PROOF OF COROLLARY 3.3. Inversion of $(\mathbf{I} - \mathbf{V}_{(S \setminus A_3)(S \setminus A_3)}(\mathbf{z}))$ by the method of partitioning yields (3.9) from (3.2); see Graybill [GRA], Section 8.2. ∎

NOTE. $(\mathbf{I} - \Phi_{A_i A_{c(i)}}\, \Phi_{A_{c(i)} A_i}) = \mathbf{I} - \mathbf{H}(A_i, A_{c(i)})$ where $\mathbf{H}$ is defined by (2.30). It was shown in the proof of Corollary 2.7 that $\mathbf{H}(A_i, A_{c(i)})^m \to \mathbf{0}$ as $m \to +\infty$ and thus $(\mathbf{I} - \Phi_{A_i A_{c(i)}}\, \Phi_{A_{c(i)} A_i})$ is invertible. Consequently, (3.9) can be used to calculate the moments of **M**.

A Taylor series expansion can be used again to obtain the probability mass function of **M** from (3.9); this will not be pursued here, however. Instead, the first marginal distribution of **M**, already considered in Corollary 2.6, will now be rederived. From (3.9), the probability mass function of M_{A_1} is easily seen to be given by

$$\{ \Pr\{ M_{A_1} = m \,|\, Z_0 = s \} : s \in A_1 \} =$$

$$\begin{cases} \mathbf{0} & \text{for } m = 0, \\[2mm] (\Phi_{A_1 A_2}\, \Phi_{A_2 A_1})^{m-1} \left\{ \Phi_{A_1 A_3} + \Phi_{A_1 A_2}\, \Phi_{A_2 A_3} \right\} & \text{for } m \geq 1, \end{cases} \tag{3.10}$$

$$\{ \Pr\{ M_{A_1} = m \,|\, Z_0 = s \} : s \in A_2 \} =$$

$$\begin{cases} \Phi_{A_2 A_3} & \text{for } m = 0, \\[2mm] \Phi_{A_2 A_1}(\Phi_{A_1 A_2}\, \Phi_{A_2 A_1})^{m-1} \left\{ \Phi_{A_1 A_3} + \Phi_{A_1 A_2}\, \Phi_{A_2 A_3} \right\} & \text{for } m \geq 1. \end{cases} \tag{3.11}$$

58

The form chosen to represent the probability mass function of M_{A_1} in Corollary 2.6 differs somewhat from (3.10) and (3.11). The representation in Corollary 2.6 is arrived at by noting that

$$\Phi_{A_1A_3} + \Phi_{A_1A_2} \Phi_{A_2A_3} = \left\{ I - \Phi_{A_1A_2} \Phi_{A_2A_1} \right\} 1, \tag{3.12}$$

$$\Phi_{A_2A_3} = 1 - \Phi_{A_2A_1} 1. \tag{3.13}$$

((3.12) and (3.13) are obtained from $Q_{A_iA_1} 1 + Q_{A_iA_2} 1 + Q_{A_iA_3} = 1$.) Then, with the initial probability vector ($\alpha_{A_1}{}^T, \alpha_{A_2}{}^T, 0$)^T, Corollay 2.6 is again seen to hold.

For the case n = 3, the following alternative to (3.2) holds.

COROLLARY 3.4. *The* A_i-*entries* (i = 1, 2, 3) *of the probability generating function of* **M** *are given by*

$$g_{A_i}(z) = \left\{ I - \sum_{\substack{j=1 \\ j \neq i}}^{3} z_i z_j \Phi_{A_iA_j} (I - z_j z_{c(i,j)} \Phi_{A_jA_{c(i,j)}} \Phi_{A_{c(i,j)}A_j})^{-1} \Phi_{A_jA_i} \right\}^{-1} \times$$

$$\left\{ \sum_{\substack{j=1 \\ j \neq i}}^{3} z_1 z_2 z_3 \Phi_{A_iA_j} (I - z_j z_{c(i,j)} \Phi_{A_jA_{c(i,j)}} \Phi_{A_{c(i,j)}A_j})^{-1} \Phi_{A_jA_{c(i,j)}} \Phi_{A_{c(i,j)}A_4} + \right.$$

$$\left. \sum_{\substack{j=1 \\ j \neq i}}^{3} z_i z_j \Phi_{A_iA_j} \Phi_{A_jA_4} + z_i \Phi_{A_iA_4} \right\}, \tag{3.14}$$

where c(i, j) *is defined for* i, j $\in$ { 1, 2, 3 }, i $\neq$ j, *by* {c(i, j)} = { 1, 2, 3 }\{ i, j }.

PROOF OF COROLLARY 3.4. For i = 3, (3.7) reads as follows

$$g_{A_3}(z) = V_{A_3A_1}(z) g_{A_1}(z) + V_{A_3A_2}(z) g_{A_2}(z) + V_{A_3A_4}(z). \tag{3.15}$$

Substitute (3.15) into (3.7) (with i = 1, 2) to get

$$\left\{ I - V_{A_1A_3}(z) V_{A_3A_1}(z) \right\} g_{A_1}(z) =$$

$$\left\{ V_{A_1A_2}(z) + V_{A_1A_3}(z) V_{A_3A_2}(z) \right\} g_{A_2}(z) +$$

$$V_{A_1A_3}(z) V_{A_3A_4}(z) + V_{A_1A_4}(z), \tag{3.16}$$

$$\left\{ \mathbf{I} - \mathbf{V}_{A_2A_3}(\mathbf{z}) \ \mathbf{V}_{A_3A_2}(\mathbf{z}) \right\} \ \mathbf{g}_{A_2}(\mathbf{z}) =$$

$$\left\{ \mathbf{V}_{A_2A_1}(\mathbf{z}) + \mathbf{V}_{A_2A_3}(\mathbf{z}) \ \mathbf{V}_{A_3A_1}(\mathbf{z}) \right\} \ \mathbf{g}_{A_1}(\mathbf{z}) +$$

$$\mathbf{V}_{A_2A_3}(\mathbf{z}) \ \mathbf{V}_{A_3A_4}(\mathbf{z}) + \mathbf{V}_{A_2A_4}(\mathbf{z}). \tag{3.17}$$

The transition from (3.16) and (3.17) to an equation in terms of $\mathbf{g}_{A_1}(\mathbf{z})$ alone is straightforward. It is as follows

$$\left\{ \mathbf{I} - \mathbf{V}_{A_1A_3}(\mathbf{z}) \ \mathbf{V}_{A_3A_1}(\mathbf{z}) \right\} \ \mathbf{g}_{A_1}(\mathbf{z}) =$$

$$\left\{ \mathbf{V}_{A_1A_2}(\mathbf{z}) + \mathbf{V}_{A_1A_3}(\mathbf{z}) \ \mathbf{V}_{A_3A_2}(\mathbf{z}) \right\} \left\{ \mathbf{I} - \mathbf{V}_{A_2A_3}(\mathbf{z}) \ \mathbf{V}_{A_3A_2}(\mathbf{z}) \right\}^{-1} \times$$

$$\left\{ \left\{ \mathbf{V}_{A_2A_1}(\mathbf{z}) + \mathbf{V}_{A_2A_3}(\mathbf{z}) \ \mathbf{V}_{A_3A_1}(\mathbf{z}) \right\} \ \mathbf{g}_{A_1}(\mathbf{z}) + \mathbf{V}_{A_2A_4}(\mathbf{z}) + \right.$$

$$\left. \mathbf{V}_{A_2A_3}(\mathbf{z}) \ \mathbf{V}_{A_3A_4}(\mathbf{z}) \right\} + \mathbf{V}_{A_1A_4}(\mathbf{z}) + \mathbf{V}_{A_1A_3}(\mathbf{z}) \ \mathbf{V}_{A_3A_4}(\mathbf{z}). \tag{3.18}$$

However, the structure of (3.18) is *not* symmetric in the indices 2 and 3. (From a formal point of view, it is desirable to establish an expression for $\mathbf{g}_{A_1}(\mathbf{z})$ which has this property; (3.14) is indeed symmetric in j, c(i,j).) It is easily seen though, that

$$\mathbf{V}_{A_1A_3}(\mathbf{z}) \ \mathbf{V}_{A_3A_2}(\mathbf{z}) \left\{ \mathbf{I} - \mathbf{V}_{A_2A_3}(\mathbf{z}) \ \mathbf{V}_{A_3A_2}(\mathbf{z}) \right\}^{-1} =$$

$$\mathbf{V}_{A_1A_3}(\mathbf{z}) \left\{ \mathbf{I} - \mathbf{V}_{A_3A_2}(\mathbf{z}) \ \mathbf{V}_{A_2A_3}(\mathbf{z}) \right\}^{-1} \mathbf{V})_{A_3A_2}(\mathbf{z}). \tag{3.19}$$

After some algebra, (3.18) and (3.19) give (3.14) for i = 1. The other cases are treated in a similar fashion. ■

The remainder of this section is devoted to the evaluation of the probability mass function of the joint distribution of M_{A_1} and M_{A_2} with n = 3. In view of the application in Section 3.3, Z_0, the initial state of Z, is assumed to be in A_1; the vector of probability generating functions under consideration is thus $\mathbf{h}(z_1, z_2) = \mathbf{g}_{A_1}(z_1, z_2, 1)$. (3.14) (for i = 1 and $z_3 = 1$) is easily rewritten in the form

$$\mathbf{h}(z_1, z_2) = \left\{ z_1 \ \boldsymbol{\Phi}_{A_1A_3} (\mathbf{I} - z_2 \ \boldsymbol{\Phi}_{A_3A_2} \ \boldsymbol{\Phi}_{A_2A_3})^{-1} \ \boldsymbol{\Phi}_{A_3A_1} + \right.$$

$$\left. z_1 \ z_2 \ \boldsymbol{\Phi}_{A_1A_2} (\mathbf{I} - z_2 \ \boldsymbol{\Phi}_{A_2A_3} \ \boldsymbol{\Phi}_{A_3A_2})^{-1} \ \boldsymbol{\Phi}_{A_2A_1} \right\} \ \mathbf{h}(z_1, z_2) +$$

$$z_1\, z_2\, \Phi_{A_1A_3} \left(I - z_2\, \Phi_{A_3A_2}\, \Phi_{A_2A_3}\right)^{-1} \Phi_{A_3A_2}\, \Phi_{A_2A_4} +$$

$$z_1\, z_2\, \Phi_{A_1A_2} \left(I - z_2\, \Phi_{A_2A_3}\, \Phi_{A_3A_2}\right)^{-1} \Phi_{A_2A_3}\, \Phi_{A_3A_4} +$$

$$z_1\, z_2\, \Phi_{A_1A_2}\, \Phi_{A_2A_4} + z_1 \left\{ \Phi_{A_1A_3}\, \Phi_{A_3A_4} + \Phi_{A_1A_4} \right\}. \tag{3.20}$$

A power series expansion of the appropriate terms in (3.20) gives

$$h(z_1, z_2) = \sum_{\ell=0}^{\infty} z_1 z_2^{\ell}\, \Psi_\ell\, h(z_1, z_2) + \sum_{\ell=0}^{\infty} z_1 z_2^{\ell}\, \mathbf{y}_\ell, \tag{3.21}$$

where

$$\Psi_0 = \Phi_{A_1A_3}\, \Phi_{A_3A_1}, \tag{3.22}$$

and for $\ell \geq 1$,

$$\Psi_\ell = \Phi_{A_1A_3} \left(\Phi_{A_3A_2}\, \Phi_{A_2A_3}\right)^{\ell} \Phi_{A_3A_1} + \Phi_{A_1A_2} \left(\Phi_{A_2A_3}\, \Phi_{A_3A_2}\right)^{\ell-1} \Phi_{A_2A_1}, \tag{3.23}$$

and

$$\mathbf{y}_0 = \Phi_{A_1A_4} + \Phi_{A_1A_3}\, \Phi_{A_3A_4}, \tag{3.24}$$

$$\mathbf{y}_1 = \Phi_{A_1A_3}\, \Phi_{A_3A_2}\, \Phi_{A_2A_4} + \Phi_{A_1A_2}\, \Phi_{A_2A_3}\, \Phi_{A_3A_4} + \Phi_{A_1A_2}\, \Phi_{A_2A_4}, \tag{3.25}$$

and for $\ell \geq 2$,

$$\mathbf{y}_\ell = \Phi_{A_1A_3} \left(\Phi_{A_3A_2}\, \Phi_{A_2A_3}\right)^{\ell-1} \Phi_{A_3A_2}\, \Phi_{A_2A_4} +$$

$$\Phi_{A_1A_2} \left(\Phi_{A_2A_3}\, \Phi_{A_3A_2}\right)^{\ell-1} \Phi_{A_2A_3}\, \Phi_{A_3A_4}. \tag{3.26}$$

We are now in a position to derive explicit expressions for the family of functions $\mathbf{u}(m_1, m_2) = \{\Pr\{ M_{A_1} = m_1, M_{A_2} = m_2 \mid Z_0 = s \} : s \in A_1 \}$, $m_1, m_2 \geq 0$, where $n = 3$, i.e., where S is partitioned as $S = A_1 \cup A_2 \cup A_3 \cup \{\omega\}$. For a concise representation of $\mathbf{u}$, the right direct product (Kronecker product) of matrices will be used, as defined in (1.5).

THEOREM 3.5. *The probability mass functions* $\mathbf{u}$ *are given in terms of the matrices in (3.22) - (3.26) for* $m_1, m_2 \geq 0$ *by*

$$(\mathbf{u}(0, 0)^T, ..., \mathbf{u}(0, m_2)^T)^T = \mathbf{0}, \tag{3.27}$$

$$(\mathbf{u}(m_1, 0)^T, ..., \mathbf{u}(m_1, m_2)^T)^T = \Omega(m_2)^{m_1-1} (\mathbf{y}_0^T, ..., \mathbf{y}_{m_2}^T)^T, \quad m_1 \geq 1, \tag{3.28}$$

where $\Omega(m_2)$ is defined by

$$\Omega(m_2) = \sum_{\ell=1}^{m_2+1} \Psi_{\ell-1} \otimes \Theta (m_2+1,\ell), \tag{3.29}$$

with the square matrix $\Theta(i, j)$ *of size* i *whose components are given by*

$$\Theta(i,j)_{p\,r} = \begin{cases} 1 \ for \ p - r + 1 = j, \\ 0 \ otherwise. \end{cases}$$

(Notice that $\Theta(i, j) = E(i, j)^T$ *with* $E(i, j)$ *defined in (2.40).)*

PROOF OF THEOREM 3.5. The Taylor series representation of $\mathbf{h}$ about $\mathbf{0} \in \mathbb{C}^2$ is given by

$$\mathbf{h} = \sum_{m_1,m_2=0}^{+\infty} \mathbf{u}(m_1, m_2) \, z_1^{m_1} \, z_2^{m_2}. \tag{3.30}$$

Substitute (3.30) into (3.21) and compare coefficients to get

$$\mathbf{u}(0, m_2) = \mathbf{0}, \ \ m_2 \geq 0, \tag{3.31}$$

and

$$\mathbf{u}(m_1, m_2) = \begin{cases} \displaystyle\sum_{\ell=0}^{m_2} \Psi_\ell \, \mathbf{u}(m_1-1, m_2-\ell) + \mathcal{Y}_{m_2} & for \ m_1 = 1, \ m_2 \geq 0, \\[4mm] \displaystyle\sum_{\ell=0}^{m_2} \Psi_\ell \, \mathbf{u}(m_1-1, m_2-\ell) & for \ m_1 \geq 2, \ m_2 \geq 0. \end{cases} \tag{3.32}$$

(3.31) implies (3.27). (3.31) and (3.32) (with $m_1 = 1$) gives (3.28) for $m_1 = 1$. (3.28) for $m_1 \geq 2$ is obtained by rewriting the second line in (3.32) as

$$(\mathbf{u}(m_1, 0)^T, \mathbf{u}(m_1, 1)^T, \mathbf{u}(m_1, 2)^T, ..., \mathbf{u}(m_1, m_2)^T)^T =$$

$$
\begin{bmatrix}
\boldsymbol{\Psi}_0 & 0 & & 0 \\
\boldsymbol{\Psi}_1 & \boldsymbol{\Psi}_0 & & 0 \\
\vdots & \vdots & \ddots & \vdots \\
\boldsymbol{\Psi}_{m_2} & \boldsymbol{\Psi}_{m_2-1} & & \boldsymbol{\Psi}_0
\end{bmatrix}
\begin{bmatrix}
\mathbf{u}(m_1-1, 0) \\
\mathbf{u}(m_1-1, 1) \\
\vdots \\
\mathbf{u}(m_1-1, m_2)
\end{bmatrix}.
\tag{3.33}
$$

By (3.29), (3.33) can be rewritten in the form

$$
(\mathbf{u}(m_1, 0)^T, \mathbf{u}(m_1, 1)^T, \mathbf{u}(m_1, 2)^T, ..., \mathbf{u}(m_1, m_2)^T)^T =
$$

$$
\boldsymbol{\Omega}(m_2) \, (\mathbf{u}(m_1-1, 0)^T, \mathbf{u}(m_1-1, 1)^T, \mathbf{u}(m_1-1, 2)^T, ..., \mathbf{u}(m_1-1, m_2)^T)^T.
\tag{3.34}
$$

(3.28) with $m_1 = 1$, in conjunction with (3.34) gives (3.28) for general m_1. ∎

We conclude this section with the cumulative distribution function of (M_{A_1}, M_{A_2}).

COROLLARY 3.6. *Assume that the initial probability vector of* Y *is* $\boldsymbol{\alpha}_{A_1}$. *Then, the cumulative distribution function of* (M_{A_1}, M_{A_2}) *is given for* m_1, $m_2 \geq 1$ *by*

$$
\Pr\{\, M_{A_1} \leq m_1, M_{A_2} \leq m_2 \,\} =
$$

$$
(\boldsymbol{\alpha}_{A_1}{}^T, ..., \boldsymbol{\alpha}_{A_1}{}^T) \left(\mathbf{I} - \boldsymbol{\Omega}(m_2)^{m_1} \right) \left(\mathbf{I} - \boldsymbol{\Omega}(m_2) \right)^{-1} (\boldsymbol{\gamma}_0{}^T, ..., \boldsymbol{\gamma}_{m_2}{}^T)^T.
\tag{3.35}
$$

∎

3.3 Tabular summary of results in Sections 3.1 and 3.2

In Table 3.1 a guide is shown to the results in Sections 3.1 and 3.2.

Variable	Characteristic	Reference
$(M_{A_1}, ..., M_{A_n})$	pgf	Theorem 3.1
$M_{A_1} + ... + M_{A_n}$	pmf	Corollary 3.2
(M_{A_1}, M_{A_2}) for $n = 2$	pgf	Corollary 3.3
$(M_{A_1}, M_{A_1}, M_{A_3})$ for $n = 3$	pgf	Corollary 3.4
(M_{A_1}, M_{A_2}) for $n = 3$	pmf	Theorem 3.5
(M_{A_1}, M_{A_2}) for $n = 3$	cdf	Corollary 3.6

Table 3.1 A guide to the distribution results in Sections 3.1 and 3.2

3.4 A power transmission reliability application

3.4.1 Numerical results

The example considered here is Model 4 from Section 1.2.1. The following parameter values have been assumed for the numerical computations:

Normal weather component failure rate:	$p_0 = 5.0$ / year;
Stormy weather component failure rate:	$p_1 = 250.0$ / year;
Component repair rate:	$\mu = 11.68$ / year;
Rate of change from normal to stormy weather:	$v_0 = 43.8$ / year;
Rate of change from stormy to normal weather:	$v_1 = 5840$ / year.

(The means of the normal and stormy periods and the mean repair times, are thus respectively given by $1 / v_0 = 200$ hours, $1 / v_1 = 1.5$ hours, and $1 / \mu = 750$ hours.) Notice, however, that the failure and repair rates in the above parameter set are not quite typical for the power transmission application; they have been selected with a view of accomodating the implementation within the system available at the time of coding this problem; the version of MATLAB for this example was what is called the 'educational version' on the Apple Macintosh SE/30. This will not, for example, admit square matrices whose size exceeds 90. This and other (mainly computing-time related) limitations made us to select the above parameter values. No doubt, the full version of MATLAB will allow any realistic parameter set to be adopted. The probability mass function of (M_{A_1}, M_{A_2}) is shown overleaf in Table 3.2 for $Y_0 = (0, 1, 1, 1)$. By direct summation (or by (3.35)), the total probability thus covered is 0.931. From Table 3.2 it is seen that for fixed m_2, the distribution $\Pr\{ M_{A_1} \in . \mid M_{A_2} = m_2 \}$ is unimodal; the same holds for the distribution $\Pr\{ M_{A_2} \in . \mid M_{A_1} = m_1 \}$ for fixed m_1. The computing (CPU-)times were as follows. Calculation of $\Omega(m_2)$: 33 sec; evaluation of (3.35): 59 sec; generation of Table 3.2: 92 sec. The size of the MATLAB code is 161 lines (comment lines excluded). The MATLAB code is shown in Section 3.4.2. The only new feature of MATLAB used here is the formatted output by fprintf which is adopted from the C language.

m_1	$m_2 = 1$	$m_2 = 2$	$m_2 = 3$	$m_2 = 4$	$m_2 = 5$	$m_2 = 6$	$m_2 = 7$	$m_2 = 8$	$m_2 = 9$	$m_2 = 10$
1	1.26_1	5.15_2	2.11_2	8.66_3	3.55_3	1.46_3	5.97_4	2.45_4	1.00_4	4.11_5
2	0.00	5.78_2	4.81_2	2.97_2	1.63_2	8.35_3	4.11_3	1.97_3	9.23_4	4.26_4
3	0.00	0.00	2.64_2	3.32_2	2.75_2	1.89_2	1.17_2	6.73_3	3.68_3	1.95_3
4	0.00	0.00	0.00	1.21_2	2.03_2	2.11_2	1.75_2	1.26_2	8.34_3	5.15_3
5	0.00	0.00	0.00	0.00	5.54_3	1.16_2	1.46_2	1.41_2	1.17_2	8.69_3
6	0.00	0.00	0.00	0.00	0.00	2.53_3	6.40_3	9.37_3	1.04_2	9.70_3
7	0.00	0.00	0.00	0.00	0.00	0.00	1.16_3	3.42_3	5.73_3	7.17_3
8	0.00	0.00	0.00	0.00	0.00	0.00	0.00	5.30_4	1.79_3	3.38_3
9	0.00	0.00	0.00	0.00	0.00	0.00	0.00	0.00	2.43_4	9.22_4
10	0.00	0.00	0.00	0.00	0.00	0.00	0.00	0.00	0.00	1.11_4
11	0.00	0.00	0.00	0.00	0.00	0.00	0.00	0.00	0.00	0.00

m_1	$m_2 = 11$	$m_2 = 12$	$m_2 = 13$	$m_2 = 14$	$m_2 = 15$	$m_2 = 16$	$m_2 = 17$	$m_2 = 18$	$m_2 = 19$	$m_2 = 20$
1	1.69_5	6.91_6	2.83_6	1.16_6	4.76_7	1.95_7	8.00_8	3.28_8	1.34_8	5.51_9
2	1.94_4	8.75_5	3.91_5	1.74_5	7.67_6	3.37_6	1.48_6	6.43_7	2.79_7	1.21_7
3	9.98_4	5.01_4	2.47_4	1.20_4	5.72_5	2.71_5	1.27_5	5.90_6	2.72_6	1.24_6
4	3.02_3	1.71_3	9.36_4	5.00_4	2.61_4	1.34_4	6.77_5	3.37_5	1.66_5	8.10_6
5	5.97_3	3.86_3	2.39_3	1.42_3	8.16_4	4.57_4	2.50_4	1.35_4	7.10_5	3.69_5
6	8.03_3	6.08_3	4.30_3	2.88_3	1.84_3	1.14_3	6.81_4	3.97_4	2.26_4	1.26_4
7	7.44_3	6.79_3	5.62_3	4.31_3	3.11_3	2.14_3	1.41_3	8.97_4	5.54_4	3.33_4
8	4.70_3	5.38_3	5.36_3	4.81_3	3.98_3	3.08_3	2.26_3	1.59_3	1.07_3	6.98_4
9	1.94_3	2.97_3	3.71_3	4.01_3	3.88_3	3.44_3	2.85_3	2.22_3	1.65_3	1.18_3
10	4.69_4	1.08_3	1.81_3	2.46_3	2.87_3	2.98_3	2.82_3	2.48_3	2.05_3	1.61_3
11	5.08_5	2.36_4	5.96_4	1.08_3	1.58_3	1.98_3	2.19_3	2.21_3	2.06_3	1.80_3
12	0.00	2.33_5	1.18_4	3.23_4	6.31_4	9.89_4	1.32_3	1.56_3	1.66_3	1.64_3
13	0.00	0.00	1.06_5	5.85_5	1.72_4	3.61_4	6.05_4	8.58_4	1.07_3	1.21_3
14	0.00	0.00	0.00	4.87_6	2.88_5	9.11_5	2.04_4	3.63_4	5.45_4	7.20_4
15	0.00	0.00	0.00	0.00	2.23_6	1.41_5	4.76_5	1.13_4	2.14_4	3.39_4
16	0.00	0.00	0.00	0.00	0.00	1.02_6	6.90_6	2.47_5	6.23_5	1.24_4
17	0.00	0.00	0.00	0.00	0.00	0.00	4.66_7	3.36_6	1.27_5	3.39_5
18	0.00	0.00	0.00	0.00	0.00	0.00	0.00	2.13_7	1.63_6	6.51_6
19	0.00	0.00	0.00	0.00	0.00	0.00	0.00	0.00	9.77_8	7.85_7
20	0.00	0.00	0.00	0.00	0.00	0.00	0.00	0.00	0.00	4.47_8

Table 3.2 Values of $\Pr\{ M_1 = m_1, M_2 = m_2 \mid Y_0 = (0, 1, 1, 1) \}$
(Negative powers of 10 are indicated by a subscript, e.g., 5.08_5 stands for 5.08×10^{-5}.)

3.4.2 MATLAB code

```
% MATLAB IMPLEMENTATION OF (3.28)

clear
disp('Parameter values assumed ...');
rho0 = 5.0
rho1 = 250.0
mu = 11.68
nu0 = 43.8
nu1 = 5840.0

disp('Now calculating Q...');

J0 = [-mu    mu; rho0  -rho0];
J1 = [-mu    mu; rho1  -rho1];

LE0E0 = kron(kron(eye(2),J0)+kron(J0,eye(2)),eye(2))+kron(eye(4),J0)-...
                                             nu0*eye(8);
LE1E1 = kron(kron(eye(2),J1)+kron(J1,eye(2)),eye(2))+kron(eye(4),J1)-...
                                             nu1*eye(8);

% State enumeration for Y:
%     1              2            3            4            5            6
% (0,0,0,0)->(0,0,0,1)->(0,0,1,0)->(0,0,1,1),->(0,1,0,0)->(0,1,0,1)->
%     7              8
% (0,1,1,0)->(0,1,1,1)->
%     9              10           11           12           13           14
% (1,0,0,0)->(1,0,0,1)->(1,0,1,0)->(1,0,1,1),->(1,1,0,0)->(1,1,0,1)->
%     15             16
% (1,1,1,0)->(1,1,1,1);

L = [LE0E0 nu0*eye(8); nu1*eye(8) LE1E1];

% State enumeration for absorbing chains:
%     1            2            3            4            5            6
% omega->(0,0,0,1)->(0,0,1,0)->(0,0,1,1),->(0,1,0,0)->(0,1,0,1)->
%              7            8
%        (0,1,1,0)->(0,1,1,1)->
%              9            10           11           12           13
%        (1,0,0,1)->(1,0,1,0)->(1,0,1,1),->(1,1,0,0)->(1,1,0,1)->
%              14           15
%        (1,1,1,0)->(1,1,1,1);

% PARTITIONING OF S: A1 = {8, 15},
%                    A2 = {4, 6 & 7, 11, 13 & 14},
%                    A3 = {2, 3 & 5, 9 & 10, 12},
%                    A4 = {1};

disp('Calculating the transition-rate matrix of the abs. Markov
process...')
LABS =[zeros(1:15)
       L(2:8,1:8)  L(2:8,10:16)
       L(10:16,9:9)  L(10:16,2:8)  L(10:16,10:16)];

for i=1:15
    if i==1
        Q(1,i)=1.0;
    else
        Q(1,i)=0.0;
```

```
        end
end

for i=2:15
    for j=1:15
        if i==j
            Q(i,j)=0.0;
        else
            Q(i,j)=-LABS(i,j)/LABS(i,i);
        end
    end
end

disp('The transition probability matrix Q of X is available.')

QA1A1 = [Q(8,8)   Q(8,15)
         Q(15,8)  Q(15,15)];
QA1A2 = [Q(8,4)       Q(8,6:7)      Q(8,11)     Q(8,13:14)
         Q(15,4)      Q(15,6:7)     Q(15,11)    Q(15,13:14)];
QA1A3 = [Q(8,2:3)  Q(8,5)  Q(8,9:10)  Q(8,12)
         Q(15,2:3) Q(15,5) Q(15,9:10) Q(15,12)];
QA1A4 = [Q(8,1)
         Q(15,1)];
QA2A1 = [Q(4,8)        Q(4,15)
         Q(6:7,8)      Q(6:7,15)
         Q(11,8)       Q(11,15)
         Q(13:14,8)    Q(13:14,15)];
QA2A2 = [Q(4,4)        Q(4,6:7)      Q(4,11)      Q(4,13:14)
         Q(6:7,4)      Q(6:7,6:7)    Q(6:7,11)    Q(6:7,13:14)
         Q(11,4)       Q(11,6:7)     Q(11,11)     Q(11,13:14)
         Q(13:14,4)    Q(13:14,6:7)  Q(13:14,11)  Q(13:14,13:14)];
QA2A3 = [Q(4,2:3)      Q(4,5)        Q(4,9:10)      Q(4,12)
         Q(6:7,2:3)    Q(6:7,5)      Q(6:7,9:10)    Q(6:7,12)
         Q(11,2:3)     Q(11,5)       Q(11,9:10)     Q(11,12)
         Q(13:14,2:3)  Q(13:14,5)    Q(13:14,9:10)  Q(13:14,12)];
QA2A4 = [Q(4,1)
         Q(6:7,1)
         Q(11,1)
         Q(13:14,1)];
QA3A1 = [Q(2:3,8)   Q(2:3,15)
         Q(5,8)     Q(5,15)
         Q(9:10,8)  Q(9:10,15)
         Q(12,8)    Q(12,15)];
QA3A2 = [Q(2:3,4)   Q(2:3,6:7)    Q(2:3,11)   Q(2:3,13:14)
         Q(5,4)     Q(5,6:7)      Q(5,11)     Q(5,13:14)
         Q(9:10,4)  Q(9:10,6:7)   Q(9:10,11)  Q(9:10,13:14)
         Q(12,4)    Q(12,6:7)     Q(12,11)    Q(12,13:14)];
QA3A3 = [Q(2:3,2:3)    Q(2:3,5)   Q(2:3,9:10)   Q(2:3,12)
         Q(5,2:3)      Q(5,5)     Q(5,9:10)     Q(5,12)
         Q(9:10,2:3)   Q(9:10,5)  Q(9:10,9:10)  Q(9:10,12)
         Q(12,2:3)     Q(12,5)    Q(12,9:10)    Q(12,12)];
QA3A4 = [Q(2:3,1)
         Q(5,1)
         Q(9:10,1)
         Q(12,1)];

PHIA1A1 = inv(eye(2)-QA1A1)*QA1A1;
PHIA1A2 = inv(eye(2)-QA1A1)*QA1A2;
PHIA1A3 = inv(eye(2)-QA1A1)*QA1A3;
PHIA1A4 = inv(eye(2)-QA1A1)*QA1A4;

PHIA2A1 = inv(eye(6)-QA2A2)*QA2A1;
PHIA2A2 = inv(eye(6)-QA2A2)*QA2A2;
PHIA2A3 = inv(eye(6)-QA2A2)*QA2A3;
PHIA2A4 = inv(eye(6)-QA2A2)*QA2A4;
```

```
PHIA3A1 = inv(eye(6)-QA3A3)*QA3A1;
PHIA3A2 = inv(eye(6)-QA3A3)*QA3A2;
PHIA3A3 = inv(eye(6)-QA3A3)*QA3A3;
PHIA3A4 = inv(eye(6)-QA3A3)*QA3A4;

indicator = 1;
indicator2 = 1;
while indicator==1
while indicator2==1
m2 = input('Enter (2 ≤) m2 :');
fix(clock)

gam = [];

for el=0:m2
  if el==0
    gam = [gam (PHIA1A4 + PHIA1A3*PHIA3A4)];
  elseif el==1
    gam = [gam (PHIA1A3*PHIA3A2*PHIA2A4 + PHIA1A2*PHIA2A3*PHIA3A4 +
PHIA1A2*PHIA2A4)];
  else
    gam = [gam (PHIA1A3*((PHIA3A2*PHIA2A3)^(el-1))*PHIA3A2*PHIA2A4 + ...
          PHIA1A2*((PHIA2A3*PHIA3A2)^(el-1))*PHIA2A3*PHIA3A4)];
  end
end
disp('The vectors gamma0, ..., gammam2 are now available.');

PSI = [];
for el=0:m2
    if el==0
        PSI = [PSI; PHIA1A3*PHIA3A1];
    else
        PSI = [PSI; (PHIA1A3*((PHIA3A2*PHIA2A3)^el)*PHIA3A1 + ...
              PHIA1A2*((PHIA2A3*PHIA3A2)^(el-1))*PHIA2A1)];
    end
end

disp('The matrices PSI0, ..., PSIm2 are now available.' );

[row col] = size(PSI);
TEMP = PSI;
OMEGA = [];
for el=0:m2
    if el>=1
        TEMP = [zeros(col,row); [eye(row-col) zeros(row-col,col)]]*TEMP;
    end
    OMEGA = [OMEGA TEMP];
end

disp('The matrix OMEGA is now available.');
fix(clock)

disp('The total probabilities covered will be as follows:');
alphaA1 = kron(ones(1,(m2+1)),[1 0]);
totprob = alphaA1*(eye(OMEGA) - OMEGA^(m2+1))*inv(eye(OMEGA) -...
                                            OMEGA)*gam(:);
disp(fprintf(...
'\n Pr[M1, M2 = 0, ...,%3.0f | s = (0,1,1,1)] = %17.10e\n',m2,totprob));
alphaA1 = kron(ones(1,(m2+1)),[0 1]);
totprob = alphaA1*inv(eye(OMEGA) - OMEGA)*gam(:);
disp(fprintf(...
'\n Pr[M1, M2 = 0, ...,%3.0f | s = (1,1,1,1)] = %17.10e\n',m2,totprob));
indicator2 = input('Chose new m2 ? (type ''1'' for "yes") ');
end
indicator2 = 1;
```

```
fix(clock)
probsum0 = 0.0;
probsum1 = 0.0;
OMEGAPOW = eye(OMEGA);
for m1=1:(m2+1)
    temp = OMEGAPOW*gam(:);
    for m=1:(m2+1)
        x = temp(2*m-1,1);
        if x>0.0
        disp(fprintf(...
        '\n Pr[M1=%3.0f,M2=%3.0f | s = (0,1,1,1)] = %17.10e\n',m1,m-1,x));
        end
        probsum0 = probsum0 + x;
        x = temp(2*m,1);
        if x>0.0
        disp(fprintf(...
        '\n Pr[M1=%3.0f,M2=%3.0f | s = (1,1,1,1)] = %17.10e\n',m1,m-1,x));
        end
        probsum1 = probsum1 + x;
    end
OMEGAPOW = OMEGAPOW*OMEGA;
end

disp('The total probabilities covered are as follows:');
disp(fprintf(...
'\n Pr[M1, M2 = 0, ...,%3.0f | s = (0,1,1,1)] = %17.10e\n',m2,probsum0));
disp(fprintf(...
'\n Pr[M1, M2 = 0, ...,%3.0f | s = (1,1,1,1)] = %17.10e\n',m2,probsum1));
fix(clock)

indicator = input('Continue ? (type ''1'' for "yes") ');
end
```

CHAPTER 4

SOJOURN TIMES FOR CONTINUOUS-PARAMETER MARKOV CHAINS

Let $Y = \{\, Y_t : t \geq 0 \,\}$ be a continuous-parameter Markov process with finite state space S. Y is assumed either irreducible, in which case S is partitioned as $S = A_1 \cup A_2$, or absorbing, in which case it has a single absorbing state ω, and then S is written as $S = A_1 \cup A_2 \cup \{\omega\}$, again with disjoint and non-empty A_1 and A_2. It is assumed that Y starts in one of the states in $A_1 \cup A_2$ at time $t = 0$. The transition rate matrix Λ and the initial probability vector α of Y are written in a block partitioned form as

$$
\Lambda = \begin{array}{c} \\ A_1 \\ A_2 \end{array}
\overset{\displaystyle A_1 \qquad\quad A_2}{
\left[\begin{array}{c|c} \Lambda_{A_1 A_1} & \Lambda_{A_1 A_2} \\ \hline \Lambda_{A_2 A_1} & \Lambda_{A_2 A_2} \end{array}\right]},
\quad
\alpha = \begin{array}{c} A_1 \\ A_2 \end{array}
\left[\begin{array}{c} \alpha_{A_1} \\ \hline \alpha_{A_2} \end{array}\right],
$$

and, for absorbing Y,

$$
\Lambda = \begin{array}{c} A_1 \\ A_2 \\ \{\omega\} \end{array}
\overset{\displaystyle A_1 \qquad\; A_2 \qquad \{\omega\}}{
\left[\begin{array}{c|c|c} \Lambda_{A_1 A_1} & \Lambda_{A_1 A_2} & \Lambda_{A_1 \{\omega\}} \\ \hline \Lambda_{A_2 A_1} & \Lambda_{A_2 A_2} & \Lambda_{A_2 \{\omega\}} \\ \hline 0 & 0 & 0 \end{array}\right]},
\quad
\alpha = \begin{array}{c} A_1 \\ A_2 \\ \{\omega\} \end{array}
\left[\begin{array}{c} \alpha_{A_1} \\ \hline \alpha_{A_2} \\ \hline \alpha_\omega \end{array}\right].
$$

The nth sojourn time of Y in A_i $(i = 1, 2)$ is the time spent by Y in A_i upon its nth visit to A_i; it will be denoted by $T_{A_i,n}$. This definition is adequate for irreducible Y since in this case Y alternates between A_1 and A_2 indefinitely and hence the nth visit to A_i is certain to take place at some point in time. However, if Y is absorbing then it may be absorbed into ω before its nth visit to A_i. Then, we define $T_{A_i,n} = 0$.

In Section 4.1, the joint distribution of any finite collection of the sojourn times in A_1 and A_2 will be derived. In Section 4.2, a Laplace transform formulation of the same result is used to arrive at the distribution of $TS_{A_1,m} = T_{A_1,1} + \dots + T_{A_1,m}$, the mth cumulative sojourn time of Y in A_1. Furthermore, for absorbing Y, the variables $\max\{T_{A_1,1}, T_{A_1,2}, \dots\}$ and $TS_{A_1,\infty} = T_{A_1,1} + T_{A_1,2} + \dots$ will also be considered. Section 4.3 gives a tabular guide to the distribution results in Chapter 4. In Section 4.4, the theory is applied to analyse some aspects of the transient behaviour of the three-component power transmission system modelled by Model 2 and Model 3 from Section 1.2.1. The core material in this chapter is based on research first reported in [CSE4] and [CSE6].

4.1 Distribution theory for sojourn times

In what follows the matrix exponential will play a vital role. This is defined for any square matrix $\mathbf{V}$, say, by the matrix power series

$$\exp\{\,\mathbf{V}\,\} = \sum_{n=0}^{\infty} \frac{\mathbf{V}^n}{n!}\,.$$

The matrix exponential is most commonly encountered in Probability Theory in the solution of what is known as the Forward Kolmogorov Equations

$$\mathbf{P}'(t) = \mathbf{P}(t)\,\mathbf{\Lambda}, \tag{4.1}$$

where $\mathbf{P}(t) = \{\,P_{s_1 s_2}(t) : s_1, s_2 \in S\,\}$ is the transition probability matrix of Y, i.e., $P_{s_1 s_2}(t) = \Pr\{\,Y_t = s_2 \mid Y_0 = s_1\,\}$. (See, for example, Bhat [BHA2], Section 7.7.) The solution of (4.1) is, with the self-evident initial condition $\mathbf{P}(0) = \mathbf{I}$, $\mathbf{P}(t) = \exp\{\,\mathbf{\Lambda}\,t\,\}$. Let us add for later reference that if Y is started according to the initial probability vector $\boldsymbol{\alpha}$, then the column vector $\mathbf{p}(t) = \{\,\Pr\{\,Y_t = s\,\} : s \in S\,\}$ obeys the equation $\mathbf{p}^T(t) = \boldsymbol{\alpha}^T\,\mathbf{P}(t)$. The Forward Kolmogorov Equations with this initial probability vector thus read as follows

$$\mathbf{p}'^T(t) = \mathbf{p}^T(t)\,\mathbf{\Lambda}. \tag{4.2}$$

The solution of (4.2) is of course $\mathbf{p}^T(t) = \boldsymbol{\alpha}^T \exp\{\,\mathbf{\Lambda}\,t\,\}$. Moler and Van Loan [MOL] have discussed several numerical techniques for the evaluation of the matrix exponential.

It is convenient at this stage to deal with a lemma about the asymptotic behaviour of the matrix exponential $\exp\{\,\mathbf{\Lambda}_{A_1 A_1}\,t\,\}$ as $t \to +\infty$.

LEMMA 4.1. *The matrix* $\exp\{\,\mathbf{\Lambda}_{A_1 A_1}\,t\,\}$ *tends to zero element-wise as* $t \to +\infty$.

PROOF OF LEMMA 4.1. We consider a continuous-time Markov chain $W = \{\,W_t : t \geq 0\,\}$ with state space $S = A_1 \cup A_2 \cup \{\omega\}$ and the following transition rate matrix

$$
\begin{array}{c}
\begin{array}{ccc} A_1 & A_2 & \{\omega\} \end{array} \\
\begin{array}{c} A_1 \\ A_2 \\ \{\omega\} \end{array}
\left[
\begin{array}{c|c|c}
\mathbf{\Lambda}_{A_1 A_1} & \mathbf{\Lambda}_{A_1 A_2} & \mathbf{\Lambda}_{A_1 \{\omega\}} \\
\hline
\mathbf{0} & \mathbf{0} & \mathbf{0} \\
\hline
\mathbf{0} & \mathbf{0} & \mathbf{0}
\end{array}
\right].
\end{array}
$$

(We may of course assume this same state space also if Y is irreducible. In that case we put $\mathbf{\Lambda}_{A_1\{\omega\}} = \mathbf{0}$ and $\mathbf{\Lambda}_{A_2\{\omega\}} = \mathbf{0}$ and Y is started in one of the states in $A_1 \cup A_2$.) The states of W in $A = A_2 \cup \{\omega\}$ are absorbing and those in A_1 are transient. Assume that W is started in A_1 according to some initial probability vector $\boldsymbol{\beta}$, i.e.,

$$\Pr\{\, W_0 = a_1 \,\} = \beta_{a_1}, a_1 \in A_1; \quad \sum_{a_1 \in A_1} \beta_{a_1} = 1. \tag{4.3}$$

The Forward Kolmogorov Equations (4.2) for W can be written in a partitioned form as

$$\left(\mathbf{p}_{A_1}{'}^T(t), \mathbf{p}_{A}{'}^T(t)\right) = \left(\mathbf{p}_{A_1}{}^T(t), \mathbf{p}_{A}{}^T(t)\right)
\begin{bmatrix}
\boldsymbol{\Lambda}_{A_1A_1} & \boldsymbol{\Lambda}_{A_1A_2} & \boldsymbol{\Lambda}_{A_1\{\omega\}} \\
0 & 0 & 0 \\
0 & 0 & 0
\end{bmatrix} =$$

$$\left(\mathbf{p}_{A_1}{}^T(t)\, \boldsymbol{\Lambda}_{A_1A_1}, \ \mathbf{p}_{A_1}{}^T(t)\, \boldsymbol{\Lambda}_{A_1A}\right). \tag{4.4}$$

Given the initial condition (4.3), the solution for the first set of equations in (4.4) (i.e., those corresponding to A_1) is given by

$$\mathbf{p}_{A_1}{}^T(t) = \boldsymbol{\beta}^T \exp\{\, \boldsymbol{\Lambda}_{A_1A_1}\, t \,\}.$$

But $\mathbf{p}_{A_1}(t)$ tends to zero element-wise as $t \to +\infty$ because A_1 is transient. The above holds for *any* stochastic vector $\boldsymbol{\beta}$, from which it follows that all row vectors of $\exp\{\, \boldsymbol{\Lambda}_{A_1A_1}\, t \,\}$ tend to zero as $t \to +\infty$. Thus, $\exp\{\, \boldsymbol{\Lambda}_{A_1A_1}\, t \,\}$ tends to zero as $t \to +\infty$. ∎

From the discrete-parameter theory we know that the distribution of the sojourn time vector in the irreducible case can be deduced from that in the absorbing case; cf. Corollary 2.3. A similar approach will be taken here: we first establish the result for the absorbing case from which the corresponding statement for irreducible Y readily follows. Let now Y be an absorbing chain on $A_1 \cup A_2 \cup \{\omega\}$. Then we have the following for the vector variable $(T_{A_1,1}, ..., T_{A_1,m})$.

THEOREM 4.2. *(a)* $\boldsymbol{\Lambda}_{A_1A_1}$ *and* $\boldsymbol{\Lambda}_{A_2A_2}$ *are invertible. (b) The joint distribution function of the first* m *sojourn times in* A_1 *is given by*

$$\Pr\{\, T_{A_1,1} \le t_1, ..., T_{A_1,m} \le t_m \,\} =$$

$$- \mathbf{w}^T \boldsymbol{\Lambda}_{A_1A_1}{}^{-1} \left\{ \mathbf{I} - \exp\{\, t_1\, \boldsymbol{\Lambda}_{A_1A_1} \,\} \right\} \times$$

$$\prod_{i=2}^{m} \left\{ \boldsymbol{\Lambda}_{A_1A_2}\, \boldsymbol{\Lambda}_{A_2A_2}{}^{-1}\, \boldsymbol{\Lambda}_{A_2A_1}\, \boldsymbol{\Lambda}_{A_1A_1}{}^{-1} \left\{ \mathbf{I} - \exp\{\, t_i\, \boldsymbol{\Lambda}_{A_1A_1} \,\} \right\} \right\} \times$$

$$\boldsymbol{\Lambda}_{A_1A_1}\, \mathbf{1} + \mathbf{1} + \mathbf{w}^T \mathbf{1} - \mathbf{w}^T \boldsymbol{\Lambda}_{A_1A_1}{}^{-1} \left\{ \mathbf{I} - \exp\{\, t_1\, \boldsymbol{\Lambda}_{A_1A_1} \,\} \right\} \times$$

$$\sum_{k=1}^{m-1} \prod_{i=2}^{k} \left\{ \boldsymbol{\Lambda}_{A_1A_2}\, \boldsymbol{\Lambda}_{A_2A_2}{}^{-1}\, \boldsymbol{\Lambda}_{A_2A_1}\, \boldsymbol{\Lambda}_{A_1A_1}{}^{-1} \left\{ \mathbf{I} - \exp\{\, t_i\, \boldsymbol{\Lambda}_{A_1A_1} \,\} \right\} \right\} \times$$

$$\{ \Lambda_{A_1A_1} - \Lambda_{A_1A_2} \Lambda_{A_2A_2}^{-1} \Lambda_{A_2A_1} \} \mathbf{1}, \tag{4.5}$$

where

$$\mathbf{w}^T = \boldsymbol{\alpha}_{A_2}^T \Lambda_{A_2A_2}^{-1} \Lambda_{A_2A_1} - \boldsymbol{\alpha}_{A_1}^T. \tag{4.6}$$

PROOF OF THEOREM 4.2. *Proof of (a).* The embedded Markov chain of Y is absorbing by definition; its transition probability matrix will be denoted by $\mathbf{P}$. The transition rate $\lambda_{s,s'}$ from s $\in S \setminus \{\omega\}$ to $s' \in S \setminus \{ s \}$ can be written as

$$\lambda_{s\,s'} = p_{s\,s'} / \mu_s \tag{4.7}$$

with μ_s defined as the mean time spent in s at any given visit to s:

$$\mu_s = \left\{ \sum_{s' \neq s} \lambda_{s\,s'} \right\}^{-1} = -1 / \lambda_{s\,s} \neq 0; \tag{4.8}$$

see Ross [ROS]. Since $p_{s\,s} = 0$, (4.7) and (4.8) can be written in matrix form to give

$$- \mathbf{D}_{A_1 \cup A_2\, A_1 \cup A_2} \Lambda_{A_1 \cup A_2\, A_1 \cup A_2} = \mathbf{I} - \mathbf{P}_{A_1 \cup A_2\, A_1 \cup A_2}, \tag{4.9}$$

where $\mathbf{D}_{A_1 \cup A_2\, A_1 \cup A_2}$ stands for the diagonal matrix with elements $\{ \mu_s : s \in A_1 \cup A_2 \}$. From (4.9) it follows that, say, $\Lambda_{A_2A_2}^{-1}$ exists since

$$- \mathbf{D}_{A_2A_2} \Lambda_{A_2A_2} = \mathbf{I} - \mathbf{P}_{A_2A_2},$$

and $\mathbf{D}_{A_2A_2}$ and $\{ \mathbf{I} - \mathbf{P}_{A_2A_2} \}$ are invertible; for the latter see Theorem 2.1 (a).

Proof of (b). To establish (4.5), we view the process through a sequence of "snapshots" taken at time instants Δt apart. To this end, we assume that the times $t_i > 0$ can be (approximately) represented as $t_i = k_i \Delta t$ with some positive integers k_i, i = 1, ..., m. The process observed at the time instances 0, Δt, $2\Delta t$, ... can be approximated by a discrete-time Markov chain with transition probability matrix

$$\mathbf{Q} = \mathbf{I} + \Delta t\, \Lambda. \tag{4.10}$$

For small Δt, (4.10) follows from the usual equations for the probability of a single transition to any of the other states when the (homogeneous) Markov assumption is made. (Incidentally, $\mathbf{Q}$ is the transition probability matrix of a uniformized discrete-parameter Markov chain for small enough Δt; see Kohlas [KOH] pp 100 and Ross [ROS] pp 174. We shall not make any use of this in the sequel in a formal sense, however.) It is assumed that Δt (the length of the "time slice") is sufficiently small to assure that the process occupies the same state throughout each time interval; this can be achieved by allowing transitions to take place at the instances Δt, $2\Delta t$, ... only. Let $N_{A_1,i}$ denote the *ith* sojourn time in A_1 of this discrete-time Markov chain.

Now, for $t_1, ..., t_m > 0$, the event $\{ T_{A_1,1} \leq t_1, ..., T_{A_1,m} \leq t_m \}$ is partitioned as

$$\{ T_{A_1,1} \leq t_1, ..., T_{A_1,m} \leq t_m \} = \{ 0 < T_{A_1,1} \leq t_1, ..., 0 < T_{A_1,m} \leq t_m \} \cup$$

$$\{ T_{A_1,1} = 0 \} \cup \bigcup_{i=2}^{m} \{ 0 < T_{A_1,1} \leq t_1, ..., 0 < T_{A_1,i-1} \leq t_{i-1}, 0 = T_{A_1,i} \},$$

from which the left hand side of (4.5) is written as

$$\Pr\{ T_{A_1,1} \leq t_1, ..., T_{A_1,m} \leq t_m \} =$$

$$\Pr\{ 0 < T_{A_1,1} \leq t_1, ..., 0 < T_{A_1,m} \leq t_m \} +$$

$$\Pr\{ T_{A_1,1} = 0 \} + \sum_{i=2}^{m} \Pr\{ 0 < T_{A_1,1} \leq t_1, ..., 0 < T_{A_1,i-1} \leq t_{i-1}, 0 = T_{A_1,i} \}. \qquad (4.11)$$

The first probability on the right hand side of (4.11) is approximated as follows

$$\Pr\{ 0 < T_{A_1,1} \leq t_1, ..., 0 < T_{A_1,m} \leq t_m \} \approx$$

$$\Pr\{ 1 \leq N_{A_1,1} \leq k_1, ..., 1 \leq N_{A_1,m} \leq k_m \} =$$

$$\sum_{n_1=1}^{k_1} \cdots \sum_{n_m=1}^{k_m} \Pr\{ N_{A_1,1} = n_1, ..., N_{A_1,m} = n_m \}. \qquad (4.12)$$

By Theorem 2.1, equation (2.2), we get for the right hand side of (4.12) the following expression

$$\mathbf{v}^T (\mathbf{I} - \mathbf{Q}_{A_1A_1})^{-1} (\mathbf{I} - \mathbf{Q}_{A_1A_1}^{k_1}) \times$$

$$\prod_{i=2}^{m} \left\{ \mathbf{Q}_{A_1A_2} (\mathbf{I} - \mathbf{Q}_{A_2A_2})^{-1} \mathbf{Q}_{A_2A_1} (\mathbf{I} - \mathbf{Q}_{A_1A_1})^{-1} (\mathbf{I} - \mathbf{Q}_{A_1A_1}^{k_1}) \right\} (\mathbf{I} - \mathbf{Q}_{A_1A_1}) \mathbf{1}, \quad (4.13)$$

with $\mathbf{v}$ defined by (2.1). To express (4.13) in terms of the submatrices of $\mathbf{\Lambda}$, we note that (4.10) implies the equations

$$\mathbf{Q}_{A_2A_2} = \mathbf{I} + \Delta t \, \mathbf{\Lambda}_{A_2A_2},$$

$$\mathbf{Q}_{A_2A_1} = \Delta t \, \mathbf{\Lambda}_{A_2A_1},$$

$$\mathbf{Q}_{A_1A_2} = \Delta t \, \mathbf{\Lambda}_{A_1A_2},$$

$$Q_{A_1A_1} = I + \Delta t \, \Lambda_{A_1A_1} = I + \frac{t_i}{k_i} \, \Lambda_{A_1A_1},$$

which, in conjunction with (4.12) and (4.13), give

$$\Pr\{\, 0 < T_{A_1,1} \le t_1, \, ..., \, 0 < T_{A_1,m} \le t_m \,\} \simeq$$

$$\left\{\boldsymbol{\alpha}_{A_2}^{\ T} (-\Lambda_{A_2A_2})^{-1} \Lambda_{A_2A_1} + \boldsymbol{\alpha}_{A_1}^{\ T}\right\} (-\Lambda_{A_1A_1})^{-1} \left\{ I - \left(I + \frac{t_1}{k_1} \Lambda_{A_1A_1}\right)^{k_1}\right\} \times$$

$$\prod_{i=2}^{m} \left\{ \Lambda_{A_1A_2} (-\Lambda_{A_2A_2})^{-1} \Lambda_{A_2A_1} (-\Lambda_{A_1A_1})^{-1} \times \right.$$

$$\left. \left\{ I - \left(I + \frac{t_i}{k_i} \Lambda_{A_1A_1}\right)^{k_i}\right\}\right\} (-\Lambda_{A_1A_1}) \, \mathbf{1} \simeq$$

$$- \mathbf{w}^T \Lambda_{A_1A_1}^{\ -1} \left\{ I - \exp\{\, t_1 \, \Lambda_{A_1A_1} \,\}\right\} \times$$

$$\prod_{i=2}^{m} \left\{ \Lambda_{A_1A_2} \Lambda_{A_2A_2}^{\ -1} \Lambda_{A_2A_1} \Lambda_{A_1A_1}^{\ -1} \left\{ I - \exp\{\, t_i \, \Lambda_{A_1A_1} \,\}\right\}\right\} \Lambda_{A_1A_1} \, \mathbf{1}. \qquad (4.14)$$

For $\Delta t \to 0$, we have $k_1, \, ..., \, k_m \to \infty$ and thus in the limit (4.14) becomes

$$\Pr\{\, 0 < T_{A_1,1} \le t_1, \, ..., \, 0 < T_{A_1,m} \le t_m \,\} =$$

$$- \mathbf{w}^T \Lambda_{A_1A_1}^{\ -1} \left\{ I - \exp\{\, t_1 \, \Lambda_{A_1A_1} \,\}\right\} \times$$

$$\prod_{i=2}^{m} \left\{ \Lambda_{A_1A_2} \Lambda_{A_2A_2}^{\ -1} \Lambda_{A_2A_1} \Lambda_{A_1A_1}^{\ -1} \left\{ I - \exp\{\, t_i \, \Lambda_{A_1A_1} \,\}\right\}\right\} \Lambda_{A_1A_1} \, \mathbf{1}. \qquad (4.15)$$

Putting in (4.15) $m = 1$ and letting $t_1 \to +\infty$, we get by Lemma 4.1

$$\Pr\{\, T_{A_1,1} = 0 \,\} = 1 - P(\, 0 < T_{A_1,1}) = 1 + \mathbf{w}^T \, \mathbf{1}. \qquad (4.16)$$

(4.5) now follows for $m = 1$ from (4.15) and (4.16) by

$$\Pr\{\, T_{A_1,1} \le t_1 \,\} = \Pr\{\, 0 < T_{A_1,1} \le t_1 \,\} + \Pr\{\, T_{A_1,1} = 0 \,\}.$$

In what follows, (4.5) will be shown for $m \ge 2$. First notice that the first two terms on the right hand side of (4.11) have been catered for by (4.15) and (4.16). The remaining probabilities on the right hand side of (4.11) (i.e., those under the summation sign) will now be considered. Letting $t_m \to +\infty$ in (4.15), we get by Lemma 4.1

$$\Pr\{\ 0 < T_{A_1,1} \le t_1,\ ...,\ 0 < T_{A_1,m\text{-}1} \le t_{m\text{-}1},\ 0 < T_{A_1,m}\ \} =$$

$$-\ \mathbf{w}^T\ \mathbf{\Lambda}_{A_1A_1}^{\ -1}\left\{\mathbf{I} - \exp\{\ t_1\ \mathbf{\Lambda}_{A_1A_1}\ \}\right\} \times$$

$$\prod_{i=2}^{m\text{-}1}\left\{\mathbf{\Lambda}_{A_1A_2}\ \mathbf{\Lambda}_{A_2A_2}^{\ -1}\ \mathbf{\Lambda}_{A_2A_1}\ \mathbf{\Lambda}_{A_1A_1}^{\ -1}\left\{\mathbf{I} - \exp\{\ t_i\ \mathbf{\Lambda}_{A_1A_1}\ \}\right\}\right\} \times$$

$$\mathbf{\Lambda}_{A_1A_2}\ \mathbf{\Lambda}_{A_2A_2}^{\ -1}\ \mathbf{\Lambda}_{A_2A_1}\ \mathbf{1}. \tag{4.17}$$

Combining (4.15) and (4.17) gives for $i \ge 2$

$$\Pr\{\ 0 < T_{A_1,1} \le t_1,\ ...,\ 0 < T_{A_1,i\text{-}1} \le t_{i\text{-}1},\ 0 = T_{A_1,i}\ \} =$$

$$\Pr\{\ 0 < T_{A_1,1} \le t_1,\ ...,\ 0 < T_{A_1,i\text{-}1} \le t_{i\text{-}1}\ \} -$$

$$\Pr\{\ 0 < T_{A_1,1} \le t_1,\ ...,\ 0 < T_{A_1,i\text{-}1} \le t_{i\text{-}1},\ 0 < T_{A_1,i}\ \} =$$

$$-\ \mathbf{w}^T\ \mathbf{\Lambda}_{A_1A_1}^{\ -1}\left\{\mathbf{I} - \exp\{\ t_1\ \mathbf{\Lambda}_{A_1A_1}\ \}\right\} \times$$

$$\prod_{j=2}^{i\text{-}1}\left\{\mathbf{\Lambda}_{A_1A_2}\ \mathbf{\Lambda}_{A_2A_2}^{\ -1}\ \mathbf{\Lambda}_{A_2A_1}\ \mathbf{\Lambda}_{A_1A_1}^{\ -1}\left\{\mathbf{I} - \exp\{\ t_j\ \mathbf{\Lambda}_{A_1A_1}\ \}\right\}\right\} \times$$

$$\left\{\mathbf{\Lambda}_{A_1A_1} - \mathbf{\Lambda}_{A_1A_2}\ \mathbf{\Lambda}_{A_2A_2}^{\ -1}\ \mathbf{\Lambda}_{A_2A_1}\right\}\mathbf{1}. \tag{4.18}$$

(4.5) now follows for $m \ge 2$ from (4.11), (4.15), (4.16), and (4.18).　∎

The method of proof chosen here to deduce Theorem 4.2 from its discrete-parameter counterpart Theorem 2.1 is that of *time discretisation*. This technique is well-known; see, e.g., Darroch and Morris [DAR].

The joint cumulative distribution function of $T_{A_1,1}, ..., T_{A_1,m}$ is obtained for irreducible Y by noting that in the irreducible case $\mathbf{\Lambda}_{A_1\{\omega\}} = \mathbf{0}$ and $\mathbf{\Lambda}_{A_2\{\omega\}} = \mathbf{0}$ and therefore

$$\mathbf{\Lambda}_{A_1A_1}\ \mathbf{1} = -\ \mathbf{\Lambda}_{A_1A_2}\ \mathbf{1},$$

$$\mathbf{\Lambda}_{A_2A_2}\ \mathbf{1} = -\ \mathbf{\Lambda}_{A_2A_1}\ \mathbf{1},$$

$$1 + \mathbf{w}^T\ \mathbf{1} = 1 + \left\{\ \boldsymbol{\alpha}_{A_2}^{\ T}\ \mathbf{\Lambda}_{A_2A_2}^{\ -1}\ \mathbf{\Lambda}_{A_2A_1} - \boldsymbol{\alpha}_{A_1}^{\ T}\right\}\mathbf{1} =$$

$$1 - \boldsymbol{\alpha}_{A_2}^{\ T}\ \mathbf{1} - \boldsymbol{\alpha}_{A_1}^{\ T}\ \mathbf{1} = 0,$$

$$\left\{\ \mathbf{\Lambda}_{A_1A_1} - \mathbf{\Lambda}_{A_1A_2}\ \mathbf{\Lambda}_{A_2A_2}^{\ -1}\ \mathbf{\Lambda}_{A_2A_1}\right\}\mathbf{1} = 0.$$

We thus have the following corollary.

COROLLARY 4.3. *For irreducible* Y, *the joint distribution function of the first* m *sojourn times in* A_1 *is given by*

$$\Pr\{ T_{A_1,1} \le t_1, ..., T_{A_1,m} \le t_m \} =$$

$$\mathbf{w}^T \mathbf{\Lambda}_{A_1 A_1}^{-1} \{\mathbf{I} - \exp\{ t_1 \mathbf{\Lambda}_{A_1 A_1} \}\} \mathbf{\Lambda}_{A_1 A_2} \times$$

$$\prod_{i=2}^{m} \{ \mathbf{\Lambda}_{A_2 A_2}^{-1} \mathbf{\Lambda}_{A_2 A_1} \mathbf{\Lambda}_{A_1 A_1}^{-1} \{\mathbf{I} - \exp\{ t_i \mathbf{\Lambda}_{A_1 A_1} \}\} \mathbf{\Lambda}_{A_1 A_2} \} \mathbf{1}, \tag{4.19}$$

where **w** *is defined by (4.6).* ∎

For the sake of completeness, let us also consider the cumulative distribution function of the *m*th sojourn time $T_{A_1,m}$ for irreducible Y. This is a result first obtained by Rubino and Sericola [RUB3].

COROLLARY 4.4. *For irreducible* Y, *the cumulative distribution function of* $T_{A_1,m}$ *is given by*

$$\Pr\{ T_{A_1,m} \le t \} =$$

$$1 + \mathbf{w}^T \{ \mathbf{\Lambda}_{A_1 A_1}^{-1} \mathbf{\Lambda}_{A_1 A_2} \mathbf{\Lambda}_{A_2 A_2}^{-1} \mathbf{\Lambda}_{A_2 A_1} \}^{m-1} \exp\{ t \mathbf{\Lambda}_{A_1 A_1} \} \mathbf{1}, \tag{4.20}$$

where **w** *is defined by (4.6).*

PROOF OF COROLLARY 4.4. For m = 1, we have by (4.19)

$$\Pr\{ T_{A_1,1} \le t \} = \mathbf{w}^T \mathbf{\Lambda}_{A_1 A_1}^{-1} \{\mathbf{I} - \exp\{ t \mathbf{\Lambda}_{A_1 A_1} \}\} \mathbf{\Lambda}_{A_1 A_2} \mathbf{1} =$$

$$1 - \mathbf{w}^T \mathbf{\Lambda}_{A_1 A_1}^{-1} \exp\{ t \mathbf{\Lambda}_{A_1 A_1} \} \mathbf{\Lambda}_{A_1 A_2} \mathbf{1} = 1 + \mathbf{w}^T \exp\{ t \mathbf{\Lambda}_{A_1 A_1} \} \mathbf{1}. \tag{4.21}$$

(The second equality in (4.21) is established by Lemma 4.1 and by noting that $\Pr\{ T_{A_1,1} \le t \}$ tends to unity as $t \to +\infty$. We use $\mathbf{\Lambda}_{A_1 A_1} \mathbf{1} = - \mathbf{\Lambda}_{A_1 A_1} \mathbf{1}$ and $\mathbf{\Lambda}_{A_1 A_1}^{-1} \exp\{ t \mathbf{\Lambda}_{A_1 A_1} \} \mathbf{\Lambda}_{A_1 A_1} = \exp\{ t \mathbf{\Lambda}_{A_1 A_1} \}$ to justify the third equality in (4.21).) For $m \ge 2$, we get the following from (4.19) by letting $t_1, ..., t_{m-1} \to +\infty$,

$$\Pr\{ T_{A_1,m} \le t \} =$$

$$\mathbf{w}^T \mathbf{\Lambda}_{A_1 A_1}^{-1} \mathbf{\Lambda}_{A_1 A_2} \{ \mathbf{\Lambda}_{A_2 A_2}^{-1} \mathbf{\Lambda}_{A_2 A_1} \mathbf{\Lambda}_{A_1 A_1}^{-1} \mathbf{\Lambda}_{A_1 A_2} \}^{m-2} \times$$

$$\mathbf{\Lambda}_{A_2 A_2}^{-1} \mathbf{\Lambda}_{A_2 A_1} \mathbf{\Lambda}_{A_1 A_1}^{-1} \{\mathbf{I} - \exp\{ t \mathbf{\Lambda}_{A_1 A_1} \}\} \mathbf{\Lambda}_{A_1 A_2} \mathbf{1}. \tag{4.22}$$

By noting that (4.22) tends to unity as $t \to +\infty$ and by Lemma 4.1 we get from (4.22)

$$\Pr\{ T_{A_1,m} \le t \} =$$

$$1 - \mathbf{w}^T \, \mathbf{\Lambda}_{A_1A_1}^{-1} \, \mathbf{\Lambda}_{A_1A_2} \left\{ \mathbf{\Lambda}_{A_2A_2}^{-1} \, \mathbf{\Lambda}_{A_2A_1} \, \mathbf{\Lambda}_{A_1A_1}^{-1} \, \mathbf{\Lambda}_{A_1A_2} \right\}^{m-2} \times$$

$$\mathbf{\Lambda}_{A_2A_2}^{-1} \, \mathbf{\Lambda}_{A_2A_1} \, \mathbf{\Lambda}_{A_1A_1}^{-1} \, \exp\{ t \, \mathbf{\Lambda}_{A_1A_1} \} \, \mathbf{\Lambda}_{A_1A_2} \, \mathbf{1} =$$

$$1 + \mathbf{w}^T \left\{ \mathbf{\Lambda}_{A_1A_1}^{-1} \, \mathbf{\Lambda}_{A_1A_2} \, \mathbf{\Lambda}_{A_2A_2}^{-1} \, \mathbf{\Lambda}_{A_2A_1} \right\}^{m-1} \times$$

$$\mathbf{\Lambda}_{A_1A_1}^{-1} \, \exp\{ t \, \mathbf{\Lambda}_{A_1A_1} \} \, \mathbf{\Lambda}_{A_1A_1} \, \mathbf{1}. \tag{4.23}$$

Using now $\mathbf{\Lambda}_{A_1A_1}^{-1} \, \exp\{ t \, \mathbf{\Lambda}_{A_1A_1} \} \, \mathbf{\Lambda}_{A_1A_1} = \exp\{ t \, \mathbf{\Lambda}_{A_1A_1} \}$, we get (4.20) from (4.23). ∎

Notice that (4.20) is identical to the formula for $\Pr\{ T_{A_1,m} \leq t \}$ as given in [RUB3], except that there the result is expressed in terms of the transition probability matrix of a uniformized Markov chain (4.10). Let us also add that (4.20) also holds if Y is absorbing; this was first established by Rubino and Sericola in [RUB4]. In our present framework, this result can be deduced from Theorem 4.2. Even though we will not elaborate on this, let us note that the technique required is similar to the one used in the last proof: the last term (the sum) on the right hand side of (4.5) for $t_1 = ... = t_{m-1} = +\infty$ is re-expressed by putting $t_1 = ... = t_m = +\infty$ in (4.5).

In the next section, the cumulative distribution function of $TS_{A_1,m}$, the m*th* cumulative sojourn time in A_1, will be derived by using Laplace transforms. As a preparatory step, in the next corollary we provide the Laplace transform of the vector variable $(T_{A_1,1}, ..., T_{A_1,m})$; this is defined for $\boldsymbol{\tau} = (\tau_1, ..., \tau_m)^T \in \mathbb{C}^{+m}, \mathbb{C}^+ = \{ z \in \mathbb{C} : \text{Re}(z) \geq 0 \}$, by

$$(T_{A_1,1}, ..., T_{A_1,m})^*(\boldsymbol{\tau}) = E\{\exp\{ -\tau_1 T_{A_1,1} - ... - \tau_m T_{A_1,m} \}\}.$$

COROLLARY 4.5. *If Y is irreducible or absorbing, the Laplace transform of the first* m *sojourn times in* A_1 *is given by*

$$(T_{A_1,1}, ..., T_{A_1,m})^*(\boldsymbol{\tau}) =$$

$$- \mathbf{w}^T \, (\mathbf{\Lambda}_{A_1A_1} - \tau_1 \, I)^{-1} \times$$

$$\prod_{i=2}^{m} \left\{ \mathbf{\Lambda}_{A_1A_2} \, \mathbf{\Lambda}_{A_2A_2}^{-1} \, \mathbf{\Lambda}_{A_2A_1} \, (\mathbf{\Lambda}_{A_1A_1} - \tau_i \, I)^{-1} \right\} \, \mathbf{\Lambda}_{A_1A_1} \, \mathbf{1} + 1 + \mathbf{w}^T \, \mathbf{1} -$$

$$- \sum_{k=1}^{m-1} \mathbf{w}^T \, (\mathbf{\Lambda}_{A_1A_1} - \tau_1 \, I)^{-1} \times$$

$$\prod_{i=2}^{k} \left\{ \mathbf{\Lambda}_{A_1A_2} \, \mathbf{\Lambda}_{A_2A_2}^{-1} \, \mathbf{\Lambda}_{A_2A_1} \, (\mathbf{\Lambda}_{A_1A_1} - \tau_i \, I)^{-1} \right\} \times$$

$$\left\{ \boldsymbol{\Lambda}_{A_1A_1} - \boldsymbol{\Lambda}_{A_1A_2} \, \boldsymbol{\Lambda}_{A_2A_2}^{-1} \, \boldsymbol{\Lambda}_{A_2A_1} \right\} \mathbf{1}, \tag{4.24}$$

where **w** *is defined by (4.6).*

PROOF OF COROLLARY 4.5. $[0,+\infty)^m$ is the support of the distribution of $(T_{A_1,1}, ..., T_{A_1,m})$. Let μ be the measure on $[0,+\infty)^m$ which is defined for any Borel-measurable set $B \subseteq [0,+\infty)^m$ by

$$\mu(B) =$$

$$\varepsilon_0(B) + \sum_{k=1}^{m} \lambda_k(\{ (t_1, ..., t_k)^T \in [0,+\infty)^k : (t_1, ..., t_k, 0, ..., 0)^T \in B \}),$$

where ε_0 is the probability measure which assigns unity to the origin and λ_k stands for the Lebesgue measure on $\mathbb{R}^k$. Define the function $h(\mathbf{t})$ on $[0,+\infty)^m$ by

$$h(\mathbf{t}) =$$

$$(1 + \mathbf{w}^T \mathbf{1}) \, I_{\{ t_1 = 0, ..., t_m = 0 \}} +$$

$$\mathbf{w}^T \exp\{ t_1 \boldsymbol{\Lambda}_{A_1A_1} \} \prod_{i=2}^{m} \left\{ - \boldsymbol{\Lambda}_{A_1A_2} \, \boldsymbol{\Lambda}_{A_2A_2}^{-1} \, \boldsymbol{\Lambda}_{A_2A_1} \exp\{ t_i \boldsymbol{\Lambda}_{A_1A_1} \} \right\} \times$$

$$\boldsymbol{\Lambda}_{A_1A_1} \mathbf{1} \, I_{\{ t_1 > 0, ..., t_m > 0 \}} +$$

$$\mathbf{w}^T \exp\{ t_1 \boldsymbol{\Lambda}_{A_1A_1} \} \sum_{k=1}^{m-1} \prod_{i=2}^{k} \left\{ - \boldsymbol{\Lambda}_{A_1A_2} \, \boldsymbol{\Lambda}_{A_2A_2}^{-1} \, \boldsymbol{\Lambda}_{A_2A_1} \exp\{ t_i \boldsymbol{\Lambda}_{A_1A_1} \} \right\} \times$$

$$\left\{ \boldsymbol{\Lambda}_{A_1A_1} - \boldsymbol{\Lambda}_{A_1A_2} \, \boldsymbol{\Lambda}_{A_2A_2}^{-1} \, \boldsymbol{\Lambda}_{A_2A_1} \right\} \mathbf{1} \, I_{\{ t_1 > 0, ..., t_k > 0, t_{k+1} = 0, ..., t_m = 0 \}}.$$

Then, by integrating h over $[0, t_1] \times ... \times [0, t_m]$, it is easily verified from (4.5) that h is a μ - density of $(T_{A_1,1}, ..., T_{A_1,m})$. The Laplace tansform of $(T_{A_1,1}, ..., T_{A_1,m})$ is therefore given by

$$(T_{A_1,1}, ..., T_{A_1,m})^*(\tau) = \int_{[0, +\infty)^m} \exp\{ - \tau_1 t_1 - ... - \tau_m t_m \} \, h(\mathbf{t}) \, \mu\{d\mathbf{t}\} =$$

$$1 + \mathbf{w}^T \mathbf{1} +$$

$$\int\limits_{(0,\,+\infty)^m} \left\{ \exp\{-\tau_1 t_1 - ... - \tau_m t_m\} \, \mathbf{w}^T \exp\{t_1 \, \mathbf{\Lambda}_{A_1A_1}\} \times \right.$$

$$\left. \prod_{i=2}^{m} \left\{ -\mathbf{\Lambda}_{A_1A_2} \mathbf{\Lambda}_{A_2A_2}^{-1} \mathbf{\Lambda}_{A_2A_1} \exp\{t_i \, \mathbf{\Lambda}_{A_1A_1}\} \right\} \mathbf{\Lambda}_{A_1A_1} \mathbf{1} \right\} dt_1...dt_m +$$

$$\sum_{k=1}^{m-1} \int\limits_{(0,\,+\infty)^k} \left\{ \exp\{-\tau_1 t_1 - ... - \tau_k t_k\} \, \mathbf{w}^T \exp\{t_1 \, \mathbf{\Lambda}_{A_1A_1}\} \times \right.$$

$$\prod_{i=2}^{k} \left\{ -\mathbf{\Lambda}_{A_1A_2} \mathbf{\Lambda}_{A_2A_2}^{-1} \mathbf{\Lambda}_{A_2A_1} \exp\{t_i \, \mathbf{\Lambda}_{A_1A_1}\} \right\} \times$$

$$\left. \left\{ \mathbf{\Lambda}_{A_1A_1} - \mathbf{\Lambda}_{A_1A_2} \mathbf{\Lambda}_{A_2A_2}^{-1} \mathbf{\Lambda}_{A_2A_1} \right\} \mathbf{1} \right\} dt_1 ... dt_k. \tag{4.25}$$

Integration by parts and Lemma 4.1 show that for $s \geq 0$

$$\int\limits_{0}^{+\infty} \exp\{-st\} \, \mathbf{\Lambda}_{A_1A_1} \exp\{t \, \mathbf{\Lambda}_{A_1A_1}\} \, dt = -\mathbf{I} + s \int\limits_{0}^{+\infty} \exp\{-st\} \exp\{t \, \mathbf{\Lambda}_{A_1A_1}\} \, dt,$$

from which it follows that $(s \, \mathbf{I} - \mathbf{\Lambda}_{A_1A_1})$ is invertible and

$$\int\limits_{0}^{+\infty} \exp\{-st\} \exp\{t \, \mathbf{\Lambda}_{A_1A_1}\} \, dt = (s \, \mathbf{I} - \mathbf{\Lambda}_{A_1A_1})^{-1}. \tag{4.26}$$

Substituting (4.26) into the right hand side of (4.25), we get (4.24). ■

The next result, Theorem 4.6, will be concerned with the joint distribution of the $2m$ sojourn times $T_{A_1,1}$, ..., $T_{A_1,m}$, $T_{A_2,1}$, ..., $T_{A_2,m}$. It is a direct extension of Theorem 4.2 in that it can be used to rederive the joint distribution of the A_1-sojourns as an m-dimensional marginal distribution. On the other hand, Theorem 4.2 is *needed* in the course of the proof of Theorem 4.6; in this sense, the two are equivalent.

THEOREM 4.6. *The joint distribution function of the first $2m$ sojourn times of Y in A_1 and A_2 is given for* $s_i, t_i \geq 0$ $(i = 1, ...,m)$ *by*

$$\Pr\{ T_{A_2,1} \leq s_1, ..., T_{A_2,m} \leq s_m, T_{A_1,1} \leq t_1, ..., T_{A_1,m} \leq t_m \} =$$

$$- \boldsymbol{\alpha}_{A_2}{}^T \, \boldsymbol{\Lambda}_{A_2 A_2}{}^{-1} \left\{ \mathbf{I} - \exp\{ s_1 \, \boldsymbol{\Lambda}_{A_2 A_2} \} \right\} \times$$

$$\boldsymbol{\Gamma}(t_1; A_2, A_1) \prod_{i=2}^{m} \left\{ \boldsymbol{\Gamma}(s_i; A_1, A_2) \, \boldsymbol{\Gamma}(t_i; A_2, A_1) \right\} \boldsymbol{\Lambda}_{A_1 A_1} \mathbf{1} +$$

$$+ \sum_{k=1}^{m} \boldsymbol{\alpha}_{A_2}{}^T \, \boldsymbol{\Lambda}_{A_2 A_2}{}^{-1} \left\{ \mathbf{I} - \exp\{ s_1 \, \boldsymbol{\Lambda}_{A_2 A_2} \} \right\} \times$$

$$\prod_{i=2}^{k} \left\{ \boldsymbol{\Gamma}(t_{i-1}; A_2, A_1) \, \boldsymbol{\Gamma}(s_i; A_1, A_2) \right\} \left(\boldsymbol{\Lambda}_{A_2 A_2} \mathbf{1} + \boldsymbol{\Lambda}_{A_2 A_1} \mathbf{1} \right) -$$

$$- \sum_{k=1}^{m-1} \boldsymbol{\alpha}_{A_2}{}^T \, \boldsymbol{\Lambda}_{A_2 A_2}{}^{-1} \left\{ \mathbf{I} - \exp\{ s_1 \, \boldsymbol{\Lambda}_{A_2 A_2} \} \right\} \times$$

$$\boldsymbol{\Gamma}(t_1; A_2, A_1) \prod_{i=2}^{k} \left\{ \boldsymbol{\Gamma}(s_i; A_1, A_2) \, \boldsymbol{\Gamma}(t_i; A_2, A_1) \right\} \left(\boldsymbol{\Lambda}_{A_1 A_2} \mathbf{1} + \boldsymbol{\Lambda}_{A_1 A_1} \mathbf{1} \right) -$$

$$- \boldsymbol{\alpha}_{A_1}{}^T \, \boldsymbol{\Lambda}_{A_1 A_1}{}^{-1} \left\{ \mathbf{I} - \exp\{ t_1 \, \boldsymbol{\Lambda}_{A_1 A_1} \} \right\} \times$$

$$\boldsymbol{\Gamma}(s_1; A_1, A_2) \prod_{i=2}^{m} \left\{ \boldsymbol{\Gamma}(t_i; A_2, A_1) \, \boldsymbol{\Gamma}(s_i; A_1, A_2) \right\} \boldsymbol{\Lambda}_{A_2 A_2} \mathbf{1} +$$

$$+ \sum_{k=1}^{m} \boldsymbol{\alpha}_{A_1}{}^T \, \boldsymbol{\Lambda}_{A_1 A_1}{}^{-1} \left\{ \mathbf{I} - \exp\{ t_1 \, \boldsymbol{\Lambda}_{A_1 A_1} \} \right\} \times$$

$$\prod_{i=2}^{k} \left\{ \boldsymbol{\Gamma}(s_{i-1}; A_1, A_2) \, \boldsymbol{\Gamma}(t_i; A_2, A_1) \right\} \left(\boldsymbol{\Lambda}_{A_1 A_1} \mathbf{1} + \boldsymbol{\Lambda}_{A_1 A_2} \mathbf{1} \right) -$$

$$- \sum_{k=1}^{m-1} \boldsymbol{\alpha}_{A_1}{}^T \, \boldsymbol{\Lambda}_{A_1 A_1}{}^{-1} \left\{ \mathbf{I} - \exp\{ t_1 \, \boldsymbol{\Lambda}_{A_1 A_1} \} \right\} \times$$

$$\boldsymbol{\Gamma}(s_1; A_1, A_2) \prod_{i=2}^{k} \left\{ \boldsymbol{\Gamma}(t_i; A_2, A_1) \, \boldsymbol{\Gamma}(s_i; A_1, A_2) \right\} \left(\boldsymbol{\Lambda}_{A_2 A_1} \mathbf{1} + \boldsymbol{\Lambda}_{A_2 A_2} \mathbf{1} \right), \tag{4.27}$$

with $\boldsymbol{\Gamma}$ *defined for non-empty, disjoint subsets* A *and* B *of* $S \setminus \{\omega\}$ *by*

$$\boldsymbol{\Gamma}(t; A, B) = \boldsymbol{\Lambda}_{AB} \, \boldsymbol{\Lambda}_{BB}{}^{-1} \left\{ \mathbf{I} - \exp\{ t \, \boldsymbol{\Lambda}_{BB} \} \right\}, \, t \geq 0. \tag{4.28}$$

PROOF OF THEOREM 4.6. Let us consider the auxiliary absorbing Markov chain $Y' = \{ Y'_t :$

$t \in [0, +\infty)$ } shown in Figure 4.1.

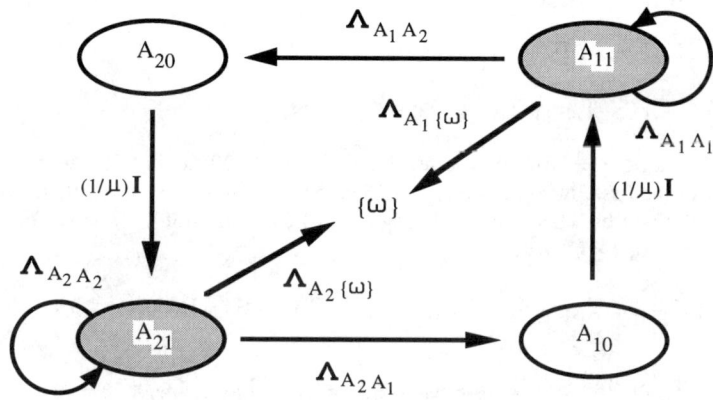

Figure 4.1 State-transition-rate diagram of the auxiliary Markov chain Y'.
(A_2' = unshaded sets of states, A_1' = shaded sets of states, ω = absorbing state.)

The state space of Y' is $S' = A_{20} \cup A_{10} \cup A_{21} \cup A_{11} \cup \{\omega\}$ where A_{2i} and A_{1i} (i = 0, 1) are disjoint instances of A_2 and A_1 respectively. Its transition-rate matrix Λ' is given by

$$
\Lambda' = \begin{array}{c} \\ A_{20} \\ A_{10} \\ A_{21} \\ A_{11} \\ \{\omega\} \end{array}
\begin{array}{c} \begin{array}{ccccc} A_{20} & A_{10} & A_{21} & A_{11} & \{\omega\} \end{array} \\
\left[\begin{array}{c|c|c|c|c}
-\mu^{-1}\mathbf{I} & 0 & \mu^{-1}\mathbf{I} & 0 & 0 \\ \hline
0 & -\mu^{-1}\mathbf{I} & 0 & \mu^{-1}\mathbf{I} & 0 \\ \hline
0 & \Lambda_{A_2 A_1} & \Lambda_{A_2 A_2} & 0 & \Lambda_{A_2\{\omega\}} \\ \hline
\Lambda_{A_1 A_2} & 0 & 0 & \Lambda_{A_1 A_1} & \Lambda_{A_1\{\omega\}} \\ \hline
0 & 0 & 0 & 0 & 0
\end{array} \right]
\end{array}
$$

μ (> 0, chosen arbitrarily) is the mean time spent in any of the states in $A_2' = A_{20} \cup A_{10}$. The initial probability vector assumed for Y' is

$$\alpha'^T = (\, 0, 0, \alpha_{A_2}^T, \alpha_{A_1}^T, 0 \,).$$

Y' can be viewed as the Markov chain Y on $A_1' = A_{21} \cup A_{11}$, supplemented by the additional set of states A_2' each visit of which is only there to mark the end of the most recent sojourn of Y in A_2 or A_1. More precisely, the following *equalities in distribution* between sojourn times of Y' and Y hold for m = 1, 2, ...

$$\mathcal{L}(T'_{A_1',1}, T'_{A_1',2}, ..., T'_{A_1',2m} \mid Y'_0 = a_2) =$$

$$\mathcal{L}(T_{A_2,1}, T_{A_1,1}, ..., T_{A_2,m}, T_{A_1,m} \mid Y_0 = a_2), \tag{4.29}$$

$$\mathcal{L}(T'_{A_1',1}, T'_{A_1',2}, ..., T'_{A_1',2m} \mid Y'_0 = a_1) =$$

$$\mathcal{L}(T_{A_1,1}, T_{A_2,1}, ..., T_{A_1,m}, T_{A_2,m} \mid Y_0 = a_1), \tag{4.30}$$

with $a_2 \in A_{21}$, $a_1 \in A_{11}$. But, on the left hand side of (4.29) and (4.30) we have the joint distributions of the first 2m sojourn times of Y' in the *same* subset A_1' of S'; their distributions are therefore known by Theorem 4.2. Using now (4.29) and (4.30), we have for any s_1, ..., s_m, t_1, ..., $t_m \in [\,0, +\infty\,)$

$$\Pr\{\ T_{A_2,1} \le s_1, ..., T_{A_2,m} \le s_m, T_{A_1,1} \le t_1, ..., T_{A_1,m} \le t_m\ \} =$$

$$\sum_{a_2 \in A_2} \Pr\{\ T_{A_2,1} \le s_1, ..., T_{A_2,m} \le s_m, T_{A_1,1} \le t_1, ..., T_{A_1,m} \le t_m \mid Y_0 = a_2\ \}$$
$$\times\ \Pr\{\ Y_0 = a_2\ \} +$$

$$\sum_{a_1 \in A_1} \Pr\{\ T_{A_2,1} \le s_1, ..., T_{A_2,m} \le s_m, T_{A_1,1} \le t_1, ..., T_{A_1,m} \le t_m \mid Y_0 = a_1\ \}$$
$$\times\ \Pr\{\ Y_0 = a_1\ \} =$$

$$\sum_{a_2 \in A_{21}} \Pr\{\ \bigcap_{i=1}^{m} \{\ T'_{A_1',2i-1} \le s_i\ \} \cap \bigcap_{i=1}^{m} \{\ T'_{A_1',2i} \le t_i\ \} \mid Y'_0 = a_2\ \}\ \alpha_{a_2} +$$

$$\sum_{a_1 \in A_{11}} \Pr\{\ \bigcap_{i=1}^{m} \{\ T'_{A_1',2i-1} \le t_i\ \} \cap \bigcap_{i=1}^{m} \{\ T'_{A_1',2i} \le s_i\ \} \mid Y'_0 = a_1\ \}\ \alpha_{a_1}. \tag{4.31}$$

Define for $\tau \ge 0$, $\mathbf{\Delta}(\tau)$ by

$$\mathbf{\Delta}(\tau) = \mathbf{\Lambda}'_{A_1'A_2'}\, \mathbf{\Lambda}'^{-1}_{A_2'A_2'}\, \mathbf{\Lambda}'_{A_2'A_1'}\, \mathbf{\Lambda}'^{-1}_{A_1'A_1'}\, \{\mathbf{I} - \exp\{\tau\, \mathbf{\Lambda}'_{A_1'A_1'}\}\}\ .$$

Then, by Theorem 4.2, the first probability on the right hand side of (4.31) is seen to be

$$-\,\mathbf{w}'(a_2)^T\, \mathbf{\Lambda}'^{-1}_{A_1'A_1'}\, \{\mathbf{I} - \exp\{\ s_1\, \mathbf{\Lambda}'_{A_1'A_1'}\ \}\}\ \times$$

$$\mathbf{\Delta}(t_1)\ \prod_{i=2}^{m}\ \{\ \mathbf{\Delta}(s_i)\, \mathbf{\Delta}(t_i)\ \}\ \mathbf{\Lambda}'_{A_1'A_1'}\, \mathbf{1} + 1 + \mathbf{w}'(a_2)^T\, \mathbf{1}\ -$$

$$-\,\mathbf{w}'(a_2)^T\, \mathbf{\Lambda}'^{-1}_{A_1'A_1'}\, \{\mathbf{I} - \exp\{\ s_1\, \mathbf{\Lambda}'_{A_1'A_1'}\ \}\}\ \times$$

$$\left\{ \sum_{k=1}^{m} \prod_{i=2}^{k} \left\{ \Delta(t_{i-1}) \Delta(s_i) \right\} + \sum_{k=1}^{m-1} \Delta(t_1) \prod_{i=2}^{k} \left\{ \Delta(s_i) \Delta(t_i) \right\} \right\} \times$$

$$\times \left\{ \mathbf{\Lambda}'_{A_1'A_1'} - \mathbf{\Lambda}'_{A_1'A_2'} \mathbf{\Lambda}'^{-1}_{A_2'A_2'} \mathbf{\Lambda}'_{A_2'A_1'} \right\} \mathbf{1},$$

where $- \mathbf{w}'(a_1')$ is the stochastic vector on A_1' which assigns unity to $a_1' \in A_1'$. The first term (sum) on the right hand side of (4.31) is therefore given by

$$- \mathbf{w}'^T \mathbf{\Lambda}'^{-1}_{A_1'A_1'} \left\{ \mathbf{I} - \exp\{ s_1 \mathbf{\Lambda}'_{A_1'A_1'} \} \right\} \Delta(t_1) \prod_{i=2}^{m} \left\{ \Delta(s_i) \Delta(t_i) \right\} \mathbf{\Lambda}'_{A_1'A_1'} \mathbf{1} -$$

$$- \mathbf{w}'^T \mathbf{\Lambda}'^{-1}_{A_1'A_1'} \left\{ \mathbf{I} - \exp\{ s_1 \mathbf{\Lambda}'_{A_1'A_1'} \} \right\} \times$$

$$\left\{ \sum_{k=1}^{m} \prod_{i=2}^{k} \left\{ \Delta(t_{i-1}) \Delta(s_i) \right\} + \sum_{k=1}^{m-1} \Delta(t_1) \prod_{i=2}^{k} \left\{ \Delta(s_i) \Delta(t_i) \right\} \right\} \times$$

$$\times \left\{ \mathbf{\Lambda}'_{A_1'A_1'} - \mathbf{\Lambda}'_{A_1'A_2'} \mathbf{\Lambda}'^{-1}_{A_2'A_2'} \mathbf{\Lambda}'_{A_2'A_1'} \right\} \mathbf{1}, \tag{4.32}$$

where

$$\mathbf{w}'^T = - (\mathbf{\alpha}_{A_2}^T, \mathbf{0}). \tag{4.33}$$

$\Delta(\tau)$ can be expressed in terms of $\Gamma(\tau; ., .)$ as

$$\Delta(\tau) = \begin{bmatrix} \mathbf{0} & - \Gamma(\tau; A_2, A_1) \\ - \Gamma(\tau; A_1, A_2) & \mathbf{0} \end{bmatrix}. \tag{4.34}$$

Using (4.34), it is easily established that for $k = 1, ..., m$

$$\prod_{i=2}^{k} \left\{ \Delta(t_{i-1}) \Delta(s_i) \right\} =$$

$$\begin{bmatrix} \prod_{i=2}^{k} \left\{ \Gamma(t_{i-1}; A_2, A_1) \Gamma(s_i; A_1, A_2) \right\} & \mathbf{0} \\ \mathbf{0} & \prod_{i=2}^{k} \left\{ \Gamma(t_{i-1}; A_1, A_2) \Gamma(s_i; A_2, A_1) \right\} \end{bmatrix}, \tag{4.35}$$

84

and

$$\Delta(t_1) \prod_{i=2}^{k} \left\{ \Delta(s_i)\,\Delta(t_i) \right\} =$$

$$\left[\begin{array}{c} \mathbf{0} \\ \hline -\,\boldsymbol{\Gamma}(t_1;\,A_1,\,A_2) \prod_{i=2}^{k} \left\{ \boldsymbol{\Gamma}(s_i;\,A_2,\,A_1)\boldsymbol{\Gamma}(t_i;\,A_1,\,A_2) \right\} \\[4pt] -\,\boldsymbol{\Gamma}(t_1;\,A_2,\,A_1) \prod_{i=2}^{k} \left\{ \boldsymbol{\Gamma}(s_i;\,A_1,\,A_2)\boldsymbol{\Gamma}(t_i;\,A_2,\,A_1) \right\} \\ \hline \mathbf{0} \end{array} \right]. \tag{4.36}$$

(4.35) and (4.36) imply that

$$\sum_{k=1}^{m} \prod_{i=2}^{k} \left\{ \Delta(t_{i-1})\,\Delta(s_i) \right\} + \sum_{k=1}^{m-1} \Delta(t_1) \prod_{i=2}^{k} \left\{ \Delta(s_i)\,\Delta(t_i) \right\} =$$

$$\left[\begin{array}{c} \sum_{k=1}^{m} \prod_{i=2}^{k} \left\{ \boldsymbol{\Gamma}(t_{i-1};\,A_2,\,A_1)\boldsymbol{\Gamma}(s_i;\,A_1,\,A_2) \right\} \\ \hline -\,\boldsymbol{\Gamma}(t_1;\,A_1,\,A_2) \sum_{k=1}^{m-1} \prod_{i=2}^{k} \left\{ \boldsymbol{\Gamma}(s_i;\,A_2,\,A_1)\boldsymbol{\Gamma}(t_i;\,A_1,\,A_2) \right\} \\[4pt] -\,\boldsymbol{\Gamma}(t_1;\,A_2,\,A_1) \sum_{k=1}^{m-1} \prod_{i=2}^{k} \left\{ \boldsymbol{\Gamma}(s_i;\,A_1,\,A_2)\boldsymbol{\Gamma}(t_i;\,A_2,\,A_1) \right\} \\ \hline \sum_{k=1}^{m} \prod_{i=2}^{k} \left\{ \boldsymbol{\Gamma}(t_{i-1};\,A_1,\,A_2)\boldsymbol{\Gamma}(s_i;\,A_2,\,A_1) \right\} \end{array} \right]. \tag{4.37}$$

Using now (4.33), (4.36) (with k = m), and (4.37) we get

$$\mathbf{w}'^{T} \boldsymbol{\Lambda}'_{A_1'A_1'}{}^{-1} \left\{ \mathbf{I} - \exp\{\,s_1\,\boldsymbol{\Lambda}'_{A_1'A_1'}\,\} \right\} \Delta(t_1) \prod_{i=2}^{m} \left\{ \Delta(s_i)\,\Delta(t_i) \right\} =$$

$$\left[\, \mathbf{0},\, \boldsymbol{\alpha}_{A_2}{}^{T} \boldsymbol{\Lambda}_{A_2A_2}{}^{-1} \left\{ \mathbf{I} - \exp\{\,s_1\,\boldsymbol{\Lambda}_{A_2A_2}\,\} \right\} \right] \times$$

$$\Gamma(t_1; A_2, A_1) \prod_{i=2}^{m} \left\{ \Gamma(s_i; A_1, A_2)\Gamma(t_i; A_2, A_1) \right\} \,],$$ (4.38)

and

$$\mathbf{w}'^{\mathrm{T}} \, \boldsymbol{\Lambda}'_{A_1'A_1'}{}^{-1} \left\{ \mathbf{I} - \exp\{ s_1 \, \boldsymbol{\Lambda}'_{A_1'A_1'} \} \right\} \times$$

$$\left\{ \sum_{k=1}^{m} \prod_{i=2}^{k} \left\{ \boldsymbol{\Delta}(t_{i-1}) \, \boldsymbol{\Delta}(s_i) \right\} + \sum_{k=1}^{m-1} \boldsymbol{\Delta}(t_1) \prod_{i=2}^{k} \left\{ \boldsymbol{\Delta}(s_i) \, \boldsymbol{\Delta}(t_i) \right\} \right\} =$$

$$\boldsymbol{\alpha}_{A_2}{}^{\mathrm{T}} \, \boldsymbol{\Lambda}_{A_2A_2}{}^{-1} \left\{ \mathbf{I} - \exp\{ s_1 \, \boldsymbol{\Lambda}_{A_2A_2} \} \right\} \times$$

$$\left[- \sum_{k=1}^{m} \prod_{i=2}^{k} \left\{ \Gamma(t_{i-1}; A_2, A_1)\Gamma(s_i; A_1, A_2) \right\} \right.$$

$$\left. \Gamma(t_1; A_2, A_1) \sum_{k=1}^{m-1} \prod_{i=2}^{k} \left\{ \Gamma(s_i; A_1, A_2)\Gamma(t_i; A_2, A_1) \right\} \right].$$ (4.39)

Finally, it is easily verified that

$$\boldsymbol{\Lambda}'_{A_1'A_1'} - \boldsymbol{\Lambda}'_{A_1'A_2'} \, \boldsymbol{\Lambda}'_{A_2'A_2'}{}^{-1} \, \boldsymbol{\Lambda}'_{A_2'A_1'} = \left[\begin{array}{c|c} \boldsymbol{\Lambda}_{A_2A_2} & \boldsymbol{\Lambda}_{A_2A_1} \\ \hline \boldsymbol{\Lambda}_{A_1A_2} & \boldsymbol{\Lambda}_{A_1A_1} \end{array} \right].$$ (4.40)

Using (4.38) - (4.40), (4.32) can be re-expressed as

$$- \boldsymbol{\alpha}_{A_2}{}^{\mathrm{T}} \, \boldsymbol{\Lambda}_{A_2A_2}{}^{-1} \left\{ \mathbf{I} - \exp\{ s_1 \, \boldsymbol{\Lambda}_{A_2A_2} \} \right\} \times$$

$$\Gamma(t_1; A_2, A_1) \prod_{i=2}^{m} \left\{ \Gamma(s_i; A_1, A_2) \, \Gamma(t_i; A_2, A_1) \right\} \boldsymbol{\Lambda}_{A_1A_1} \mathbf{1} +$$

$$+ \sum_{k=1}^{m} \boldsymbol{\alpha}_{A_2}{}^{\mathrm{T}} \, \boldsymbol{\Lambda}_{A_2A_2}{}^{-1} \left\{ \mathbf{I} - \exp\{ s_1 \, \boldsymbol{\Lambda}_{A_2A_2} \} \right\} \times$$

$$\prod_{i=2}^{k} \left\{ \Gamma(t_{i-1}; A_2, A_1) \, \Gamma(s_i; A_1, A_2) \right\} (\boldsymbol{\Lambda}_{A_2A_2} \mathbf{1} + \boldsymbol{\Lambda}_{A_2A_1} \mathbf{1}) -$$

$$- \sum_{k=1}^{m-1} \alpha_{A_2}^T \Lambda_{A_2A_2}^{-1} \left\{ \mathbf{I} - \exp\{ s_1 \Lambda_{A_2A_2} \} \right\} \times$$

$$\Gamma(t_1; A_2, A_1) \prod_{i=2}^{k} \left\{ \Gamma(s_i; A_1, A_2) \Gamma(t_i; A_2, A_1) \right\} \left(\Lambda_{A_1A_2} \mathbf{1} + \Lambda_{A_1A_1} \mathbf{1} \right). \tag{4.41}$$

The corresponding expression for the second term on the right hand side of (4.31) is obtained from (4.41) by interchanging the roles of A_2 and A_1. Summing this and (4.41) gives (4.27). ∎

For *irreducible* Y, we immediately deduce the following.

COROLLARY 4.7. *For irreducible* Y, *the joint distribution function of the first* 2m *sojourn times of* Y *in* A_1 *and* A_2 *is given for* $s_i, t_i \geq 0$ (i = 1, ...,m) *by*

$$\Pr\{ T_{A_2,1} \leq s_1, ..., T_{A_2,m} \leq s_m, T_{A_1,1} \leq t_1, ..., T_{A_1,m} \leq t_m \} =$$

$$- \alpha_{A_2}^T \Lambda_{A_2A_2}^{-1} \left\{ \mathbf{I} - \exp\{ s_1 \Lambda_{A_2A_2} \} \right\} \times$$

$$\Gamma(t_1; A_2, A_1) \prod_{i=2}^{m} \left\{ \Gamma(s_i; A_1, A_2) \Gamma(t_i; A_2, A_1) \right\} \Lambda_{A_1A_1} \mathbf{1} -$$

$$- \alpha_{A_1}^T \Lambda_{A_1A_1}^{-1} \left\{ \mathbf{I} - \exp\{ t_1 \Lambda_{A_1A_1} \} \right\} \times$$

$$\Gamma(s_1; A_1, A_2) \prod_{i=2}^{m} \left\{ \Gamma(t_i; A_2, A_1) \Gamma(s_i; A_1, A_2) \right\} \Lambda_{A_2A_2} \mathbf{1},$$

with Γ *defined by (4.28).* ∎

In the next chapter, the Laplace transform formulation of Theorem 4.6. will be required. The Laplace transform of the random vector $(T_{A_1,1}, ..., T_{A_1,m}, T_{A_2,1}, ..., T_{A_2,m})$ is defined for τ, $\sigma \in \mathbb{C}^{+m}$ by

$$(T_{A_1,1}, ..., T_{A_1,m}, T_{A_2,1}, ..., T_{A_2,m})^*(\tau; \sigma) =$$

$$E\{\exp\{ - \tau_1 T_{A_1,1} - ... - \tau_m T_{A_1,m} - \sigma_1 T_{A_2,1} - ... - \sigma_m T_{A_2,m} \}\}.$$

With this notation then, we have the following corollary.

COROLLARY 4.8. *If* Y *is irreducible or absorbig, the Laplace transform of* $(T_{A_1,1}, ..., T_{A_1,m},$ $T_{A_2,1}, ..., T_{A_2,m})$ *is given by*

$$(T_{A_1,1}, ..., T_{A_1,m}, T_{A_2,1}, ..., T_{A_2,m})^*(\tau; \sigma) =$$

$$f_{A_2A_1,m}(\boldsymbol{\sigma};\boldsymbol{\tau}) + f_{A_1A_2,m}(\boldsymbol{\tau};\boldsymbol{\sigma}),\tag{4.42}$$

where $f_{AB,m}(\boldsymbol{\sigma};\boldsymbol{\tau})$ *is defined for any disjoint and non-empty* $A, B \subset S \setminus \{\omega\}$
by

$$f_{AB,m}(\boldsymbol{\sigma};\boldsymbol{\tau}) =$$

$$-\boldsymbol{\alpha}_A^T (\boldsymbol{\Lambda}_{AA} - \sigma_1 I)^{-1} \boldsymbol{\Lambda}_{AB} (\boldsymbol{\Lambda}_{BB} - \tau_1 I)^{-1} \times$$

$$\prod_{i=2}^{m} \left\{ \boldsymbol{\Lambda}_{BA} (\boldsymbol{\Lambda}_{AA} - \sigma_i I)^{-1} \boldsymbol{\Lambda}_{AB} (\boldsymbol{\Lambda}_{BB} - \tau_i I)^{-1} \right\} \boldsymbol{\Lambda}_{BB} \, \mathbf{1} +$$

$$+ \sum_{k=1}^{m} \boldsymbol{\alpha}_A^T (\boldsymbol{\Lambda}_{AA} - \sigma_1 I)^{-1} \times$$

$$\prod_{i=2}^{k} \left\{ \boldsymbol{\Lambda}_{AB} (\boldsymbol{\Lambda}_{BB} - \tau_{i-1} I)^{-1} \boldsymbol{\Lambda}_{BA} (\boldsymbol{\Lambda}_{AA} - \sigma_i I)^{-1} \right\} \left\{ \boldsymbol{\Lambda}_{AA} \, \mathbf{1} + \boldsymbol{\Lambda}_{AB} \, \mathbf{1} \right\} -$$

$$- \sum_{k=1}^{m-1} \boldsymbol{\alpha}_A^T (\boldsymbol{\Lambda}_{AA} - \sigma_1 I)^{-1} \boldsymbol{\Lambda}_{AB} (\boldsymbol{\Lambda}_{BB} - \tau_1 I)^{-1} \times$$

$$\prod_{i=2}^{k} \left\{ \boldsymbol{\Lambda}_{BA} (\boldsymbol{\Lambda}_{AA} - \sigma_i I)^{-1} \boldsymbol{\Lambda}_{AB} (\boldsymbol{\Lambda}_{BB} - \tau_i I)^{-1} \right\} \left\{ \boldsymbol{\Lambda}_{BA} \, \mathbf{1} + \boldsymbol{\Lambda}_{BB} \, \mathbf{1} \right\}.\tag{4.43}$$

PROOF OF COROLLARY 4.8. The reasoning is along the lines of the proof of Corollary 4.5. $[0, +\infty)^{2m}$ is the support of the distribution of the random vector $(T_{A_1,1}, ..., T_{A_1,m}, T_{A_2,1}, ..., T_{A_2,m})$. Let μ be the measure on $[0,+\infty)^{2m}$ which is defined for any Borel-measurable set B $\subseteq [0,+\infty)^{2m}$ by

$$\mu(B) = \lambda_{2m}(B) +$$

$$+ \sum_{k=1}^{m} \lambda_{2k-1}(\{ (t_1, ..., t_{k-1}, s_1, ..., s_k)^T \in [0,+\infty)^{2k-1} :$$

$$(t_1, ..., t_{k-1}, 0, ..., 0, s_1, ..., s_k, 0, ..., 0)^T \in B \}) +$$

$$+ \sum_{k=1}^{m} \lambda_{2k-1}(\{ (t_1, ..., t_k, s_1, ..., s_{k-1})^T \in [0,+\infty)^{2k-1} :$$

$$(t_1, ..., t_k, 0, ..., 0, s_1, ..., s_{k-1}, 0, ..., 0)^T \in B \}) +$$

$$+ \sum_{k=1}^{m-1} \lambda_{2k}(\{ (t_1, ..., t_k, s_1, ..., s_k)^T \in [0, +\infty)^{2k} :$$

$$(t_1, ..., t_k, 0, ..., 0, s_1, ..., s_k, 0, ..., 0)^T \in B \}),$$

where λ_k is the Lebesgue measure on $\mathbb{R}^k$. Put for $\mathbf{s}, \mathbf{t} \in [0, +\infty)^m$,

$$h(\mathbf{s}; \mathbf{t}) = h_{A_2 A_1, m}(\mathbf{s}; \mathbf{t}) + h_{A_1 A_2, m}(\mathbf{t}; \mathbf{s}), \qquad (4.44)$$

where $h_{AB,m}(\mathbf{s}; \mathbf{t})$ is defined by

$$h_{AB,m}(\mathbf{s}; \mathbf{t}) =$$

$$- \boldsymbol{\alpha}_A^T \exp\{ s_1 \boldsymbol{\Lambda}_{AA} \} \boldsymbol{\Lambda}_{AB} \exp\{ t_1 \boldsymbol{\Lambda}_{BB} \} \times$$

$$\prod_{i=2}^{m} \left\{ \boldsymbol{\Lambda}_{BA} \exp\{ s_i \boldsymbol{\Lambda}_{AA} \} \boldsymbol{\Lambda}_{AB} \exp\{ t_i \boldsymbol{\Lambda}_{BB} \} \right\} \boldsymbol{\Lambda}_{BB} \mathbf{1} \times$$

$$I_{\{ t_1 > 0, ..., t_m > 0, s_1 > 0, ..., s_m > 0 \}} +$$

$$+ \sum_{k=1}^{m} \boldsymbol{\alpha}_A^T \exp\{ s_1 \boldsymbol{\Lambda}_{AA} \} \times$$

$$\prod_{i=2}^{k} \left\{ \boldsymbol{\Lambda}_{AB} \exp\{ t_{i-1} \boldsymbol{\Lambda}_{BB} \} \boldsymbol{\Lambda}_{BA} \exp\{ s_i \boldsymbol{\Lambda}_{AA} \} \right\} \left\{ \boldsymbol{\Lambda}_{AA} \mathbf{1} + \boldsymbol{\Lambda}_{AB} \mathbf{1} \right\} \times$$

$$I_{\{ t_1, ..., t_{k-1} > 0, t_k = ... = t_m = 0, s_1, ..., s_k > 0, s_{k+1} = ... = s_m = 0 \}} -$$

$$- \sum_{k=1}^{m-1} \boldsymbol{\alpha}_A^T \exp\{ s_1 \boldsymbol{\Lambda}_{AA} \} \boldsymbol{\Lambda}_{AB} \exp\{ t_1 \boldsymbol{\Lambda}_{BB} \} \times$$

$$\prod_{i=2}^{k} \left\{ \boldsymbol{\Lambda}_{BA} \exp\{ s_i \boldsymbol{\Lambda}_{AA} \} \boldsymbol{\Lambda}_{AB} \exp\{ t_i \boldsymbol{\Lambda}_{BB} \} \right\} \left\{ \boldsymbol{\Lambda}_{BA} \mathbf{1} + \boldsymbol{\Lambda}_{BB} \mathbf{1} \right\} \times$$

$$I_{\{ t_1, ..., t_k > 0, t_{k+1} = ... = t_m = 0, s_1, ..., s_k > 0, s_{k+1} = ... = s_m = 0 \}}. \qquad (4.45)$$

By integrating h over $[0, t_1] \times ... \times [0, t_m] \times [0, s_1] \times ... \times [0, s_m]$, it is easily established by (4.27) that h is a μ - density of $(T_{A_1,1}, ..., T_{A_1,m}, T_{A_2,1}, ..., T_{A_2,m})$. By (4.44), the Laplace transform in question is thus given by

$$(T_{A_1,1}, ..., T_{A_1,m}, T_{A_2,1}, ..., T_{A_2,m})^*(\boldsymbol{\tau}; \boldsymbol{\sigma}) =$$

$$\int_{[0, +\infty)^{2m}} \exp\{ - \boldsymbol{\sigma}^T \mathbf{s} - \boldsymbol{\tau}^T \mathbf{t} \} \left(h_{A_2A_1,m}(\mathbf{s}; \mathbf{t}) + h_{A_1A_2,m}(\mathbf{t}; \mathbf{s}) \right) \mu\{d\mathbf{s}, d\mathbf{t}\}. \qquad (4.46)$$

Using (4.45), it is seen that

$$\int_{[0, +\infty)^{2m}} \exp\{ - \boldsymbol{\sigma}^T \mathbf{s} - \boldsymbol{\tau}^T \mathbf{t} \} \, h_{AB,m}(\mathbf{s}; \mathbf{t}) \, \mu\{d\mathbf{s}, d\mathbf{t}\} = f_{AB,m}(\boldsymbol{\sigma}; \boldsymbol{\tau}),$$

which shows (4.42) by (4.44) and (4.46). ∎

We conclude this section with some remarks on *Markov reward processes*. (Two recent references on such processes are [TRI1] and [REI2].) This tool is used to evaluate the performance of a system modelled by a continuous-time Markov chain Y. In its simplest form, the reward process comprises Y and a set of positive, constant reward rates $\{ r_s : s \in S \}$. Each visit of length Δt by Y to $s \in S$ is deemed to result in the accumulation of a reward $r_s \Delta t$. Let **R** denote the diagonal matrix of the reward rates associated with visits to $A_1 \cup A_2$ and assume that the conditions of Theorem 4.2 hold. Then, accumulated rewards due to visits in A_1 may be expressed as sojourn times in A_1 of a new Markov chain whose transition rate matrix is given by

$$\boldsymbol{\Lambda}^* = \left[\begin{array}{c|c} \mathbf{R}^{-1} & \mathbf{0} \\ \hline \mathbf{0} & \mathbf{0} \end{array} \right] \boldsymbol{\Lambda} =$$

$$\left[\begin{array}{c|c|c} \mathbf{R}_{A_1A_1}^{-1} \boldsymbol{\Lambda}_{A_1A_1} & \mathbf{R}_{A_1A_1}^{-1} \boldsymbol{\Lambda}_{A_1A_2} & \mathbf{R}_{A_1A_1}^{-1} \boldsymbol{\Lambda}_{A_1\{\omega\}} \\ \hline \mathbf{R}_{A_2A_2}^{-1} \boldsymbol{\Lambda}_{A_2A_1} & \mathbf{R}_{A_2A_2}^{-1} \boldsymbol{\Lambda}_{A_2A_2} & \mathbf{R}_{A_2A_2}^{-1} \boldsymbol{\Lambda}_{A_2\{\omega\}} \\ \hline \mathbf{0} & \mathbf{0} & \mathbf{0} \end{array} \right].$$

It is seen in particular that the choice of the reward rates associated with the states in A_2 will not affect the joint distribution of accumulated rewards due to the first m visits in A_1; for instance, $\boldsymbol{\Lambda}^*_{A_1A_2} \boldsymbol{\Lambda}^*_{A_2A_2}{}^{-1} \boldsymbol{\Lambda}^*_{A_2A_1} \boldsymbol{\Lambda}^*_{A_1A_1}{}^{-1}$ evaluates to

$$(\mathbf{R}_{A_1A_1}^{-1}\boldsymbol{\Lambda}_{A_1A_2})(\mathbf{R}_{A_2A_2}^{-1}\boldsymbol{\Lambda}_{A_2A_2})^{-1}(\mathbf{R}_{A_2A_2}^{-1}\boldsymbol{\Lambda}_{A_2A_1})(\mathbf{R}_{A_1A_1}^{-1}\boldsymbol{\Lambda}_{A_1A_1})^{-1} =$$

$$\mathbf{R}_{A_1A_1}^{-1} \boldsymbol{\Lambda}_{A_1A_2} \boldsymbol{\Lambda}_{A_2A_2}^{-1} \boldsymbol{\Lambda}_{A_2A_1} \boldsymbol{\Lambda}_{A_1A_1}^{-1} \mathbf{R}_{A_1A_1},$$

which is obviously not dependent on $\mathbf{R}_{A_2A_2}$.

4.2 Some further distribution results related to sojourn times

This section is devoted to the analysis of the variables $TS_{A_1,m} = T_{A_1,1} + ... + T_{A_1,m}$, $\max\{T_{A_1,1}, T_{A_1,2}, ...\}$, and $TS_{A_1,\infty} = T_{A_1,1} + T_{A_1,2} + ...$. (The latter two are finite for absorbing Y only.) The cumulative distribution functions of $TS_{A_1,m}$ and $TS_{A_1,\infty}$ are known from [RUB2] where a different method of proof is employed: there, no use is made of other results on sojourn times, but a separate (univariate) renewal argument is applied. The reasonings in [RUB2], [RUB3], and [RUB4] are in fact such that no one presentation is dependent upon, or makes use of, the results in the other two. In this section, the full information inherent in the joint distribution of sojourn times is exploited to rederive those results in a unified fashion. The distribution of $\max\{T_{A_1,1}, T_{A_1,2}, ...\}$ was first considered in [CSE6]. We start with $TS_{A_1,m}$.

THEOREM 4.9. *The cumulative distribution function of the sum of the lengths of the first m sojourns in* A_1, $TS_{A_1,m}$, *is given by*

$$\Pr\{\ TS_{A_1,m} \le t\ \} = 1 + (\mathbf{w}^T, \mathbf{0}, ..., \mathbf{0}) \exp\{\ t\ \mathbf{M}_m\ \} \mathbf{1}, t \ge 0, \tag{4.47}$$

where the block partitioned square matrix $\mathbf{M}_m$ *is defined via (2.13) as a matrix of type* $\boldsymbol{\Delta}_m$,

$$\mathbf{M}_m = \boldsymbol{\Delta}_m(\boldsymbol{\Phi}, ..., \boldsymbol{\Phi}; \boldsymbol{\Psi}, ..., \boldsymbol{\Psi}),$$

with

$$\boldsymbol{\Phi} = \boldsymbol{\Lambda}_{A_1 A_1}, \quad \boldsymbol{\Psi} = - \boldsymbol{\Lambda}_{A_1 A_2} \boldsymbol{\Lambda}_{A_2 A_2}^{-1} \boldsymbol{\Lambda}_{A_2 A_1}.$$

w *is defined by (4.6).*

PROOF OF THEOREM 4.9. Before entering the proof proper, let us state a lemma which is easily seen to hold.

LEMMA 4.10. *The matrix* $\boldsymbol{\Delta}_n(\boldsymbol{\Phi}_1, ..., \boldsymbol{\Phi}_n; \boldsymbol{\Psi}_1, ..., \boldsymbol{\Psi}_{n-1})$ *in (2.13) is invertible if* $\boldsymbol{\Phi}_1^{-1}, ...,$ $\boldsymbol{\Phi}_n^{-1}$ *exist and then*

$$(\boldsymbol{\Delta}_n(\boldsymbol{\Phi}_1, ..., \boldsymbol{\Phi}_n; \boldsymbol{\Psi}_1, ..., \boldsymbol{\Psi}_{n-1}))^{-1} = \boldsymbol{\Xi}_n(\boldsymbol{\Phi}_1, ..., \boldsymbol{\Phi}_n; \boldsymbol{\Psi}_1, ..., \boldsymbol{\Psi}_{n-1}) =$$

$$\begin{bmatrix} \boldsymbol{\Xi}_{11} & \boldsymbol{\Xi}_{12} & \boldsymbol{\Xi}_{13} & \cdots & \boldsymbol{\Xi}_{1n} \\ \mathbf{0} & \boldsymbol{\Xi}_{22} & \boldsymbol{\Xi}_{23} & \cdots & \boldsymbol{\Xi}_{2n} \\ \mathbf{0} & \mathbf{0} & \boldsymbol{\Xi}_{33} & \cdots & \boldsymbol{\Xi}_{3n} \\ \vdots & \vdots & \vdots & & \vdots \\ \mathbf{0} & \mathbf{0} & \mathbf{0} & \cdots & \boldsymbol{\Xi}_{n-1\,n} \\ \mathbf{0} & \mathbf{0} & \mathbf{0} & \cdots & \boldsymbol{\Xi}_{nn} \end{bmatrix},$$

where

$$\Xi_{ij} = \Phi_i^{-1} \prod_{k=i}^{j-1} \left\{ - \Psi_k \Phi_{k+1}^{-1} \right\}, \ 1 \le i \le j \le n. \qquad \blacksquare$$

Proof of Theorem 4.9. From Corollary 4.5 it follows that the Laplace transform of $TS_{A_1,m}$ is given by

$$(TS_{A_1,m})^*(\sigma) =$$

$$- \mathbf{w}^T (\mathbf{\Lambda}_{A_1A_1} - \sigma \, \mathbf{I})^{-1} \left\{ \mathbf{\Lambda}_{A_1A_2} \mathbf{\Lambda}_{A_2A_2}^{-1} \mathbf{\Lambda}_{A_2A_1} (\mathbf{\Lambda}_{A_1A_1} - \sigma \, \mathbf{I})^{-1} \right\}^{m-1} \mathbf{\Lambda}_{A_1A_1} \mathbf{1} +$$

$$+ \mathbf{1} + \mathbf{w}^T \mathbf{1} -$$

$$- \sum_{k=1}^{m-1} \mathbf{w}^T (\mathbf{\Lambda}_{A_1A_1} - \sigma \, \mathbf{I})^{-1} \left\{ \mathbf{\Lambda}_{A_1A_2} \mathbf{\Lambda}_{A_2A_2}^{-1} \mathbf{\Lambda}_{A_2A_1} (\mathbf{\Lambda}_{A_1A_1} - \sigma \, \mathbf{I})^{-1} \right\}^{k-1} \times$$

$$\left\{ \mathbf{\Lambda}_{A_1A_1} - \mathbf{\Lambda}_{A_1A_2} \mathbf{\Lambda}_{A_2A_2}^{-1} \mathbf{\Lambda}_{A_2A_1} \right\} \mathbf{1}, \qquad (4.48)$$

Next, it will be shown that (4.48) can be rewritten in terms of $\mathbf{M}_m$ as

$$(TS_{A_1,m})^*(\sigma) = \mathbf{1} + \mathbf{w}^T \mathbf{1} - (\mathbf{w}^T, \mathbf{0}, ..., \mathbf{0}) (\mathbf{M}_m - \sigma \, \mathbf{I})^{-1} \mathbf{M}_m \mathbf{1}. \qquad (4.49)$$

To see this, we note that

$$(\mathbf{M}_m - \sigma \, \mathbf{I}) = \mathbf{\Delta}_m(\mathbf{\Phi} - \sigma \, \mathbf{I}, ..., \mathbf{\Phi} - \sigma \, \mathbf{I}; \mathbf{\Psi}, ..., \mathbf{\Psi}).$$

But, $(\mathbf{\Phi} - \sigma \, \mathbf{I}) = (\mathbf{\Lambda}_{A_1A_1} - \sigma \, \mathbf{I})$ is invertible by (4.26), and thus $(\mathbf{M}_m - \sigma \, \mathbf{I})^{-1}$ exists by Lemma 4.10. The right hand side of (4.49) is therefore equal to

$$\mathbf{1} + \mathbf{w}^T \mathbf{1} - (\mathbf{w}^T, \mathbf{0}, ..., \mathbf{0}) \, \Xi_m(\mathbf{\Phi} - \sigma \, \mathbf{I}, ..., \mathbf{\Phi} - \sigma \, \mathbf{I}; \mathbf{\Psi}, ..., \mathbf{\Psi}) \begin{bmatrix} (\mathbf{\Phi} + \mathbf{\Psi}) \, \mathbf{1} \\ \hline \vdots \\ \hline (\mathbf{\Phi} + \mathbf{\Psi}) \, \mathbf{1} \\ \hline \mathbf{\Phi} \, \mathbf{1} \end{bmatrix} =$$

$$\mathbf{1} + \mathbf{w}^T \mathbf{1} - \sum_{k=1}^{m-1} \mathbf{w}^T (\mathbf{\Phi} - \sigma \, \mathbf{I})^{-1} \left\{ - \mathbf{\Psi} (\mathbf{\Phi} - \sigma \, \mathbf{I})^{-1} \right\}^{k-1} (\mathbf{\Phi} + \mathbf{\Psi}) \, \mathbf{1} -$$

$$- w^T (\Phi - \sigma I)^{-1} \left\{ - \Psi (\Phi - \sigma I)^{-1} \right\}^{m-1} \Phi \ 1,$$

which by (4.48) is indeed the Laplace transform of $TS_{A_1,m}$. (4.47) now follows from (4.49) by Laplace transform inversion since

$$(\sigma I - M_m)^{-1} = \int_0^{+\infty} \exp\{-\sigma t\} \exp\{t M_m\} \ dt. \tag{4.50}$$

((4.50) can be shown along the lines of (4.26) by using in the proof of Lemma 4.1 a suitably defined auxiliary Markov chain.)

(NOTE. We want to elaborate on this process of 'inversion' in some more detail for the sake of clarity and also since the same type of reasoning will surface in the proof of Theorem 4.12, and, in Chapter 5, in the proofs of Corollary 5.4 and Proposition 5.5. Let μ be the measure on $[0, +\infty)$ which is defined for Borel-measurable $B \subseteq [0, +\infty)$ by

$$\mu(B) = \varepsilon_0(B) + \lambda(B \cap (0, +\infty)),$$

where ε_0 is the probability measure which assigns unity to the origin and λ stands for the Lebesgue measure on $\mathbb{R}$. Then, it is readily seen that for the function h, defined on $[0, +\infty)$ by

$$h(t) = (1 + w^T 1) I_{\{0\}}(t) + (w^T, 0, ..., 0) \exp\{t M_m\} M_m 1 I_{(0, +\infty)}(t),$$

the following holds by (4.50)

$$\int_{[0, +\infty)} h(t) \exp\{-\sigma t\} \mu\{dt\} = 1 + w^T 1 - (w^T, 0, ..., 0)(M_m - \sigma I)^{-1} M_m 1.$$

It is therefore h(t) a μ-density of the distribution of $TS_{A_1,m}$. (4.47) is now obtained by integrating h with respect to μ over $[0, t]$.) ∎

Before turning our attention to $TS_{A_1,\infty}$, we establish a lemma which will be used in the sequel. For the rest of this section, Y is absorbing.

LEMMA 4.11. *(a) All the entries of* $\Lambda_{A_1A_1}^{-1} \Lambda_{A_1A_2} \Lambda_{A_2A_2}^{-1} \Lambda_{A_2A_1}$ *are non-negative. (b) For any* $\sigma \geq 0$, *we have element-wise*

$$\left\{ \Lambda_{A_1A_2} \Lambda_{A_2A_2}^{-1} \Lambda_{A_2A_1} (\Lambda_{A_1A_1} - \sigma I)^{-1} \right\}^k \to 0, \ as \ k \to +\infty. \tag{4.51}$$

PROOF OF LEMMA 4.11. *Proof of (a).* From the proof, and with the notation, of Theorem 4.2.(a), we have the following set of equations which is equivalent to (4.9)

$$D_{A_1A_1} \Lambda_{A_1A_2} = P_{A_1A_2}, \qquad D_{A_2A_2} \Lambda_{A_2A_1} = P_{A_2A_1}, \tag{4.52}$$

$$- D_{A_1A_1} \Lambda_{A_1A_1} = I - P_{A_1A_1}, \quad - D_{A_2A_2} \Lambda_{A_2A_2} = I - P_{A_2A_2}. \tag{4.53}$$

From (4.52) and (4.53) it follows that

$$\Lambda_{A_1A_1}^{-1} \Lambda_{A_1A_2} \Lambda_{A_2A_2}^{-1} \Lambda_{A_2A_1} =$$

$$(I - P_{A_1A_1})^{-1} P_{A_1A_2} (I - P_{A_2A_2})^{-1} P_{A_2A_1}, \tag{4.54}$$

which shows the assertion by writing the inverses on the right hand side of (4.54) as infinite 'geometric sums'.

Proof of (b). First, it is easily established by (2.33) that element-wise,

$$\left\{ (I - P_{A_1A_1})^{-1} P_{A_1A_2} (I - P_{A_2A_2})^{-1} P_{A_2A_1} \right\}^k \to 0, \text{ as } k \to +\infty, \tag{4.55}$$

where, as before, P stands for the transition probability matrix of the embedded Markov chain of Y. (Notice that this reasoning has already been employed in the proof of Corollary 2.7.) We get by (4.53)

$$(\Lambda_{A_1A_1} - \sigma I) = - D_{A_1A_1}^{-1} (I - P_{A_1A_1}) - \sigma I =$$

$$- D_{A_1A_1}^{-1} \left\{ (I + \sigma D_{A_1A_1}) - P_{A_1A_1} \right\},$$

from which it is seen that

$$- (\Lambda_{A_1A_1} - \sigma I)^{-1} = \left\{ I - (I + \sigma D_{A_1A_1})^{-1} P_{A_1A_1} \right\}^{-1} (I + \sigma D_{A_1A_1})^{-1} D_{A_1A_1}. \tag{4.56}$$

But, $(I + \sigma D_{A_1A_1})^{-1} P_{A_1A_1}$ has non-negative entries only and it is element-wise not greater than $P_{A_1A_1}$; this relationship also holds, of course, for any power of these two matrices. Consequently, (4.56) shows that, elementwise,

$$0 \le - (\Lambda_{A_1A_1} - \sigma I)^{-1} \le$$

$$(I - P_{A_1A_1})^{-1} (I + \sigma D_{A_1A_1})^{-1} D_{A_1A_1} \le (I - P_{A_1A_1})^{-1} D_{A_1A_1}. \tag{4.57}$$

On the other hand, by (4.52) and (4.53) we have

$$\Lambda_{A_1A_2} \Lambda_{A_2A_2}^{-1} \Lambda_{A_2A_1} =$$

$$D_{A_1A_1}^{-1} P_{A_1A_2} \left\{ - D_{A_2A_2}^{-1} (I - P_{A_2A_2}) \right\}^{-1} D_{A_2A_2}^{-1} P_{A_2A_1} =$$

$$- D_{A_1A_1}^{-1} P_{A_1A_2} (I - P_{A_2A_2})^{-1} P_{A_2A_1}. \tag{4.58}$$

From (4.57) and (4.58) it follows that, elementwise,

$$0 \le \mathbf{\Lambda}_{A_1A_2} \mathbf{\Lambda}_{A_2A_2}^{-1} \mathbf{\Lambda}_{A_2A_1} (\mathbf{\Lambda}_{A_1A_1} - \sigma \mathbf{I})^{-1} \le$$

$$\mathbf{D}_{A_1A_1}^{-1} \mathbf{P}_{A_1A_2} (\mathbf{I} - \mathbf{P}_{A_2A_2})^{-1} \mathbf{P}_{A_2A_1} (\mathbf{I} - \mathbf{P}_{A_1A_1})^{-1} \mathbf{D}_{A_1A_1}. \qquad (4.59)$$

(4.51) now follows from (4.55) and (4.59). ∎

We are now in a position to consider $TS_{A_1,\infty}$.

THEOREM 4.12. *For absorbing* Y, *the cumulative distribution function of* $TS_{A_1,\infty}$ *is given for* t ≥ 0 *by*

$$\Pr\{ TS_{A_1,\infty} \le t \} = 1 + \mathbf{w}^T \exp\{ t (\mathbf{\Lambda}_{A_1A_1} - \mathbf{\Lambda}_{A_1A_2} \mathbf{\Lambda}_{A_2A_2}^{-1} \mathbf{\Lambda}_{A_2A_1}) \} \mathbf{1}, \qquad (4.60)$$

where **w** *is defined by (4.6).*

PROOF OF THEOREM 4.12. Two different threads of thought will be pursued here: the first one will allow the structure of the cumulative distribution function of $TS_{A_1,\infty}$ to be inferred from an informal limit consideration in the Laplace transform domain. The second one (which delivers the rigorous proof but is apparent to a lesser extent) utilizes a certain differential equation from [RUB2].

From (4.48) and (4.51) it follows that the Laplace transform of $TS_{A_1,\infty}$ is

$$(TS_{A_1,\infty})^*(\sigma) =$$

$$1 + \mathbf{w}^T \mathbf{1} - \mathbf{w}^T (\mathbf{\Lambda}_{A_1A_1} - \sigma \mathbf{I})^{-1} \left\{ \mathbf{I} - \mathbf{\Lambda}_{A_1A_2} \mathbf{\Lambda}_{A_2A_2}^{-1} \mathbf{\Lambda}_{A_2A_1} (\mathbf{\Lambda}_{A_1A_1} - \sigma \mathbf{I})^{-1} \right\}^{-1} \times$$

$$\left\{ \mathbf{\Lambda}_{A_1A_1} - \mathbf{\Lambda}_{A_1A_2} \mathbf{\Lambda}_{A_2A_2}^{-1} \mathbf{\Lambda}_{A_2A_1} \right\} \mathbf{1} =$$

$$1 + \mathbf{w}^T \mathbf{1} - \mathbf{w}^T \left\{ \mathbf{\Lambda}_{A_1A_1} - \mathbf{\Lambda}_{A_1A_2} \mathbf{\Lambda}_{A_2A_2}^{-1} \mathbf{\Lambda}_{A_2A_1} - \sigma \mathbf{I} \right\}^{-1} \times$$

$$\left\{ \mathbf{\Lambda}_{A_1A_1} - \mathbf{\Lambda}_{A_1A_2} \mathbf{\Lambda}_{A_2A_2}^{-1} \mathbf{\Lambda}_{A_2A_1} \right\} \mathbf{1}. \qquad (4.61)$$

(4.60) now follows from (4.61) by direct Laplace transform inversion (along the lines of the one in the proof of Theorem 4.9) if (4.26) can be established with $\mathbf{\Lambda}_{A_1A_1}$ replaced by $\mathbf{\Lambda}_{A_1A_1} - \mathbf{\Lambda}_{A_1A_2} \mathbf{\Lambda}_{A_2A_2}^{-1} \mathbf{\Lambda}_{A_2A_1}$. However, this step could not be successfully completed; the missing step is the assertion of Lemma 4.1 with $\mathbf{\Lambda}_{A_1A_1}$ replaced by $\mathbf{\Lambda}_{A_1A_1} - \mathbf{\Lambda}_{A_1A_2} \mathbf{\Lambda}_{A_2A_2}^{-1} \mathbf{\Lambda}_{A_2A_1}$. Let us add in passing that for the simplest case, when namely $A_1 = \{ a_1 \}$ and $A_2 = \{ a_2 \}$, the assertion holds since with, say,

$$\mathbf{\Lambda} = \begin{array}{c} \\ \{a_1\} \\ \\ \{a_2\} \\ \\ \{\omega\} \end{array} \begin{array}{ccc} \{a_1\} & \{a_2\} & \{\omega\} \\ \left[\begin{array}{ccc} -\lambda_1 - \lambda_2 & \lambda_1 & \lambda_2 \\ \\ \nu_1 & -\nu_1 - \nu_2 & \nu_2 \\ \\ 0 & 0 & 0 \end{array}\right] \end{array},$$

we have

$$\mathbf{\Lambda}_{A_1A_1} - \mathbf{\Lambda}_{A_1A_2} \mathbf{\Lambda}_{A_2A_2}^{-1} \mathbf{\Lambda}_{A_2A_1} = -\lambda_1\left(1 - \frac{\nu_1}{\nu_1 + \nu_2}\right) - \lambda_2 < 0.$$

We now embark on the alternative approach to prove (4.60). Let $\mathbf{M}_m$, $\mathbf{\Phi}$, and $\mathbf{\Psi}$ be defined as in Theorem 4.9. First, note that due to the structure of $\mathbf{M}_m$, we have

$$\mathbf{M}_m{}^k = \left[\begin{array}{c|c} \mathbf{\Phi}^k & * \; * \; * \\ \hline 0 & \mathbf{M}_{m-1}{}^k \end{array}\right], \; k = 0, 1, 2, \dots ,$$

from which

$$\exp\{\, t\, \mathbf{M}_m \,\} = \left[\begin{array}{c|c} \exp\{t\, \mathbf{\Phi}\} & * \; * \; * \\ \hline 0 & \exp\{t\, \mathbf{M}_{m-1}\} \end{array}\right].$$

Let the column vector $\mathbf{h}_m(t)$ denote the first group of components of $\exp\{\, t\, \mathbf{M}_m \,\}\,\mathbf{1}$; the number of its entries is the cardinality of A_1. For *any* $\mathbf{w}$ we have

$$\mathbf{w}^T \frac{d}{dt}\, \mathbf{h}_m(t) = \frac{d}{dt}\left\{\mathbf{w}^T\, \mathbf{h}_m(t)\right\} = \frac{d}{dt}\left\{(\mathbf{w}^T, \mathbf{0}, \dots, \mathbf{0})\, \exp\{\, t\, \mathbf{M}_m \,\}\, \mathbf{1}\right\} =$$

$$(\mathbf{w}^T, \mathbf{0}, \dots, \mathbf{0})\, \mathbf{M}_m\, \exp\{\, t\, \mathbf{M}_m \,\}\, \mathbf{1} =$$

$$(\mathbf{w}^T \mathbf{\Phi}, \mathbf{w}^T \mathbf{\Psi}, \mathbf{0}, \dots, \mathbf{0})\, \exp\{\, t\, \mathbf{M}_m \,\}\, \mathbf{1} =$$

$$(\mathbf{w}^T \mathbf{\Phi}, \mathbf{0}, \dots, \mathbf{0})\, \exp\{\, t\, \mathbf{M}_m \,\}\, \mathbf{1} \; +$$

$$(\mathbf{0}, \mathbf{w}^T \mathbf{\Psi}, \mathbf{0}, \dots, \mathbf{0}) \left[\begin{array}{c|c} \exp\{t\, \mathbf{\Phi}\} & * \; * \; * \\ \hline 0 & \exp\{t\, \mathbf{M}_{m-1}\} \end{array}\right] \mathbf{1} =$$

$\mathbf{w}^T \boldsymbol{\Phi} \, \mathbf{h}_m(t) + \mathbf{w}^T \boldsymbol{\Psi} \, \mathbf{h}_{m-1}(t).$

Thus, we have the differential equation $\mathbf{h}_m'(t) = \boldsymbol{\Phi} \, \mathbf{h}_m(t) + \boldsymbol{\Psi} \, \mathbf{h}_{m-1}(t)$. By (4.47),

$$\Pr\{\, TS_{A_1,m} \leq t \,\} = 1 + \mathbf{w}^T \, \mathbf{h}_m(t). \qquad (4.62)$$

We know that the left hand side of (4.62) converges (compact-uniformly) for $\boldsymbol{\alpha}_{A_2} = \mathbf{0}$ and *any* stochastic vector $\boldsymbol{\alpha}_{A_1}$ ($= -\mathbf{w}$), and hence, so does $\mathbf{h}_m(t)$ component-wise to some limit $\mathbf{h}_\infty(t)$ which then satisfies

$$\mathbf{h}_\infty'(t) = (\boldsymbol{\Phi} + \boldsymbol{\Psi}) \, \mathbf{h}_\infty(t). \qquad (4.63)$$

Solving (4.63), we get

$$\mathbf{h}_\infty(t) = \exp\{\, t \, (\boldsymbol{\Phi} + \boldsymbol{\Psi}) \,\} \, \mathbf{c},$$

with $\mathbf{c} = \mathbf{1}$ since by (4.47) (with $t = 0$ and $m \rightarrow +\infty$) it is $\Pr\{\, TS_{A_1,\infty} = 0 \,\} = 1 + \mathbf{w}^T \, \mathbf{1}$. ∎

We now close this section with a theorem on $\max\{T_{A_1,1}, T_{A_1,2}, \ldots\}$.

THEOREM 4.13. *For absorbing* Y, *the cumulative distribution function of* $\max\{T_{A_1,1}, T_{A_1,2}, \ldots\}$ *is given by*

$$\Pr\{\, \max\{T_{A_1,1}, T_{A_1,2}, \ldots\} \leq t \,\} = 1 + \mathbf{w}^T \, \mathbf{1} - \mathbf{w}^T \, \boldsymbol{\Lambda}_{A_1A_1}^{-1} \left\{ \mathbf{I} - \exp\{\, t \, \boldsymbol{\Lambda}_{A_1A_1} \} \right\} \times$$

$$\left\{ \mathbf{I} - \boldsymbol{\Lambda}_{A_1A_2} \, \boldsymbol{\Lambda}_{A_2A_2}^{-1} \, \boldsymbol{\Lambda}_{A_2A_1} \, \boldsymbol{\Lambda}_{A_1A_1}^{-1} \left\{ \mathbf{I} - \exp\{\, t \, \boldsymbol{\Lambda}_{A_1A_1} \} \right\} \right\}^{-1} \times$$

$$\left\{ \boldsymbol{\Lambda}_{A_1A_1} - \boldsymbol{\Lambda}_{A_1A_2} \, \boldsymbol{\Lambda}_{A_2A_2}^{-1} \, \boldsymbol{\Lambda}_{A_2A_1} \right\} \, \mathbf{1}, \qquad (4.64)$$

where $\mathbf{w}$ *is defined by (4.6).*

PROOF OF THEOREM 4.13. From (4.5), we get with $t_1 = \ldots = t_m = t \geq 0$,

$$\Pr\{\, \max\{T_{A_1,1}, \ldots, T_{A_1,m} \} \leq t \,\} = 1 + \mathbf{w}^T \, \mathbf{1} -$$

$$- \mathbf{w}^T \, \boldsymbol{\Lambda}_{A_1A_1}^{-1} \left\{ \mathbf{I} - \exp\{\, t \, \boldsymbol{\Lambda}_{A_1A_1} \} \right\} \times$$

$$\left\{ \boldsymbol{\Lambda}_{A_1A_2} \, \boldsymbol{\Lambda}_{A_2A_2}^{-1} \, \boldsymbol{\Lambda}_{A_2A_1} \, \boldsymbol{\Lambda}_{A_1A_1}^{-1} \left\{ \mathbf{I} - \exp\{\, t \, \boldsymbol{\Lambda}_{A_1A_1} \} \right\} \right\}^{m-1} \times$$

$$\boldsymbol{\Lambda}_{A_1A_1} \, \mathbf{1} - \mathbf{w}^T \, \boldsymbol{\Lambda}_{A_1A_1}^{-1} \left\{ \mathbf{I} - \exp\{\, t \, \boldsymbol{\Lambda}_{A_1A_1} \} \right\} \times$$

$$\sum_{k=1}^{m-1} \left\{ \mathbf{\Lambda}_{A_1A_2} \, \mathbf{\Lambda}_{A_2A_2}^{-1} \, \mathbf{\Lambda}_{A_2A_1} \, \mathbf{\Lambda}_{A_1A_1}^{-1} \left\{ \mathbf{I} - \exp\{ t \, \mathbf{\Lambda}_{A_1A_1} \} \right\} \right\}^{k-1} \times$$

$$\left\{ \mathbf{\Lambda}_{A_1A_1} - \mathbf{\Lambda}_{A_1A_2} \, \mathbf{\Lambda}_{A_2A_2}^{-1} \, \mathbf{\Lambda}_{A_2A_1} \right\} \mathbf{1}, \tag{4.65}$$

(4.64) now follows from (4.65) if we can show that for $k \to +\infty$,

$$\left\{ \mathbf{\Lambda}_{A_1A_2} \, \mathbf{\Lambda}_{A_2A_2}^{-1} \, \mathbf{\Lambda}_{A_2A_1} \, \mathbf{\Lambda}_{A_1A_1}^{-1} \left\{ \mathbf{I} - \exp\{ t \, \mathbf{\Lambda}_{A_1A_1} \} \right\} \right\}^{k} \to 0. \tag{4.66}$$

We know from the proof of Lemma 4.1 that for any $t \geq 0$ and *any* stochastic vector $\boldsymbol{\beta}$ on A_1, the row vector $\mathbf{p}^T = \boldsymbol{\beta}^T \exp\{ \mathbf{\Lambda}_{A_1A_1} t \}$ is substochastic. This shows that the entries of $\boldsymbol{\beta}^T \{ \mathbf{I} - \exp\{ \mathbf{\Lambda}_{A_1A_1} t \} \} = \boldsymbol{\beta}^T - \mathbf{p}^T$ are in $[-1, 1]$; Therefore, the entries of $\{ \mathbf{I} - \exp\{ \mathbf{\Lambda}_{A_1A_1} t \} \}$ lie between -1 and $+1$. (4.66) now follows by Lemma 4.11. ■

It should be noted that the distributions in (4.47) and (4.60) (but not that in (4.64)) belong to the class of phase-type distributions. A review of this class of distributions in the reliability setting and further references on this can be found in [WEB].

4.3 Tabular summary of results in Sections 4.1 and 4.2

In Table 4.1 a guide is shown to the results in Sections 4.1 and 4.2.

Variable	Characteristic	Reference
$(T_{A_1,1}, \ldots, T_{A_1,m})$	cdf	Theorem 4.2, Corollary 4.3
$T_{A_1,m}$	cdf	Corollary 4.4
$(T_{A_1,1}, \ldots, T_{A_1,m})$	LT	Corollary 4.5
$(T_{A_2,1}, \ldots, T_{A_2,m}, T_{A_1,1}, \ldots, T_{A_1,m})$	cdf	Theorem 4.6, Corollary 4.7
$(T_{A_2,1}, \ldots, T_{A_2,m}, T_{A_1,1}, \ldots, T_{A_1,m})$	LT	Corollary 4.8
$TS_{A_1,m}$	cdf	Theorem 4.9
$TS_{A_1,\infty}$	cdf	Theorem 4.12
$\max\{ T_{A_1,1}, T_{A_1,2}, \ldots \}$	cdf	Theorem 4.13

Table 4.1 A guide to the distribution results in Sections 4.1 and 4.2

4.4 An application: further dependability characteristics of the three-unit power transmission model

4.4.1 Numerical results

The theory will now be applied to obtain some more dependability characteristics of Model 2 from Section 1.2.1. Some discrete-parameter aspects of this system are already know from Section 2.2. The set of model parameters used here will be the first two shown in Table 2.2. Assume that all system components are operational at t = 0, i.e., $\alpha_{A_1}^T = (1, 0, 0)$, where A_1 and A_2 are defined respectively by $A_1 = G = \{ 1, 2, 3 \}$ and $A_2 = B = \{ 4, 5 \}$; see Figure 1.2.

The dependability characteristics from Section 4.2 have been implemented in MATLAB on an Apple Macintosh IIx; the source code is shown in Section 4.4.2. The new language feature of MATLAB not encountered thus far is the matrix exponential expm. This is obviously a most useful function for the implementation of Markov models. Use has also been made of the facility which allows new matrices to be built up in a symbolic fashion from already defined ones (this feature allowed M_m to be defined recursively). Finally, the plot function proved useful in obtaining preliminary working versions of the graphs shown below; a dedicated graphics package (cricketGRAPH) was used, however, to prepare the plots shown here.

The cumulative distribution function of $TS_{A_1,m}$, i.e., of the total system up time during the first m working periods, is shown in Figures 4.2 and 4.3 (see page 100) for the parameter sets № 1 and № 2 respectively. The time period for the plots is 120 years. The order of magnitude of this time scale reflects accepted design practice: in the US, steel structures are required to withstand, for example, wind loads which correspond to a 50-year mean recurrence interval ([KUZ], pp. 23). It is seen that an increase of the failure rates λ_T and λ_c' makes the distribution functions to be shifted to the left, i.e., the cumulative working periods become shorter in a statistical sense. The 'educational version' of MATLAB was used to perform the calculations. For any fixed t, the computing times for the five values $\{Pr\{ TS_{A_1,m} \leq t \} : m = 1, ..., 5\}$ were (in sec), respectively, 1, 1, 2, 3, and 6. A reduction in these figures is expected when using the full version of MATLAB.

The failure set A_2 comprises two rather disparate elements: the return rate from '4' into A_1 is 2190 / year whereas that from '5' is only 46 / year. The analysis thus far cannot differentiate between working periods separated by 'minor' and 'major' breakdowns (corresponding to visits to '4' and '5' respectively). In order to be able to do so, the system's behaviour will be examined until a tower failure occurs. The Markov chain will be converted into an absorbing one by assuming that $\mu_T = 0$. Then, the set of 'good' states is $A_1 = \{ 1, 2, 3 \}$, and that of the 'bad' states is $A_2 = \{ 4 \}$. $\omega = 5$ is the absorbing state and it stands for 'system failure'. What we have arrived at is Model 3 from Section 1.2.1. The system's transition rate matrix Λ is now given by

$$\Lambda = \begin{bmatrix} * & \lambda_1 & \lambda_2 & \lambda_c' & \lambda_T \\ \mu_1 & * & 0 & \lambda_2+\lambda_c' & \lambda_T \\ \mu_2 & 0 & * & \lambda_1+\lambda_c' & \lambda_T \\ 0 & \mu_2 & \mu_1 & * & \lambda_T \\ 0 & 0 & 0 & 0 & 0 \end{bmatrix}. \tag{4.67}$$

For the absorbing case, the Figures 4.4 and 4.5 (see page 101) show the results of the calculations for the parameter sets $N^{\underline{o}}$ 1 and $N^{\underline{o}}$ 2 respectively. As expected, it is seen from the figures that the distribution function of $TS_{A_1,\infty}$ is the limit of that of $TS_{A_1,m}$ as m $\to\infty$. Comparing the two figures, it is also seen that for fixed m, the probability mass of $TS_{A_1,m}$ is shifted towards smaller values of t when the hazard rate parameters λ_T and λ_c' are increased. This is a reflection of the system's decrease in dependability. Between the corresponding quantities for irreducible and absorbing Y, represented in Figures 4.2 and 4.4 respectively (and, similarly, in Figures 4.3 and 4.5 respectively), we have, with obvious notation, the following interrelationship for m < +∞

$Pr\{\ TS_{A_1,m}(abs) \le t\ \} =$

$1 - Pr\{\{\ TS_{A_1,m}(irr) > t\ \} \cap$

$\{$ no tower failure during the first t time units of operation $\}\} \ge$

$1 - Pr\{\ TS_{A_1,m}(irr) > t\ \} = Pr\{\ TS_{A_1,m}(irr) \le t\ \}.$ \hfill (4.68)

Each curve in Figure 4.2 is therefore bounded from above by its counterpart in Figure 4.4. From (4.68) it is also seen that for fixed t their difference is

$Pr\{\{\ TS_{A_1,m}(irr) > t\ \} \cap$

$\{$ at least one tower failure during the first t time units of operation $\}\}.$

The Figures 4.6 and 4.7 on page 102 show for the parameter sets $N^{\underline{o}}$ 1 and $N^{\underline{o}}$ 2 respectively, the distribution function of max$\{\ T_{A_1,1},\ ...,\ T_{A_1,m}\ \}$, as given in (4.65), and that of max$\{\ T_{A_1,1},\ T_{A_1,2},\ ...\ \}$, as given in (4.64). It is seen that the distribution function of max$\{\ T_{A_1,1},\ ...,\ T_{A_1,m}\ \}$ rapidly converges to its limit as m $\to \infty$. Comparing Figures 4.6 and 4.7, we see that the curves in Figure 4.7 are shifted to the left, reflecting the increased unreliability of the system due to higher values of λ_T and λ_c'.

The MATLAB code for the present example is shown in the next section (pages 103 - 105).

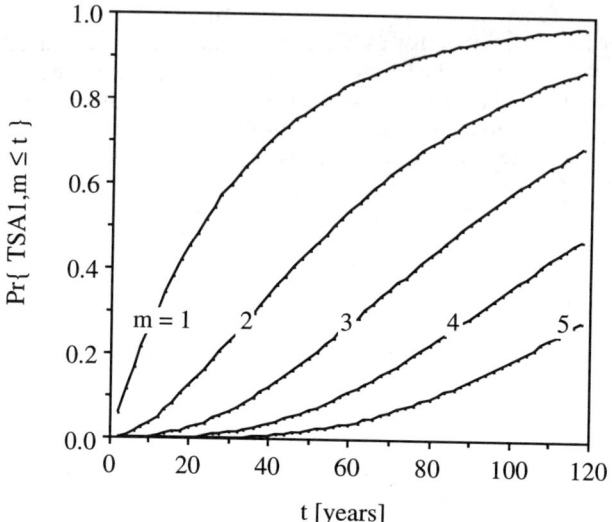

Figure 4.2 The distribution function of $TS_{A_1,m}$ for m = 1, ..., 5 for the parameter set № 1

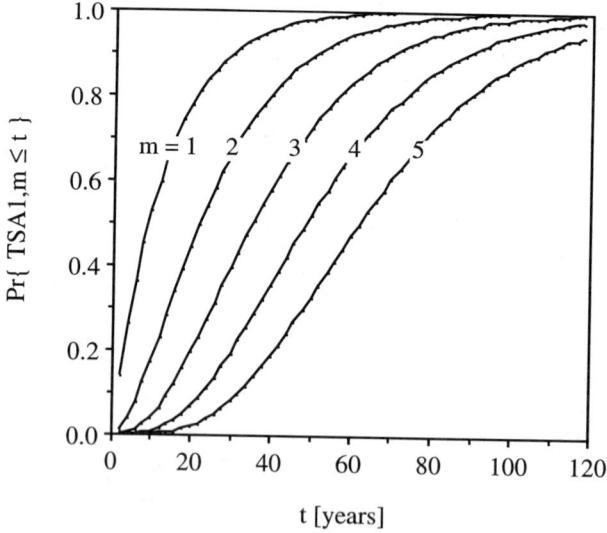

Figure 4.3 The distribution function of $TS_{A_1,m}$ for m = 1, ..., 5 for the parameter set № 2

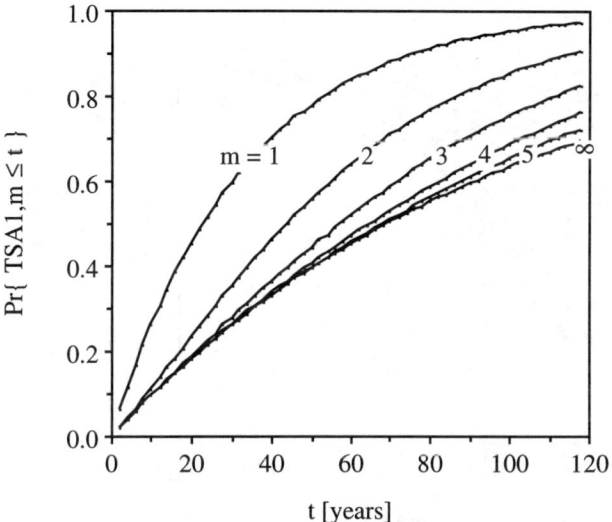

Figure 4.4 The distribution function of $TS_{A_1,m}$ for m = 1, ..., 5, ∞ for the parameter set № 1

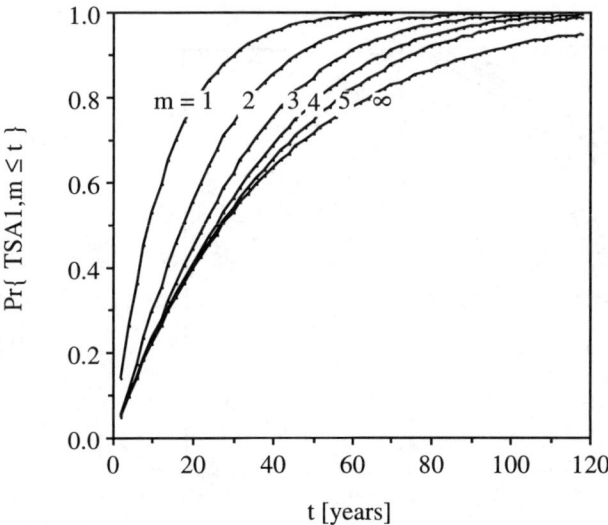

Figure 4.5 The distribution function of $TS_{A_1,m}$ for m = 1, ..., 5, ∞ for the parameter set № 2

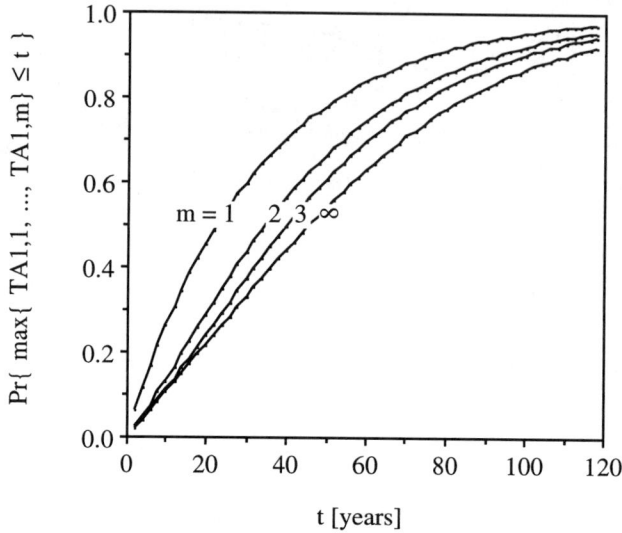

Figure 4.6 The distribution function of $\max\{\,T_{A_1,1},\,...,\,T_{A_1,m}\,\}$, $m \in \{\,1, 2, 3, \infty\,\}$, for the parameter set № 1

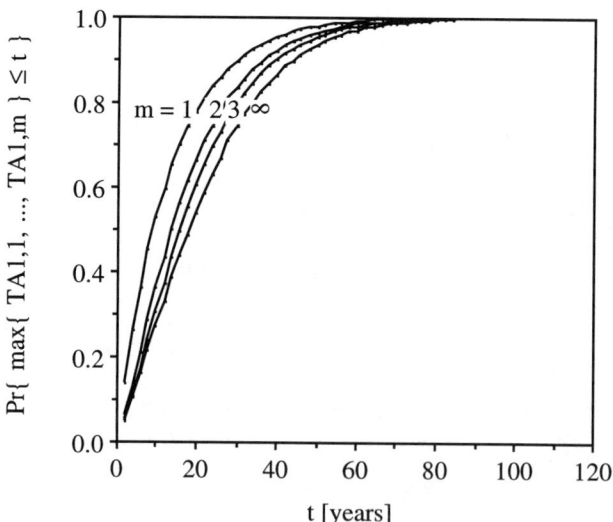

Figure 4.7 The distribution function of $\max\{\,T_{A_1,1},\,...,\,T_{A_1,m}\,\}$, $m \in \{\,1, 2, 3, \infty\,\}$, for the parameter set № 2

4.4.2 <u>MATLAB code</u>

(a) MATLAB implementation of (4.47) (irreducible case)

```
lmax = input('Enter lmax ...');    % = maximum number of working periods
tmax = input('Enter tmax ... ');
deltat = input('Enter steplength of t ... ');
x = [deltat];
lamt = input('Enter lambdat in 1/year... ');
lamc = input('Enter lambdacdash in 1/year ...');
lam1 = 0.25
lam2 = 0.25
mu1 = 1095
mu2 = 1095
mut = 46

LDD = [ -(mu1+mu2+lamt)   lamt
          0              -mut];

LUU = [-(lam1+lam2+lamc+lamt)  lam1                      lam2
         mu1             -(lam2+lamc+mu1+lamt)     0
         mu2                  0                -(lam1+lamc+mu2+lamt)];

LUD = [lamc         lamt
       lam2+lamc    lamt
       lam1+lamc    lamt];

LDU = [0    mu2    mu1
       mut  0      0];

PSI = - LUD*inv(LDD)*LDU;

[nu nd] = size(LUD);

f=[0];                    % ARRAY OF RESULTS
for l=1:lmax
  if l==1
     BIGM = LUU;
     w = [-1 0 0];
  else
     j = max(size(BIGM));
     TEMP1 = [BIGM [zeros(j - nu,nu); PSI]];
     TEMP2 = [zeros(nu,j) LUU];
     BIGM = [TEMP1; TEMP2];
     clear TEMP1 TEMP2;
     w = [w zeros(1, nu)];
  end
  i = 1;
  t = deltat;
  while (t < tmax)
     x(i) = t;
     f(i,l) = 1 + w*expm(t*BIGM)*(ones(1,max(size(BIGM))))';
     i = i+1;
     t = i*deltat;
  end
end

disp(f)
disp(x')
clg;                      % clears graph window
plot(x,f,'-');
```

(b) MATLAB implementation of (4.47) & (4.60) (absorbing case)

```
lmax = input('Enter lmax ...'); % = maximum number of working periods
tmax = input('Enter tmax ... ');
deltat = input('Enter steplength of t ... ');
x = [deltat];
lamt = input('Enter lambdat in 1/year... ');
lamc = input('Enter lambdacdash in 1/year ...');
lam1 = 0.25
lam2 = 0.25
mu1 = 1095
mu2 = 1095

LDD = [ -(mu1+mu2+lamt)];
LUU = [-(lam1+lam2+lamc+lamt)    lam1                      lam2
          mu1                  -(lam2+lamc+mu1+lamt)  0
          mu2                    0                       -(lam1+lamc+mu2+lamt)];
LUD = [lamc
        lam2+lamc
        lam1+lamc];
LDU = [0     mu2     mu1];
PSI = - LUD*inv(LDD)*LDU;
[nu nd] = size(LUD);
f=[0];                    % ARRAYS OF RESULTS
g=[0];                    % TAKES VALUES FOR THE LIMIT CDF
for l=1:lmax
  if l==1
     BIGM = LUU;
     w = [-1 0 0];
  else
     j = max(size(BIGM));
     TEMP1 = [BIGM [zeros(j - nu,nu); PSI]];
     TEMP2 = [zeros(nu,j) LUU];
     BIGM = [TEMP1; TEMP2];
     clear TEMP1 TEMP2;
     w = [w zeros(1, nu)];
  end
  i = 1;
  t = deltat;
  while (t < tmax)
     x(i) = t;
     f(i,l) = 1 + w*expm(t*BIGM)*(ones(1,max(size(BIGM))))');
     i = i+1;
     t = i*deltat;
  end
end
w0 = [-1 0 0];
M0 = LUU + PSI;
i = 1;
t = deltat;
while (t < tmax)
    x(i) = t;
    g(i) = 1 + w0*expm(t*M0)*(ones(1,nu)');
    i = i+1;
    t = i*deltat;
end
disp(g)
disp(f)
disp(x')
clg;                       % clears graph window
plot(x,f,'-',x,g,'-');
```

(c) MATLAB implementation of (4.64) & (4.65) (absorbing case)

```
lmax = input('Enter lmax ...'); % = maximum number of working periods
tmax = input('Enter tmax ... ');
deltat = input('Enter steplength of t ... ');
x = [deltat];

lamt = input('Enter lambdat in 1/year... ');
lamc = input('Enter lambdacdash in 1/year ...');

lam1 = 0.25
lam2 = 0.25
mu1 = 1095
mu2 = 1095

LDD = [ -(mu1+mu2+lamt)];
LUU = [-(lam1+lam2+lamc+lamt)   lam1                         lam2
            mu1                 -(lam2+lamc+mu1+lamt)   0
            mu2                 0                       -(lam1+lamc+mu2+lamt)];
LUD = [lamc
        lam2+lamc
        lam1+lamc];
LDU = [0    mu2    mu1];
[nu nd] = size(LUD);
LUUI = inv(LUU);
M1 = LUD*inv(LDD)*LDU*LUUI;
M2 = (LUU - LUD*inv(LDD)*LDU)*ones(nu,1);
w = [-1 0 0];
f=[0];                       % ARRAYS OF RESULTS
g=[0];                       % TAKES VALUES FOR THE LIMIT CDF

i = 1;
t = deltat;

while (t < tmax)
    x(i) = t;
    M3 = eye(nu) - expm(t*LUU);
    M4 = M1*M3;
    M5 = w*LUUI*M3;
    g(i) = - M5*inv(eye(nu) - M4)*M2;
    for l=1:lmax
        SUM = zeros(nu,nu);
        for k=1:(l-1)
            SUM = SUM + M4^(k-1);
        end
        f(i,l) = - M5*((M4^(l-1))*LUU*ones(nu,1) + SUM*M2);
    end
    i = i+1;
    t = i*deltat;
end
disp(g)
disp(f)
disp(x')
clg;                         % clears graph window
plot(x,f,'-',x,g,'-');
```

CHAPTER 5

THE NUMBER OF VISITS TO A SUBSET OF THE STATE SPACE BY A CONTINUOUS-PARAMETER IRREDUCIBLE MARKOV CHAIN DURING A FINITE TIME INTERVAL

Here, we continue the study of continuous-parameter finite Markov processes. The notation and assumptions introduced in Chapter 4 will be retained here, the only additional requirement being that Y be irreducible. In this chapter, we shall explore the variable $M_{A_1}(t)$, $t > 0$, which is defined as the number of periods spent by Y in A_1 during $[0, t]$. If A_1 stands for the set of system 'up' states then, obviously, $M_{A_1}(t)$ is the number of working periods during the first t time units. (Notice that $M_{A_1}(t)$ is the continuous-time analogue of the variable $K_{A_1,n}$ which was defined for discrete-parameter Markov chains by (2.57).) It is easily seen that $M_{A_1}(t)$ is related to R(t), the number of completed repair events during $[0, t]$, by

$$R(t) = \begin{cases} M_{A_1}(t) - 1 & \text{if } Y_0 \in A_1, \\ M_{A_1}(t) & \text{if } Y_0 \in A_2. \end{cases} \qquad (5.1)$$

R(t) is a meaningful measure of the total (undiscounted) repair cost incurred during $[0, t]$, if every repair event (modelled by transitions from A_2 to A_1) carries the same cost. In Section 5.1, a closed form expression for the probability mass function of $M_{A_1}(t)$ will be derived. In Section 5.2, the theory is applied to obtain numerical results for R(t) for Model 1 from Section 1.2.1. This chapter is based on research first reported in [CSE3].

5.1 The variable $M_{A_1}(t)$

5.1.1 The main result

It is easily seen that due to the irreducibility of Y, $M_{A_1}(t)$ and the sequences of (cumulative) sojourn times are interrelated by the following set of equations

$$\Pr\{ M_{A_1}(t) = 0 \} = \sum_{a_2 \in A_2} \Pr\{ T_{A_2,1} > t \mid Y_0 = a_2 \} \Pr\{ Y_0 = a_2 \}, \qquad (5.2)$$

and, for m = 1, 2, ...,

$$\Pr\{ M_{A_1}(t) = m \} =$$

$$\sum_{a_1 \in A_1} \Big(\Pr\{ TS_{A_1,m-1} + TS_{A_2,m-1} \le t \mid Y_0 = a_1 \} -$$

$$- \Pr\{ TS_{A_1,m} + TS_{A_2,m} \le t \mid Y_0 = a_1 \} \Big) \Pr\{ Y_0 = a_1 \} +$$

$$\sum_{a_2 \in A_2} \Big(\Pr\{ TS_{A_1,m-1} + TS_{A_2,m} \le t \mid Y_0 = a_2 \} -$$

$$- \Pr\{ TS_{A_1,m} + TS_{A_2,m+1} \le t \mid Y_0 = a_2 \} \Big) \Pr\{ Y_0 = a_2 \}. \qquad (5.3)$$

Corollary 4.8 will enable us to derive, via Laplace transforms, the joint cumulative distribution function of the variables in (5.3). This then, with (5.2) and (5.3), will yield an expression for the probability mass function of $M_{A_1}(t)$. To formulate the results, define the matrices $\Theta_m(A, B)$ and $E_m(A, B)$ for $m \ge 1$ and disjoint, non-empty $A, B \subset S$ via (2.13) by

$$\Theta_m(A, B) =$$

$$\Delta_{2m}(\Lambda_{AA}, \Lambda_{BB}, ..., \Lambda_{AA}, \Lambda_{BB}; \Lambda_{AB}, \Lambda_{BA}, \Lambda_{AB}, \Lambda_{BA}..., \Lambda_{AB}),$$

$$E_m(A, B) =$$

$$\Delta_{2m-1}(\Lambda_{AA}, \Lambda_{BB}, ..., \Lambda_{AA}, \Lambda_{BB}, \Lambda_{AA}; \Lambda_{AB}, \Lambda_{BA}, ..., \Lambda_{AB}, \Lambda_{BA}).$$

We are now in a position to state the following on $M_{A_1}(t)$.

THEOREM 5.1. *The probability mass function of* $M_{A_1}(t)$ *is given by*

$$\Pr\{ M_{A_1}(t) = 0 \} = \alpha_{A_2}{}^T \exp\{ t \, \Lambda_{A_2 A_2} \} \, \mathbf{1}, \qquad (5.4)$$

$$\Pr\{ M_{A_1}(t) = 1 \} = (\alpha_{A_1}{}^T, \mathbf{0}) \exp\{ t \, \Theta_1(A_1, A_2) \} \, \mathbf{1} -$$

$$- \alpha_{A_2}{}^T \exp\{ t \, \Lambda_{A_2 A_2} \} \, \mathbf{1} + (\alpha_{A_2}{}^T, \mathbf{0}, \mathbf{0}) \exp\{ t \, E_2(A_2, A_1) \} \, \mathbf{1}, \qquad (5.5)$$

and, for $m \ge 2$,

$$\Pr\{ M_{A_1}(t) = m \} = (\alpha_{A_2}{}^T, \mathbf{0}, ..., \mathbf{0}) \exp\{ t \, E_{m+1}(A_2, A_1) \} \, \mathbf{1} -$$

$$- (\alpha_{A_2}{}^T, \mathbf{0}, ..., \mathbf{0}) \exp\{ t \, E_m(A_2, A_1) \} \, \mathbf{1} +$$

$$+ (\alpha_{A_1}{}^T, \mathbf{0}, ..., \mathbf{0}) \exp\{ t \, \Theta_m(A_1, A_2) \} \, \mathbf{1} -$$

$$- (\alpha_{A_1}{}^T, \mathbf{0}, ..., \mathbf{0}) \exp\{ t \, \Theta_{m-1}(A_1, A_2) \} \, \mathbf{1}. \qquad (5.6)$$

The proof of Theorem 5.1 will be dealt with in the next section.

5.1.2 The proof of Theorem 5.1

The proof makes use of (5.2) and (5.3) and it is in the following four steps:

(A) Establish (5.4).
(B) Evaluate for $\ell \geq 1$ the distribution function of $TS_{A_1,\ell} + TS_{A_2,\ell}$.
(C) Evaluate for $\ell \geq 2$ the distribution function of $TS_{A_1,\ell-1} + TS_{A_2,\ell}$.
(D) Combine (B) and (C) to give (5.5) and (5.6).

The above four points will now be dealt with one by one.

(A) From Corollary 4.4 we know that for $t \geq 0$ and $a_2 \in A_2$,

$$\Pr\{ T_{A_2,1} \leq t \mid Y_0 = a_2 \} = 1 - \delta_{a_2}{}^T \exp\{ t \Lambda_{A_2A_2} \} \mathbf{1}, \tag{5.7}$$

where δ_{a_2} is the column vector on A_2 the only non-zero entry of which is at its a_2th position. Combining (5.7) with (5.2) gives (5.4).

(B) The Laplace transform of the joint distribution of the first 2ℓ sojourn times, $(T_{A_1,1}, ..., T_{A_1,\ell}, T_{A_2,1}, ..., T_{A_2,\ell})$, is available from Corollary 4.8. In the next proposition, this is recast into matrix form.

PROPOSITION 5.2. *The Laplace transform of* $(T_{A_1,1}, ..., T_{A_1,\ell}, T_{A_2,1}, ..., T_{A_2,\ell})$ *is given by*

$$(T_{A_1,1}, ..., T_{A_1,\ell}, T_{A_2,1}, ..., T_{A_2,\ell})^*(\tau; \sigma) =$$

$$(\alpha_{A_1}{}^T, 0, ..., 0) \left\{ \Theta_\ell(A_1, A_2) - \Upsilon_\ell(\tau, \sigma; A_1, A_2) \right\}^{-1} \Theta_\ell(A_1, A_2)\ \mathbf{1} +$$

$$(\alpha_{A_2}{}^T, 0, ..., 0) \left\{ \Theta_\ell(A_2, A_1) - \Upsilon_\ell(\sigma, \tau; A_2, A_1) \right\}^{-1} \Theta_\ell(A_2, A_1)\ \mathbf{1}, \tag{5.8}$$

where the matrix Υ *is defined for disjoint, non-empty* A, B $\subset$ S *by*

$$\Upsilon_\ell(\sigma, \tau; A, B) = \begin{array}{c} \\ \\ A \\ B \\ A \\ \\ \\ \\ \\ B \end{array} \begin{array}{cccccc} A & B & A & & B \\ \left[\begin{array}{c|c|c|c|c} \sigma_1 I & 0 & 0 & & 0 \\ \hline 0 & \tau_1 I & 0 & \cdots & 0 \\ \hline 0 & 0 & \sigma_2 I & \cdots & 0 \\ \hline \vdots & \vdots & \vdots & \cdots & \vdots \\ \hline 0 & 0 & 0 & \cdots & \tau_\ell I \end{array}\right] \end{array}.$$

PROOF OF PROPOSITION 5.2. By Corollary 4.8, it obviously suffices to verify that, say, the first term on the right hand side of (5.8) is equal to $f_{A_1A_2,\ell}(\tau; \sigma)$. (The latter is defined by (4.43).) It is

$$\Theta_\ell(A_1, A_2) - \Upsilon_\ell(\tau, \sigma; A_1, A_2) = \Delta_{2\ell}(\Phi_1, ..., \Phi_{2\ell}; \Psi_1, ..., \Psi_{2\ell-1}), \qquad (5.9)$$

where for $i = 1, ..., \ell$ and $j = 1, ..., \ell-1$ it is

$$\Phi_{2i-1} = \Lambda_{A_1 A_1} - \tau_i I, \quad \Phi_{2i} = \Lambda_{A_2 A_2} - \sigma_i I,$$

$$\Psi_{2i-1} = \Lambda_{A_1 A_2}, \quad \Psi_{2j} = \Lambda_{A_2 A_1}.$$

But, an expression is provided for the inverse of the matrix in (5.9) by Lemma 4.10. It is therefore, with the notation of Lemma 4.10,

$$\left\{ \Theta_\ell(A_1, A_2) - \Upsilon_\ell(\tau, \sigma; A_1, A_2) \right\}^{-1} = \Xi_{2\ell}(\Phi_1, ..., \Phi_{2\ell}; \Psi_1, ..., \Psi_{2\ell-1}).$$

All what now remains to be shown, is that

$$(\alpha_{A_1}^T, 0, ..., 0) \, \Xi_{2\ell}(\Phi_1, ..., \Phi_{2\ell}; \Psi_1, ..., \Psi_{2\ell-1}) \, \Theta_\ell(A_1, A_2) \, 1 =$$

$$f_{A_1 A_2, \ell}(\tau; \sigma). \qquad (5.10)$$

To show (5.10), first notice that

$$\Theta_\ell(A_1, A_2) = \begin{bmatrix} \Lambda_{A_1 A_1} & \Lambda_{A_1 A_2} & 0 & \cdots & 0 \\ 0 & \Lambda_{A_2 A_2} & \Lambda_{A_2 A_1} & \cdots & 0 \\ 0 & 0 & \Lambda_{A_1 A_1} & \cdots & 0 \\ \vdots & \vdots & \vdots & & \vdots \\ 0 & 0 & 0 & \cdots & \Lambda_{A_1 A_2} \\ 0 & 0 & 0 & \cdots & \Lambda_{A_2 A_2} \end{bmatrix},$$

from which it follows by the irreducibility of Y that

$$\Theta_\ell(A_1, A_2) \, 1 = \begin{bmatrix} 0 \\ \vdots \\ \vdots \\ 0 \\ \Lambda_{A_2 A_2} 1 \end{bmatrix}.$$

The left hand side of (5.10) is therefore

$$\alpha_{A_1}{}^T \, \Xi_{12\ell}(\Phi_1, ..., \Phi_{2\ell}; \Psi_1, ..., \Psi_{2\ell-1}) \, \Lambda_{A_2A_2} \, \mathbf{1} =$$

$$\alpha_{A_1}{}^T \, \Phi_1{}^{-1} \prod_{k=1}^{2\ell-1} \left\{ - \Psi_k \, \Phi_{k+1}{}^{-1} \right\} \Lambda_{A_2A_2} \, \mathbf{1} =$$

$$- \alpha_{A_1}{}^T \, \Phi_1{}^{-1} \prod_{k=1}^{2\ell-1} \left\{ \Psi_k \, \Phi_{k+1}{}^{-1} \right\} \Lambda_{A_2A_2} \, \mathbf{1}. \tag{5.11}$$

The expression on the right hand side of (5.11) is easily seen to be $f_{A_1A_2,\ell}(\tau; \sigma)$ since, again due the irreducibility of Y, all but the first term on the right hand side of (4.43) are zero. ∎

The next corollary follows immediately.

COROLLARY 5.3. *For* $\ell \geq 1$, *the Laplace transform of* $TS_{A_1,\ell} + TS_{A_2,\ell}$ *is given by*

$$(TS_{A_1,\ell} + TS_{A_2,\ell})^*(\tau) =$$

$$(\alpha_{A_1}{}^T, 0, ..., 0) \, (\Theta_\ell(A_1, A_2) - \tau \, I)^{-1} \, \Theta_\ell(A_1, A_2) \, \mathbf{1} +$$

$$(\alpha_{A_2}{}^T, 0, ..., 0) \, (\Theta_\ell(A_2, A_1) - \tau \, I)^{-1} \, \Theta_\ell(A_2, A_1) \, \mathbf{1}. \tag{5.12}$$

∎

We are now in a position to establish the cumulative distribution function of $TS_{A_1,\ell} + TS_{A_2,\ell}$.

COROLLARY 5.4. *For* $\ell \geq 1$, *the cumulative distribution function of* $TS_{A_1,\ell} + TS_{A_2,\ell}$ *is given by*

$$\Pr\{ TS_{A_1,\ell} + TS_{A_2,\ell} \leq t \} = 1 - (\alpha_{A_1}{}^T, 0, ..., 0) \exp\{ t \, \Theta_\ell(A_1, A_2) \} \, \mathbf{1} -$$

$$- (\alpha_{A_2}{}^T, 0, ..., 0) \exp\{ t \, \Theta_\ell(A_2, A_1) \} \, \mathbf{1}. \tag{5.13}$$

PROOF OF COROLLARY 5.4. (5.13) is deduced from (5.12) along the lines of the Laplace transform inverson in the proof of Theorem 4.9; we only need to prove that for any partition {A, B} of S:

$$\exp\{ t \, \Theta_\ell(A, B) \} \to 0 \text{ as } t \to \infty. \tag{5.14}$$

To justify (5.14), consider a Markov chain whose transition rate matrix is

$$
\begin{bmatrix}
\mathbf{\Theta}_\ell(A, B) & \vdots & \mathbf{R} \\
\text{-------} & \vdots & \text{------} \\
\mathbf{L} & \vdots & \mathbf{W}
\end{bmatrix},
$$

with $\mathbf{W}$, $\mathbf{R}$, and $\mathbf{L}$ respectively defined by

$$
\mathbf{W} = \begin{bmatrix}
\mathbf{\Lambda}_{AA} & \vdots & \mathbf{\Lambda}_{AB} \\
\text{------} & \vdots & \text{------} \\
\mathbf{0} & \vdots & \mathbf{\Lambda}_{BB}
\end{bmatrix}, \quad
\mathbf{R} = \begin{bmatrix}
\mathbf{0} & \vdots & & \vdots \mathbf{0} & \vdots & \mathbf{\Lambda}_{BA}{}^T \\
\text{--} & \vdots & \cdots & \vdots \text{--} & \vdots & \text{------} \\
\mathbf{0} & \vdots & & \vdots \mathbf{0} & \vdots & \mathbf{0}
\end{bmatrix}^T,
$$

and

$$
\mathbf{L} = \begin{bmatrix}
\mathbf{0} & \vdots & \mathbf{0} & \vdots & & \vdots & \mathbf{0} \\
\text{------} & \vdots & \text{--} & \vdots & \cdots & \vdots & \text{--} \\
\mathbf{\Lambda}_{BA} & \vdots & \mathbf{0} & \vdots & & \vdots & \mathbf{0}
\end{bmatrix}.
$$

It is irreducible. (5.14) therefore holds by Lemma 4.1. ∎

(C) For disjoint, non-empty A, $B \subset S$, define the matrix $\mathbf{K}_\ell(A, B)$ via (2.13) by

$$
\mathbf{K}_\ell(A, B) = \mathbf{\Delta}_{2\ell-1}(\mathbf{\Lambda}_{AA}, \mathbf{\Lambda}_{BB}, \ldots, \mathbf{\Lambda}_{AA}, \mathbf{\Lambda}_{BB};
$$

$$
\mathbf{\Lambda}_{AB}, \mathbf{\Lambda}_{BA}, \ldots, \mathbf{\Lambda}_{AB}, \mathbf{\Lambda}_{BA}\cdots, -\mathbf{\Lambda}_{BA}\,\mathbf{\Lambda}_{AA}{}^{-1}\,\mathbf{\Lambda}_{AB}) =
$$

$$
\left(= \begin{bmatrix}
\mathbf{\Lambda}_{AA} & \mathbf{\Lambda}_{AB} & \mathbf{0} & \cdots & \mathbf{0} & \mathbf{0} & \mathbf{0} \\
\mathbf{0} & \mathbf{\Lambda}_{BB} & \mathbf{\Lambda}_{BA} & \cdots & \mathbf{0} & \mathbf{0} & \mathbf{0} \\
\mathbf{0} & \mathbf{0} & \mathbf{\Lambda}_{AA} & \cdots & \mathbf{0} & \mathbf{0} & \mathbf{0} \\
\vdots & \vdots & \vdots & & \vdots & \vdots & \vdots \\
\mathbf{0} & \mathbf{0} & \mathbf{0} & \cdots & \mathbf{\Lambda}_{AA} & \mathbf{\Lambda}_{AB} & \mathbf{0} \\
\mathbf{0} & \mathbf{0} & \mathbf{0} & \cdots & \mathbf{0} & \mathbf{\Lambda}_{BB} & -\mathbf{\Lambda}_{BA}\mathbf{\Lambda}_{AA}{}^{-1}\mathbf{\Lambda}_{AB} \\
\mathbf{0} & \mathbf{0} & \mathbf{0} & \cdots & \mathbf{0} & \mathbf{0} & \mathbf{\Lambda}_{BB}
\end{bmatrix} \right)
$$

Then, the following holds.

PROPOSITION 5.5. *For $\ell \geq 2$, the cumulative distribution function of* $TS_{A_1, \ell-1} + TS_{A_2, \ell}$ *is given by*

$$\Pr\{ TS_{A_1,\ell-1} + TS_{A_2,\ell} \le t \} = 1 - (\boldsymbol{\alpha}_{A_1}^T, \mathbf{0}, ..., \mathbf{0}) \exp\{ t\, \mathbf{K}_\ell(A_1, A_2) \}\, \mathbf{1} -$$

$$- (\boldsymbol{\alpha}_{A_2}^T, \mathbf{0}, ..., \mathbf{0}) \exp\{ t\, \mathbf{E}_\ell(A_2, A_1) \}\, \mathbf{1}. \tag{5.15}$$

PROOF OF PROPOSITION 5.5. The reasoning is similar to that in the proof of (5.13). Putting in Corollary 4.8, $m = \ell$, and $\boldsymbol{\tau} = (\rho, ..., \rho, \rho, \rho, 0)$, $\boldsymbol{\sigma} = (\rho, ..., \rho) \in \mathbb{C}^{+\ell}$, the Laplace transform of $TS_{A_1,\ell-1} + TS_{A_2,\ell}$ is obtained as

$$(TS_{A_1,\ell-1} + TS_{A_2,\ell})^*(\rho) =$$

$$f_{A_2A_1,\ell}(\rho, ..., \rho; \rho, ..., \rho, \rho, 0) + f_{A_1A_2,\ell}(\rho, ..., \rho, \rho, 0; \rho, ..., \rho). \tag{5.16}$$

By (4.43) and since Y is irreducible, the first term on the right hand side of (5.16) is

$$f_{A_2A_1,\ell}(\rho, ..., \rho; \rho, ..., \rho, \rho, 0) =$$

$$- \boldsymbol{\alpha}_{A_2}^T (\boldsymbol{\Lambda}_{A_2A_2} - \rho\, \mathbf{I})^{-1} \boldsymbol{\Lambda}_{A_2A_1} (\boldsymbol{\Lambda}_{A_1A_1} - \rho\, \mathbf{I})^{-1} \times$$

$$\left\{ \boldsymbol{\Lambda}_{A_1A_2} (\boldsymbol{\Lambda}_{A_2A_2} - \rho\, \mathbf{I})^{-1} \boldsymbol{\Lambda}_{A_2A_1} (\boldsymbol{\Lambda}_{A_1A_1} - \rho\, \mathbf{I})^{-1} \right\}^{\ell-2} \times$$

$$\boldsymbol{\Lambda}_{A_1A_2} (\boldsymbol{\Lambda}_{A_2A_2} - \rho\, \mathbf{I})^{-1} \boldsymbol{\Lambda}_{A_2A_1}\, \mathbf{1}. \tag{5.17}$$

To show that

$$(\boldsymbol{\alpha}_{A_2}^T, \mathbf{0}, ..., \mathbf{0}) (\mathbf{E}_\ell(A_2, A_1) - \rho\, \mathbf{I})^{-1} \mathbf{E}_\ell(A_2, A_1)\, \mathbf{1} = f_{A_2A_1,\ell}(\rho, ..., \rho; \rho, ..., \rho, \rho, 0)$$

$$\tag{5.18}$$

first notice that

$$\mathbf{E}_\ell(A_2, A_1)\, \mathbf{1} = \begin{bmatrix} \mathbf{0} \\ \hdashline \vdots \\ \hdashline \mathbf{0} \\ \hdashline \boldsymbol{\Lambda}_{A_2A_2}\, \mathbf{1} \end{bmatrix}. \tag{5.19}$$

Furthermore, define

$$\boldsymbol{\Phi}_1 = \boldsymbol{\Phi}_3 = ... = \boldsymbol{\Phi}_{2\ell-1} = \boldsymbol{\Lambda}_{A_2A_2} - \rho\, \mathbf{I},$$

$$\boldsymbol{\Phi}_2 = \boldsymbol{\Phi}_4 = ... = \boldsymbol{\Phi}_{2\ell-2} = \boldsymbol{\Lambda}_{A_1A_1} - \rho\, \mathbf{I},$$

$$\Psi_1 = \Psi_3 = \dots = \Psi_{2\ell-3} = \Lambda_{A_2 A_1},$$

$$\Psi_2 = \Psi_4 = \dots = \Psi_{2\ell-2} = \Lambda_{A_1 A_2}.$$

It is easily seen that

$$E_\ell(A_2, A_1) - \rho I = \Delta_{2\ell-1}(\Phi_1, \Phi_1, \dots, \Phi_{2\ell-1}; \Psi_1, \Psi_2, \dots, \Psi_{2\ell-2}).$$

(5.18) now holds by (5.17), (5.19), Lemma 4.10, and the following

$$(\alpha_{A_2}^T, 0, \dots, 0) \, (E_\ell(A_2, A_1) - \rho I)^{-1} \, E_\ell(A_2, A_1) \, 1 =$$

$$\alpha_{A_2}^T \, \Xi_{1\,(2\ell-1)}(\Phi_1, \dots, \Phi_{2\ell-1}; \Psi_1, \dots, \Psi_{2\ell-2}) \, \Lambda_{A_2 A_2} \, 1 =$$

$$\alpha_{A_2}^T \, \Phi_1^{-1} \prod_{k=1}^{2\ell-2} \left\{ -\Psi_k \, \Phi_{k+1}^{-1} \right\} \Lambda_{A_2 A_2} \, 1 =$$

$$- \alpha_{A_2}^T \, \Phi_1^{-1} \prod_{k=1}^{2\ell-2} \left\{ \Psi_k \, \Phi_{k+1}^{-1} \right\} \Lambda_{A_2 A_1} (\Lambda_{A_1 A_1} - 0 \, I)^{-1} \Lambda_{A_1 A_1} \, 1 =$$

$$f_{A_2 A_1, \ell}(\rho, \dots, \rho; \rho, \dots, \rho, \rho, 0). \tag{5.20}$$

(The last step in (5.20) is seen by (4.43) and the irreducibility of Y.) We now deal with the second term on the righ hand side of (5.16). It will be shown that

$$(\alpha_{A_1}^T, 0, \dots, 0) \, (K_\ell(A_1, A_2) - \rho I)^{-1} \, K_\ell(A_1, A_2) \, 1 = f_{A_1 A_2, \ell}(\rho, \dots, \rho, \rho, 0; \rho, \dots, \rho).$$

$$\tag{5.21}$$

It is by (4.43)

$$f_{A_1 A_2, \ell}(\rho, \dots, \rho, \rho, 0; \rho, \dots, \rho) =$$

$$- \alpha_{A_1}^T \, (\Lambda_{A_1 A_1} - \rho I)^{-1} \Lambda_{A_1 A_2} (\Lambda_{A_2 A_2} - \rho I)^{-1} \times$$

$$\left\{ \Lambda_{A_2 A_1} (\Lambda_{A_1 A_1} - \rho I)^{-1} \Lambda_{A_1 A_2} (\Lambda_{A_2 A_2} - \rho I)^{-1} \right\}^{\ell-2} \times$$

$$\Lambda_{A_2 A_1} \, \Lambda_{A_1 A_1}^{-1} \, \Lambda_{A_1 A_2} (\Lambda_{A_2 A_2} - \rho I)^{-1} \Lambda_{A_2 A_2} \, 1. \tag{5.22}$$

Futhermore, notice that

$$
\mathbf{K}_\ell(A_1, A_2)\ \mathbf{1} =
\begin{bmatrix}
\mathbf{0} \\
\hdashline
\vdots \\
\hdashline
\mathbf{0} \\
\hdashline
\mathbf{\Lambda}_{A_2A_2}\mathbf{1} - \mathbf{\Lambda}_{A_2A_1}\mathbf{\Lambda}_{A_1A_1}^{-1}\mathbf{\Lambda}_{A_1A_2}\mathbf{1} \\
\hdashline
\mathbf{\Lambda}_{A_2A_2}\mathbf{1}
\end{bmatrix}
=
\begin{bmatrix}
\mathbf{0} \\
\hdashline
\vdots \\
\hdashline
\mathbf{0} \\
\hdashline
\mathbf{\Lambda}_{A_2A_2}\mathbf{1}
\end{bmatrix},
$$

$$(5.23)$$

and define

$$\mathbf{\Phi}_1 = \mathbf{\Phi}_3 = \dots = \mathbf{\Phi}_{2(\ell-2)+1} = \mathbf{\Lambda}_{A_1A_1} - \rho\,\mathbf{I},$$

$$\mathbf{\Phi}_2 = \mathbf{\Phi}_4 = \dots = \mathbf{\Phi}_{2(\ell-2)+2} = \mathbf{\Lambda}_{A_2A_2} - \rho\,\mathbf{I},$$

$$\mathbf{\Phi}_{2(\ell-2)+3} = \mathbf{\Lambda}_{A_2A_2} - \rho\,\mathbf{I},$$

$$\mathbf{\Psi}_1 = \mathbf{\Psi}_3 = \dots = \mathbf{\Psi}_{2(\ell-2)+1} = \mathbf{\Lambda}_{A_1A_2},$$

$$\mathbf{\Psi}_2 = \mathbf{\Psi}_4 = \dots = \mathbf{\Psi}_{2(\ell-2)} = \mathbf{\Lambda}_{A_2A_1},$$

$$\mathbf{\Psi}_{2(\ell-2)+2} = -\,\mathbf{\Lambda}_{A_2A_1}\,\mathbf{\Lambda}_{A_1A_1}^{-1}\,\mathbf{\Lambda}_{A_1A_2}.$$

It is easily seen that

$$\mathbf{K}_\ell(A_1, A_2) - \rho\,\mathbf{I} = \mathbf{\Delta}_{2\ell-1}(\mathbf{\Phi}_1, \dots, \mathbf{\Phi}_{2\ell-1};\ \mathbf{\Psi}_1, \dots, \mathbf{\Psi}_{2\ell-2}).$$

(5.21) now holds by (5.22), (5.23), Lemma 4.10, and the following

$$(\boldsymbol{\alpha}_{A_1}^{\mathrm{T}}, 0, \dots, 0)\,(\mathbf{K}_\ell(A_1, A_2) - \rho\,\mathbf{I})^{-1}\,\mathbf{K}_\ell(A_1, A_2)\,\mathbf{1} =$$

$$\boldsymbol{\alpha}_{A_1}^{\mathrm{T}}\,\mathbf{\Xi}_{1(2\ell-1)}(\mathbf{\Phi}_1, \dots, \mathbf{\Phi}_{2\ell-1};\ \mathbf{\Psi}_1, \dots, \mathbf{\Psi}_{2\ell-2})\,\mathbf{\Lambda}_{A_2A_2}\,\mathbf{1} =$$

$$\boldsymbol{\alpha}_{A_1}^{\mathrm{T}}\,\mathbf{\Phi}_1^{-1}\prod_{k=1}^{2\ell-2}\left\{-\,\mathbf{\Psi}_k\,\mathbf{\Phi}_{k+1}^{-1}\right\}\mathbf{\Lambda}_{A_2A_2}\,\mathbf{1} =$$

$$-\,\boldsymbol{\alpha}_{A_1}^{\mathrm{T}}\,\mathbf{\Phi}_1^{-1}\,\mathbf{\Psi}_1\,\mathbf{\Phi}_2^{-1}\prod_{j=1}^{\ell-2}\left\{\mathbf{\Psi}_{2j}\,\mathbf{\Phi}_{2j+1}^{-1}\,\mathbf{\Psi}_{2j+1}\,\mathbf{\Phi}_{2j+2}^{-1}\right\}\times$$

$$\mathbf{\Lambda}_{A_2A_1}\,\mathbf{\Lambda}_{A_1A_1}^{-1}\,\mathbf{\Lambda}_{A_1A_2}\,(\mathbf{\Lambda}_{A_2A_2} - \rho\,\mathbf{I})^{-1}\,\mathbf{\Lambda}_{A_2A_2}\,\mathbf{1} = f_{A_1A_2,\ell}(\rho, \dots, \rho, \rho, 0; \rho, \dots, \rho).$$

(5.16), (5.18), and (5.21) give the Laplace transform of $TS_{A_1,\ell-1} + TS_{A_2,\ell}$ as

$$(TS_{A_1,\ell-1} + TS_{A_2,\ell})^*(\rho) =$$

$$(\boldsymbol{\alpha}_{A_1}{}^T, \mathbf{0}, ..., \mathbf{0}) \, (\mathbf{K}_\ell(A_1, A_2) - \rho \, \mathbf{I})^{-1} \, \mathbf{K}_\ell(A_1, A_2) \, \mathbf{1} +$$

$$(\boldsymbol{\alpha}_{A_2}{}^T, \mathbf{0}, ..., \mathbf{0}) \, (\mathbf{E}_\ell(A_2, A_1) - \rho \, \mathbf{I})^{-1} \, \mathbf{E}_\ell(A_2, A_1) \, \mathbf{1},$$

from which (5.15) can be obtained by Laplace transform inversion. According to the process of inversion discussed in the proof of Theorem 4.9, we only need to show that

$$\exp\{ \, t \, \mathbf{E}_\ell(A_2, A_1) \, \} \to \mathbf{0}, t \to +\infty, \tag{5.24}$$

$$\exp\{ \, t \, \mathbf{K}_\ell(A_1, A_2) \, \} \to \mathbf{0}, t \to +\infty. \tag{5.25}$$

The proof of (5.24) is close to that of (5.14) and will be omitted. To prove (5.25), consider the auxiliary Markov chain Z whose state-transition-rate diagram is as shown in Figure 5.1 (page 116). Its state space is the union of all the shaded sets of states, i.e., $A_{11} \cup A_{21} \cup ... \cup A_{1 \, \ell-1} \cup A_{2 \, \ell-1} \cup A_{2 \, \ell} \cup A_{1 \, \ell+1} \cup A_{2 \, \ell+1}$, where A_{1i} and A_{2j} ($i \in \{ 1, ..., \ell-1, \ell+1 \}, j \in \{ 1, ..., \ell+1 \}$) are disjoint instances of A_1 and A_2 respectively. The set $A_{1 \, \ell}$ is an ℓth instance of A_1 which, however, does not form part of the state space of Z. When in $A_{1 \, \ell}$, the global clock is temporarily 'stopped' and it is restarted upon arrival of Z into $A_{2 \, \ell}$. By this device, the matrix of transition rates from $A_{2 \, \ell-1}$ into $A_{2 \, \ell}$ becomes $- \boldsymbol{\Lambda}_{A_2 A_1} \boldsymbol{\Lambda}_{A_1 A_1}{}^{-1} \boldsymbol{\Lambda}_{A_1 A_2}$.

(This is justified as follows. For $t \geq 0$, $\delta t \geq 0$, and $a_{2 \, \ell-1} \in A_{2 \, \ell-1}$, $a_{2 \, \ell} \in A_{2 \, \ell}$ we have for $\delta t \to 0$,

$$\Pr\{ \, W_{t+\delta t+0} = a_{2 \, \ell} \mid W_t = a_{2 \, \ell-1} \, \} =$$

$$\sum_{a_{1 \, \ell} \in A_{1 \, \ell}} \Pr\{ \, W_{t+\delta t+0} = a_{2 \, \ell} \mid W_{t+\delta t-0} = a_{1 \, \ell} \, \} \Pr\{ \, W_{t+\delta t-0} = a_{1 \, \ell} \mid W_t = a_{2 \, \ell-1} \, \} =$$

$$\sum_{a_{1 \, \ell} \in A_{1 \, \ell}} q_{a_{1 \, \ell} \, a_{2 \, \ell}} (\lambda_{a_{2 \, \ell-1} a_{1 \, \ell}} + o(\delta t)), \tag{5.26}$$

with $q_{a_{1 \, \ell} \, a_{2 \, \ell}}$ defined as the probability of W entering $A_{2 \, \ell}$ via $a_{2 \, \ell}$, given that it has arrived into $A_{1 \, \ell}$ through $a_{1 \, \ell}$. This latter probability depends only on the embedded Markov chain $Z = \{ Z_0, Z_1, ... \}$ of Y. More precisely, if we denote by $\mathbf{P}$ the transition probability matrix of Z, then

$$q_{a_1 a_2} = \Pr\{ \, Z_1 = a_2 \mid Z_0 = a_1 \, \} +$$

$$\Pr\{ \, Z_2 = a_2, Z_1 \in A_1 \mid Z_0 = a_1 \, \} +$$

$$\Pr\{ \, Z_3 = a_2, Z_1, Z_2 \in A_1 \mid Z_0 = a_1 \, \} + ... =$$

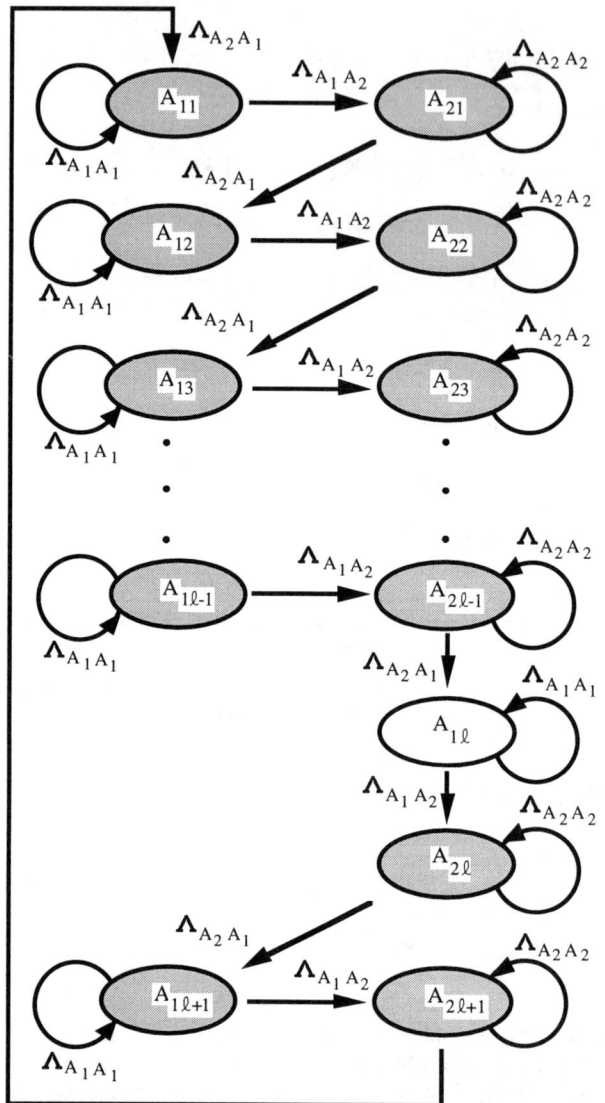

Figure 5.1 State-transition-rate diagram of the auxiliary Markov chain Z

$$p_{a_1 a_2} + \sum_{z_1 \in A_1} p_{a_1 z_1} p_{z_1 a_2} + \sum_{z_1 \in A_1} \sum_{z_2 \in A_1} p_{a_1 z_1} p_{z_1 z_2} p_{z_2 a_2} + \ldots =$$

(a_1, a_2) - entry of $\left\{ \mathbf{P}_{A_1A_2} + \mathbf{P}_{A_1A_1} \mathbf{P}_{A_1A_2} + \mathbf{P}_{A_1A_1}^2 \mathbf{P}_{A_1A_2} + \cdots \right\} =$

(a_1, a_2) - entry of $\left\{ (\mathbf{I} - \mathbf{P}_{A_1A_1})^{-1} \mathbf{P}_{A_1A_2} \right\}.$

On the other hand, we know that $\mathbf{P}$ and $\mathbf{\Lambda}$ are interrelated by (4.9); in particular,

$$\mathbf{I} - \mathbf{P}_{A_1A_1} = - \mathbf{D}_{A_1A_1} \mathbf{\Lambda}_{A_1A_1}, \quad \mathbf{P}_{A_1A_2} = \mathbf{D}_{A_1A_1} \mathbf{\Lambda}_{A_1A_2},$$

from which it follows that

$$q_{a_1 a_2} = (a_1, a_2) \text{ - entry of } \left\{ - \mathbf{\Lambda}_{A_1A_1}^{-1} \mathbf{\Lambda}_{A_1A_2} \right\}. \tag{5.27}$$

(5.26) and (5.27) show that the matrix of transition rates of W from $A_{2\ell-1}$ to $A_{2\ell}$ is $-\mathbf{\Lambda}_{A_2A_1} \mathbf{\Lambda}_{A_1A_1}^{-1} \mathbf{\Lambda}_{A_1A_2}.)$

W is irreducible and $\mathbf{K}_\ell(A_1, A_2)$ is a submatrix of its transition rate matrix which corresponds to transitions between states of a set which is not the full state space. Therefore, by Lemma 4.1, (5.25) holds. ∎

(D) The case $m = 1$. With $TS_{A_1,0} \equiv TS_{A_2,0} \equiv 0$, (5.3) now reads as follows

$\Pr\{ M_{A_1}(t) = 1 \} =$

$$\underset{a_1 \in A_1}{\Sigma} \ (1 - \Pr\{ TS_{A_1,1} + TS_{A_2,1} \le t \mid Y_0 = a_1 \}) \Pr\{ Y_0 = a_1 \} +$$

$$\underset{a_2 \in A_2}{\Sigma} \left(\Pr\{ TS_{A_2,1} \le t \mid Y_0 = a_2 \} - \right.$$

$$\left. - \Pr\{ TS_{A_1,1} + TS_{A_2,2} \le t \mid Y_0 = a_2 \} \right) \Pr\{ Y_0 = a_2 \}. \tag{5.28}$$

We have by Corollary 4.4

$$\Pr\{ TS_{A_2,1} \le t \} = 1 + \{ \boldsymbol{\alpha}_{A_1}^T \mathbf{\Lambda}_{A_1A_1}^{-1} \mathbf{\Lambda}_{A_1A_2} - \boldsymbol{\alpha}_{A_2}^T \} \exp\{ t \mathbf{\Lambda}_{A_2A_2} \} \mathbf{1},$$

from which it follows that

$$\underset{a_2 \in A_2}{\Sigma} \Pr\{ TS_{A_2,1} \le t \mid Y_0 = a_2 \} \Pr\{ Y_0 = a_2 \} =$$

$$\boldsymbol{\alpha}_{A_2}^T \mathbf{1} - \boldsymbol{\alpha}_{A_2}^T \exp\{ t \mathbf{\Lambda}_{A_2A_2} \} \mathbf{1}. \tag{5.29}$$

Corollary 5.4 shows that

$$\sum_{a_1 \in A_1} \Pr\{\, TS_{A_1,1} + TS_{A_2,1} \le t \mid Y_0 = a_1 \,\} \Pr\{\, Y_0 = a_1 \,\} =$$

$$\alpha_{A_1}{}^T \mathbf{1} - (\alpha_{A_1}{}^T, \mathbf{0}) \exp\{\, t\, \Theta_1(A_1, A_2) \,\} \mathbf{1}. \tag{5.30}$$

Finally, by Proposition 5.5,

$$\sum_{a_2 \in A_2} \Pr\{\, TS_{A_1,1} + TS_{A_2,2} \le t \mid Y_0 = a_2 \,\} \Pr\{\, Y_0 = a_2 \,\} =$$

$$\alpha_{A_2}{}^T \mathbf{1} - (\alpha_{A_2}{}^T, \mathbf{0}, \mathbf{0}) \exp\{\, t\, E_2(A_2, A_1) \,\} \mathbf{1}. \tag{5.31}$$

(5.5) now follows from (5.28) - (5.31).

The case $m \ge 2$. The reasoning is similar to the previous one. Use now (5.3), (5.13), and (5.15) to arrive at (5.6). ∎

5.2 An application: the number of repairs of a two-unit power transmission system during a finite time interval

5.2.1 Numerical results and implementation issues

The system of application is Model 1 from Section 1.2.1. The parameter values assumed are as follows

Normal weather component failure rates:	$\lambda_1 = \lambda_2 = 0.5$ / year,	(5.32)
Stormy weather component failure rates:	$\lambda_1' = \lambda_2' = 25.0$ / year,	(5.33)
Component repair rates:	$\mu_1 = \mu_2 = \mu_1' = \mu_2' = 1168$ / year,	
Rate of change from normal to stormy:	$\nu = 43.8$ / year,	
Rate of change from stormy to normal:	$\rho = 5840$ / year.	

Let us note that the latter three of the above equations are equivalent to assuming the following mean values

Mean component repair times:	7.5 hours,	
Mean duration of normal weather:	200 hours,	(5.34)
Mean duration of stormy weather:	1.5 hours.	

The values in (5.32) - (5.34) are from [BIL1], Chapter V; they are based on real- life data. From Figure 1.1, the set of 'up' states is $A_1 = \{\, 1, ..., 6 \,\}$, the set of 'down' states is $A_2 = \{\, 7, 8 \,\}$. The quantity of interest is $R(t)$, the number of completed repair events, i.e., transitions from A_2 to A_1, during [0, t]. The initial state is '1', i.e., all components are assumed to be 'up' when the system is started. Thus, from (5.1), $R(t)$ in terms of $M_{A_1}(t)$ is $R(t) = M_{A_1}(t) - 1$. The numerical values produced by a MATLAB implementation of (5.4) - (5.6) are shown in Table

5.1. The MATLAB code is shown in Section 5.2.2. The computation of each of the entries in the last row in Table 5.1 took 2 CPU minutes on the Apple Macintosh IIx.

t [yr]	r = 0	r = 1	r = 2	r = 3	r = 4
1.0	$9.979\ 677_1$	$2.027\ 544_3$	$4.674\ 673_6$	$9.959\ 086_9$	$2.013\ 667_{11}$
2.5	$9.949\ 242_1$	$5.056\ 341_3$	$1.937\ 243_5$	$6.303\ 566_8$	$1.845\ 163_{10}$
5.0	$9.898\ 723_1$	$1.006\ 328_2$	$6.413\ 865_5$	$3.215\ 138_7$	$1.380\ 894_9$
10.0	$9.798\ 451_1$	$1.992\ 462_2$	$2.282\ 907_4$	$1.927\ 943_6$	$1.332\ 344_8$
20.0	$9.600\ 946_1$	$3.904\ 789_2$	$8.444\ 488_4$	$1.287\ 719_5$	$1.550\ 824_7$
50.0	$9.031\ 997_1$	$9.183\ 749_2$	$4.787\ 547_3$	$1.704\ 478_4$	$4.658\ 338_6$
100.0	$8.157\ 682_1$	$1.658\ 965_1$	$1.708\ 264_2$	$1.187\ 282_3$	$6.264\ 472_5$

Table 5.1 The probability mass function of the number of completed repairs R(t),
$$\Pr\{\ R(t) = r \mid Y_0 = 1\ \},$$
(Negative powers of 10 are indicated by a subscript, e.g., 9.432_1 stands for 9.432×10^{-1}.)

We note in passing that this example will be recomputed in Chapter 10 using an alternative method which is based on the numerical inversion of Laplace transforms. Hence, that method will also serve for the validation of our present approach.

A third approach, the 'randomization technique' is a viable alternative to the above two for certain parameter values: an indication of how to compute the distribution function of $M_{A_1}(t)$ by randomization is given by De Souza E Silva and Gail in [DES2]. However, for the above example the randomization technique is not appropriate for the following reason. With the above parameter values, which are well within the practical range, the numerical value of η, the rate parameter of the subordinated Poisson process, will satisfy $\eta \geq 8176$ / year; we may in fact choose $\eta = 8176$ / year. Therefore, on most systems the randomization technique will deliver 'zero' as an answer since $\exp\{ -\eta t \}$ will be computed as zero, at least for values of t which are in the practical range ($t \geq 1$ year, say). References in which the randomization technique is used are numerous; see, for example, [DES1] - [DES3], [GROS1], [GROS2], and [SER]. We, too, will use this technique in Chapter 6.

The only new MATLAB language element in this example is `sprintf`. It allows the creation of strings as is usual in the C language.

5.2.2 MATLAB code

```
% Supply and show data...

disp('Normal weather failure rates in 1/year...');
lambda1=0.5;
lambda2=0.5;
disp(fprintf('lambda1 = %12.5e\n',lambda1));
disp(fprintf('lambda2 = %12.5e\n',lambda2));

disp('Stormy weather failure rates in 1/year...');
lambda1d=25.0;
lambda2d=25.0;
disp(fprintf('lambda1d = %12.5e\n',lambda1d));
```

```
disp(fprintf('lambda2d = %12.5e\n',lambda2d));

% Assumed mean repair time is 7.5 hours throughout. Thus,

disp('Repair rates in 1/year...');
mu1 =1168;
mu2 =1168;
mu1d=1168;
mu2d=1168;
disp(fprintf('mu1 = %12.5e\n',mu1));
disp(fprintf('mu2 = %12.5e\n',mu2));
disp(fprintf('mu1d = %12.5e\n',mu1d));
disp(fprintf('mu2d = %12.5e\n',mu2d));

% Assumed mean duration of normal weather is 200 hours. Thus,

disp('Rate of change from normal to stormy weather in 1/year...');
nu=43.8;
disp(fprintf('nu = %12.5e\n',nu));

% Assumed mean duration of stormy weather is 1.5 hours. Thus,

disp('Rate of change from stormy to normal weather in 1/year...');
rho=5840;
disp(fprintf('rho = %12.5e\n',rho));

RATES=[0     nu      lambda2  0        lambda1  0        0        0
       rho   0       0        lambda2d 0        lambda1d 0        0
       mu2   0       0        nu       0        0        lambda1  0
       0     mu2d    rho      0        0        0        0        lambda1d
       mu1   0       0        0        0        nu       lambda2  0
       0     mu1d    0        0        rho      0        0        lambda2d
       0     0       mu1      0        mu2      0        0        nu
       0     0       0        mu1d     0        mu2d     rho      0        ];

disp('Transition rate matrix...');
TR_RATE_MAT=RATES-diag(RATES*ones(8,1))

% Partition TR_RATE_MAT: up states    A1 = U = {1, ..., 6},
%                        down states A2 = D = {7, 8}.

LUU = TR_RATE_MAT(1:6,1:6);
LDD = TR_RATE_MAT(7:8,7:8);
LUD = TR_RATE_MAT(1:6,7:8);
LDU = TR_RATE_MAT(7:8,1:6);

t = input('Enter length of time period t in years ... ');
m = input('Enter m (≥ 1) for Pr( MA1 ≤ m ) ... ');

[nu nd] = size(LUD);

TEMP = [LUU LUD; zeros(nd, nu) LDD];
THETA = TEMP;
init = input('Enter initial state in {1, ..., 6} ... ');
E6 = eye(6);

fix(clock)
Prtemp = 0.0;
alpha = E6(init:init,:);
alpha = [alpha zeros(1, nd)];
Pr = alpha*expm(t*THETA)*(ones(1,max(size(THETA)))');
text1 = sprintf('\n Pr[ MA1(%6.2f) = %2.0f | Y0 = %1.0f] =',t,1,init);
text2 = sprintf(' %22.15e\n',Pr - Prtemp);
disp([text1,text2]);
fix(clock)
Prtemp = Pr;
```

```
for i=1:(m-1)
  j = max(size(THETA));
  TEMP1 = [THETA [zeros(j - nd,nu); LDU] zeros(j,nd)];
  TEMP2 = [zeros(nu+nd,j) TEMP];
  THETA = [TEMP1; TEMP2];
  alpha = [alpha zeros(1, nu+nd)];
  Pr = alpha*expm(t*THETA)*(ones(1,max(size(THETA)))');
  text1 = sprintf(' Pr[ MA1(%6.2f) = %2.0f | Y0 = %1.0f] =',t,i+1,init);
  text2 = sprintf(' %22.15e\n',Pr - Prtemp);
  disp([text1,text2]);
  fix(clock)
  Prtemp = Pr;
end
```

CHAPTER 6

A COMPOUND MEASURE OF DEPENDABILITY FOR CONTINUOUS-TIME MARKOV MODELS OF REPAIRABLE SYSTEMS

In the previous chapter, we were looking at the distribution of $M_{A_1}(t_0)$, the number of visits during $[0, t_0]$, $t_0 > 0$, to a subset A_1 of the state space S by an irreducible finite Markov process Y. In the reliability context, $M_{A_1}(t_0)$ will become the number of repair periods $M_B(t_0)$ if we put $A_1 = B$, the set of system 'down' states. The cumulative distribution function of $M_B(t_0)$ is of course available from Theorem 5.1. For reliability assessment it is desirable to consider several system characteristics simultaneously; in this chapter, $M_B(t_0)$ will be supplemented by $T_G(t_0)$, the total time spent by Y during $[0, t_0]$ in the set of 'good' states $G = A_2 = S \setminus A_1$. More precisely, we shall consider here $\Pr\{ T_G(t_0) > t, M_B(t_0) \le m \}$ for $t \in (0, t_0)$ and $m \in \{ 0, 1, \ldots \}$, i.e., the probability that during $[0, t_0]$ the system is 'up' for more than t units of time *and* that the number of 'down' periods does not exceed m.

$M_B(t_0)$ can be viewed as a measure of the (undiscounted) cost incurred during $[0, t_0]$; this interpretation is justified if every 'down' period is associated with the same cost irrespective of its duration. $T_G(t_0)$ measures the total amount of work delivered by the system during $[0, t_0]$. The dependability measure considered here is therefore a *compound* one, incorporating both $T_G(t_0)$ and $M_B(t_0)$. This dependability measure combines and generalizes the two hitherto unrelated performance variables $T_G(t_0)$ and $M_B(t_0)$. Notice in particular that the complementary distribution function of $T_G(t_0)$ and the distribution function of $M_B(t_0)$ (for $Y_0 \in G$) are respectively obtained in the limit by letting $m \to +\infty$ and $t \to 0$. Furthermore, given $Y_0 \in G$, the dependability measure is the system reliability if $m = 0$. Thus, several of the familiar system characteristics are expressible in terms of our dependability measure.

The system is assumed to be repairable and hence Y is irreducible. The quantity $T_G(t_0) / t_0$ is known as *interval availability* and it is the object of interest in many papers. Individually, both $T_G(t_0)$ and $M_B(t_0)$ have well-known distributions: we know $M_B(t_0)$ from Chapter 5 whereas a closed form expression for the cumulative distribution function of $T_G(t_0)$ is available from Sericola [SER]; algorithms for the computation of (as opposed to closed form expressions for) the cumulative distribution functions of $T_G(t_0)$ are known from [DES1]. Most recently, Rubino and Sericola [RUB5] presented a computational scheme for the cumulative distribution function of $T_G(t_0)$ under certain additional conditions on the process Y. In the last three papers cited above, the randomization technique is used; it is based on a representation of a continuous-time Markov process in terms of a pair of independent processes: one of them is a homogeneous Poisson process determining the time instants at which state changes occur; the other is a discrete-parameter Markov chain describing the state changes themselves.

Our approach, too, is based on the randomization technique. The event of interest, when represented in terms of the underlying discrete-parameter Markov chain, will turn out to involve essentially a discrete-parameter version of the performance measure under consideration. This issue will be discussed in the Section 6.1. In Section 6.2 then, a closed form expression is obtained for our performance measure in the discrete-parameter case. It is based on the knowledge of the joint distribution of any finite collection of sojourn times in G and B of a discrete-parameter Markov chain from Section 2.1.3. In Section 6.3, we explore the viability of our analytical result for the numerical computation of the dependability measure under consideration by the example of Model 5 from Section 1.2.1.

This chapter is based on material first reported in [CSE11].

6.1 The dependability measure and its evaluation by randomization

For the sake of completeness and for later reference, the framework for the randomization technique is briefly reviewed first. For details, see, for example, De Souza e Silva and Gail [DES1] and the references therein. As usual, let $\mathbf{\Lambda}$ stand for the transition rate matrix of Y. Choose

$$\mu \geq \max\{ \sum_{\substack{s' \in S \\ s' \neq s}} \lambda_{ss'} : s \in S \},$$

and put

$$\mathbf{P} = \mu^{-1} \mathbf{\Lambda} + \mathbf{I}.$$

Then, $\mathbf{P}$ is a suitable transition probability matrix of the subordinated discrete-parameter Markov chain $Z = \{ Z_0, Z_1, \dots \}$. The Poisson process $W = \{ W_t : t \geq 0 \}$ alluded to earlier is idependent of Z and has rate parameter μ. The randomized process $\{ Z_{W_t} : t \geq 0 \}$ is a Markov process with the same distribution as Y. To express now $\Pr\{ T_G(t_0) > t, M_B(t_0) \leq n \}$ in terms of this new representation of Y, the 'number of state changes in $[0, t_0]$', K, is introduced, which is the number of events of the Poisson process W during $[0, t_0]$. It is

$$\Pr\{ K = k \} = \frac{t_0^k \mu^k}{k!} \exp\{ -t_0 \mu \}, \quad k = 0, 1, \dots . \tag{6.1}$$

Given that $K = k$ (≥ 1), let $U_{1k} < \dots < U_{kk}$ denote the k event times of the Poisson process W. It is well-known that they have the same distribution as the order statistics of a random sample of size $k \geq 1$ from a uniform distribution on $(0, t_0)$. Thus, for $u \in (0, t_0)$,

$$\Pr\{ U_{ik} \leq u \} = \sum_{j=i}^{k} \binom{k}{j} \left(\frac{u}{t_0}\right)^j \left(1 - \frac{u}{t_0}\right)^{k-j}, \quad i = 0, 1, \dots, k+1. \tag{6.2}$$

(Note that by definition $U_{0k} \equiv 0$ and $U_{k+1k} \equiv t_0$.) It is also known that the $k+1$ interevent times

$$V_{1k} = U_{1k}, V_{2k} = U_{2k} - U_{1k}, \dots, V_{kk} = U_{kk} - U_{k-1k}, V_{k+1k} = t_0 - U_{kk}$$

are exchangeable.

To evaluate $\Pr\{ T_G(t_0) > t, M_B(t_0) \leq n \}$, we write by (6.1)

$$\Pr\{ T_G(t_0) > t, M_B(t_0) \leq n \} =$$

$$\Pr\{\, T_G(t_0) > t, M_B(t_0) \le n \mid K = 0 \,\} \exp\{\, -t_0\, \mu \,\} +$$

$$\sum_{\nu=0}^{n} \sum_{k=1}^{\infty} \Pr\{\, T_G(t_0) > t, M_B(t_0) = \nu \mid K = k \,\} \Pr\{\, K = k \,\}. \tag{6.3}$$

The first term on the right-hand side of (6.3) is, of course, $\Pr\{\, Y_0 \in G \,\} \exp\{\, -t_0\, \mu \,\}$. It is easily seen that given $K = k\ (\ge 1)$, the pair $\big(T_G(t_0), M_B(t_0)\big)$ can be replaced by

$$\Big(\, \sum_{r=1}^{k+1} V_{rk}\, I_{\{Z_{r-1} \in G\}}\, ,\, I_{\{Z_0 \in B\}} + \sum_{r=1}^{k} I_{\{Z_r \in B,\, Z_{r-1} \in G\}} \,\Big),$$

where, of course, $I_{\{...\}}$ stands for the indicator function of the index set $\{...\}$. Thus, for $k \ge 1$,

$$\Pr\{\, T_G(t_0) > t, M_B(t_0) = \nu \mid K = k \,\} =$$

$$\Pr\Big\{\, \sum_{r=1}^{k+1} V_{rk}\, I_{\{Z_{r-1} \in G\}} > t,\, I_{\{Z_0 \in B\}} + \sum_{r=1}^{k} I_{\{Z_r \in B,\, Z_{r-1} \in G\}} = \nu \,\Big\} =$$

$$\sum_{i=1}^{k+1} \Pr\Big\{\, \sum_{r=1}^{k+1} V_{rk}\, I_{\{Z_{r-1} \in G\}} > t,\, I_{\{Z_0 \in B\}} +$$

$$\sum_{r=1}^{k} I_{\{Z_r \in B,\, Z_{r-1} \in G\}} = \nu,\ \sum_{r=1}^{k+1} I_{\{Z_{r-1} \in G\}} = i \,\Big\} =$$

$$\sum_{i=1}^{k+1} \Pr\{\, U_{ik} > t \,\} \Pr\Big\{\, \sum_{r=1}^{k+1} I_{\{Z_{r-1} \in G\}} = i,\, I_{\{Z_0 \in B\}} + \sum_{r=1}^{k} I_{\{Z_r \in B,\, Z_{r-1} \in G\}} = \nu \,\Big\}. \tag{6.4}$$

(The last step in (6.4) holds by the exchangeability of $V_{1k}, ..., V_{k+1\,k}$ and by the independence of W and Z.) Summing over $\nu = 0, ..., n$ in (6.4), we get from (6.1) - (6.4)

$$\Pr\{\, T_G(t_0) > t, M_B(t_0) \le n \,\} = \Pr\{\, Y_0 \in G \,\} \exp\{\, -t_0\, \mu \,\} +$$

$$\sum_{k=1}^{\infty} \sum_{i=1}^{k+1} \sum_{j=0}^{i-1} \frac{t_0^k\, \mu^k}{k!} \exp\{\, -t_0\, \mu \,\} \binom{k}{j} \Big(\frac{t}{t_0}\Big)^j \Big(1 - \frac{t}{t_0}\Big)^{k-j} \Delta(k, i, n), \tag{6.5}$$

where

$$\Delta(k, i, n) =$$

$$\sum_{\nu=0}^{n} \Pr\{ \sum_{r=1}^{k+1} I_{\{Z_{r-1} \in G\}} = i, I_{\{Z_0 \in B\}} + \sum_{r=1}^{k} I_{\{Z_r \in B, Z_{r-1} \in G\}} = \nu \}. \tag{6.6}$$

It is seen from (6.6) that $\Delta(k, i, n)$ is a probability which can be expressed in terms of the first $k + 1$ variables of the subordinated Markov process Z as follows

$$\Delta(k, i, n) = \Pr\{ \text{ The total 'time' spent by } Z_0, ..., Z_k \text{ in G is i } and$$
$$B \text{ is visited by } Z_0, ..., Z_k \text{ no more than n times.}\}. \tag{6.7}$$

Let us add in passing that (6.7) shows that for $n \geq k + 1$,

$$\Delta(k, i, n) = \Delta(k, i, k+1) = \Pr\{ \text{ The total 'time' spent by } Z_0, ..., Z_k \text{ in G is i } \}.$$

This relationship has two important implications. First, it allows computational savings to be made if (6.5) is required for more than one value of n. Second, by letting $n \to +\infty$ in (6.5), a closed form expression for the (complementary) distribution function of $T_G(t_0)$ is obtained

$$\Pr\{ T_G(t_0) > t \} = \Pr\{ Y_0 \in G \} \exp\{ -t_0 \mu \} +$$

$$\sum_{k=1}^{\infty} \sum_{i=1}^{k+1} \sum_{j=0}^{i-1} \frac{t_0^k \mu^k}{k!} \exp\{ -t_0 \mu \} \binom{k}{j} \left(\frac{t}{t_0}\right)^j \left(1 - \frac{t}{t_0}\right)^{k-j} \Delta(k, i, k+1). \tag{6.8}$$

(6.8) is known from De Souza e Silva and Gail [DES1].

By (6.5) and (6.7), the evaluation of the continuous-time dependability measure under consideration is essentially reduced to that of its discrete-parameter counterpart. (Strictly speaking, the analogy is, of course, between $\Delta(k, i, n)$ and the *density* $\Pr\{ T_G(t_0) \in dt, M_B(t_0) \leq n \}$.) In the next section a closed form expression will be derived for $\Delta(k, i, n)$.

6.2 The evaluation of $\Delta(k, i, n)$

The purpose of this section is to furnish a closed form expression for the probability in (6.7). The following notation will be used. D stands for the closed unit disk in $\mathbb{C}$, i.e., $D = \{ z \in \mathbb{C} : |z| \leq 1 \}$. In most cases, the size of the matrices and vectors used will be obvious from the context but, if necessary, it will be indicated by an appropriate subscript or pair of subscripts, e.g., $\mathbf{1}_k = \mathbf{1} \in \mathbb{R}^k$, $\mathbf{0}_{1 \times k}^T = (0, 0, ..., 0)^T \in \mathbb{R}^k$.

The plan of obtaining $\Delta(k, i, n)$ is as follows. In Section 6.2.1, for *fixed* $k \in \{ 1, 2, ... \}$, the first $k + 1$ components of the *irreducible* Markov chain Z, i.e., $Z_0, ..., Z_k$, will be equivalently replaced by an auxiliary *absorbing* Markov chain $X^{(k)} = \{ X_0^{(k)}, X_1^{(k)}, ... \}$ with a suitably defined (enlarged) state space $S^{(k)}$ which is partitioned as $S^{(k)} = G^{(k)} \cup B^{(k)} \cup \{ \omega \}$. $\Delta(k, i, n)$ in terms of $X^{(k)}$ will be

$$\Delta(k, i, n) =$$

$$\Pr\{ \text{ Total 'time' spent by } X^{(k)} \text{ in } G^{(k)} \text{ until absorption is i } and$$
$$B^{(k)} \text{ is visited by } X^{(k)} \text{ until absorption no more than n times.}\}. \tag{6.9}$$

126

In Section 6.2.2 then, an expression will be derived for the probability in (6.9). The reasoning there is based on the joint distribution of the sojourn times of $X^{(k)}$ in both $G^{(k)}$ and $B^{(k)}$.

6.2.1 The auxiliary absorbing Markov chain $X^{(k)}$

Let $k \in \{1,2, ...\}$ be fixed. Figure 6.1 shows the state-transition diagram of an auxiliary absorbing Markov chain $X^{(k)}$ with state space $S^{(k)} = G^{(k)} \cup B^{(k)} \cup \{\omega\}$, where $G^{(k)} = G_0 \cup ... \cup G_k$ and $B^{(k)} = B_0 \cup ... \cup B_k$ (shaded in Figure 6.1); $G_0, ..., G_k$ and $B_0, ..., B_k$ are disjoint instances of G and B respectively and ω is the absorbing state.

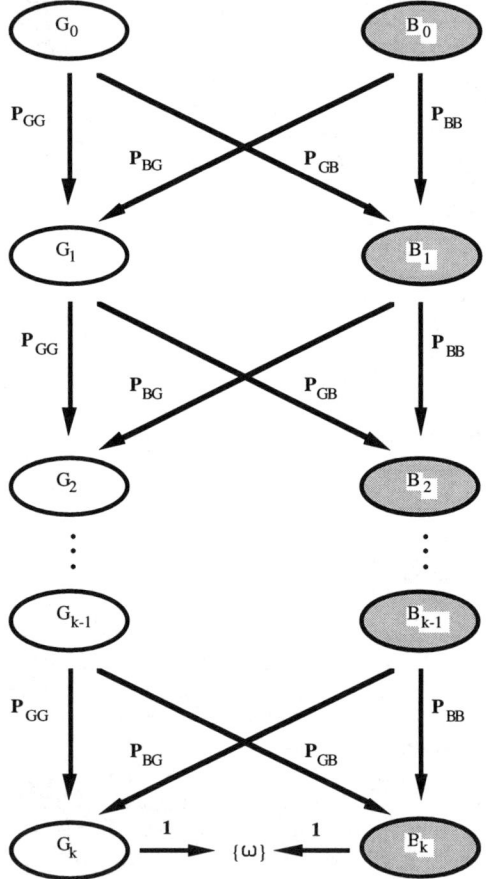

Figure 6.1 The state-transition diagram of the auxiliary Markov chain $X^{(k)}$

$X^{(k)}$ is equivalent to the finite sequence $\{ Z_0, ..., Z_k \}$ in the sense that $\Delta(k, i, n)$ from (6.7) is re-expressed by (6.9) in terms of $X^{(k)}$. For $\ell \in \{ 1, 2, ... \}$, let $E(\ell)$ denote the square matrix of size ℓ with entries

$$e(\ell)_{ij} = \begin{cases} 1 \text{ for } j - i = 1, \\ 0 \text{ otherwise.} \end{cases}$$

Then it is seen from Figure 6.1 that the transition probability matrix of $X^{(k)}$ is

$$Q = \begin{array}{c} \\ G^{(k)} \\ B^{(k)} \\ \{\omega\} \end{array} \begin{array}{ccc} G^{(k)} & B^{(k)} & \{\omega\} \\ \left[\begin{array}{ccc|ccc|cc} \mathbf{P}_{GG} \otimes \mathbf{E}(k+1) & \mathbf{P}_{GB} \otimes \mathbf{E}(k+1) & \mathbf{Q}_{G^{(k)}\{\omega\}} \\ \hline \mathbf{P}_{BG} \otimes \mathbf{E}(k+1) & \mathbf{P}_{BB} \otimes \mathbf{E}(k+1) & \mathbf{Q}_{B^{(k)}\{\omega\}} \\ \hline \mathbf{0} & \mathbf{0} & \mathbf{1} \end{array} \right], \end{array} \qquad (6.10)$$

where $\otimes$ denotes the *right* direct product of matrices as defined in (1.5). $\mathbf{Q}_{G^{(k)}\{\omega\}}$ and $\mathbf{Q}_{B^{(k)}\{\omega\}}$ are column vectors the only non-zero entries of which are unity in their last (i.e., the $(k+1)st$) group of components. The initial probability vector $\boldsymbol{\beta}^{(k)}$ of $X^{(k)}$ is given in terms of Y's initial probability vector $\boldsymbol{\alpha} = (\boldsymbol{\alpha}_G^T, \boldsymbol{\alpha}_B^T)^T$ by

$$\boldsymbol{\beta}^{(k)} = \left(\boldsymbol{\alpha}_G^T \otimes (1, \mathbf{0}_{1 \times k}), \boldsymbol{\alpha}_B^T \otimes (1, \mathbf{0}_{1 \times k}), 0 \right)^T.$$

6.2.2 The closed form expression for $\Delta(k, i, n)$

In this section, the auxiliary Markov chain $X^{(k)}$ will be used to establish a closed form expression for the probability in (6.9). For $k \in \{ 1, 2, \dots \}$ fixed, we put $X = X^{(k)}$, $\mathscr{S} = S^{(k)}$, $\mathscr{G} = G^{(k)}$, and $\mathscr{B} = B^{(k)}$. Let $\{ N_{\mathscr{G},1}, N_{\mathscr{G},2}, \dots \}$ and $\{ N_{\mathscr{B},1}, N_{\mathscr{B},2}, \dots \}$ stand for the lengths of consecutive sojourns of X in $\mathscr{G}$ and $\mathscr{B}$ respectively; $N_{\mathscr{S}',i} = 0$ if X visits $\mathscr{S}' \in \{ \mathscr{G}, \mathscr{B} \}$ less than i times. $M_{\mathscr{B}}$ stands for the number of visits of X to $\mathscr{B}$ until absorption. The n*th cumulative* sojourn time in $\mathscr{G}$ is denoted by $NS_{\mathscr{G},n} = N_{\mathscr{G},1} + \dots + N_{\mathscr{G},n}$; the total time spent by X in $\mathscr{G}$ until absorption is $NS_{\mathscr{G},\infty}$.

The generating function of $\{ \Delta(k,i,n) : i \geq 0 \}$. The next proposition expresses the generating function of $\{ \Delta(k,i,n) : i \geq 0 \}$ in terms of the sojourn time variables of X in $\mathscr{G}$ and $\mathscr{B}$.

PROPOSITION 6.1. *The generating function of* $\{ \Delta(k,i,n) : i \geq 0 \}$ *is for* $z \in D$

$$\sum_{i=0}^{\infty} \Delta(k, i, n) z^i = g_{k,n}(z, 0), \qquad (6.11)$$

where $g_{k,n}$ *is the probability generating function of* $(NS_{\mathscr{G},n+1}, N_{\mathscr{B},n+1})$, *i.e., for* $(z, w) \in D^2$,

$$g_{k,n}(z, w) = E \left[z^{N_{\mathscr{G},1} + \dots + N_{\mathscr{G},n+1}} w^{N_{\mathscr{B},n+1}} \right]. \qquad (6.12)$$

PROOF OF PROPOSITION 6.1. It is

$$\sum_{i=0}^{\infty} \Delta(k, i, n)\, z^i = \sum_{i=0}^{\infty} \Pr\{\, NS_{\mathcal{G},\infty} = i, M_{\mathcal{B}} \le n \,\}\, z^i =$$

$$\Pr\{\, M_{\mathcal{B}} \le n \,\}\, E\left[z^{NS_{\mathcal{G},\infty}} \middle| M_{\mathcal{B}} \le n \right] = E\left[z^{NS_{\mathcal{G},\infty}} I_{\{M_{\mathcal{B}} \le n\}} \right]. \tag{6.13}$$

Due to the equivalence

$$NS_{\mathcal{G},\infty} = i, M_{\mathcal{B}} \le n \Leftrightarrow N_{\mathcal{G},1} + \ldots + N_{\mathcal{G},n+1} = i, N_{\mathcal{B},n+1} = 0, \tag{6.14}$$

the right-hand side of (6.13) can be rewritten as

$$E\left[z^{NS_{\mathcal{G},\infty}} I_{\{M_{\mathcal{B}} \le n\}} \right] = E\left[z^{N_{\mathcal{G},1} + \ldots + N_{\mathcal{G},n+1}} I_{\{N_{\mathcal{B},n+1} = 0\}} \right] =$$

$$\lim_{\substack{w \to 0 \\ w \in \mathbb{C} \setminus \{0\}}} E\left[z^{N_{\mathcal{G},1} + \ldots + N_{\mathcal{G},n+1}} w^{N_{\mathcal{B},n+1}} \right] = g_{k,n}(z, 0), \tag{6.15}$$

from which (6.11) follows. ∎

For a more explicit expression for the generating function of $\{\, \Delta(k,i,n) : i \ge 0 \,\}$, the right-hand side of (6.12) will now be examined. Using Theorem 2.16 with $A_1 = \mathcal{G}$, $A_2 = \mathcal{B}$, $m = n + 1$, $\mathbf{z}_1 = (\, z, \ldots, z\,) \in D^{n+1}$, $\mathbf{z}_2 = (\, 1, \ldots, 1, w\,) \in D^{n+1}$, and $\boldsymbol{\beta}_{A_1}^T = \boldsymbol{\alpha}_G^T \otimes (\, 1, \mathbf{0}_{1 \times k}\,)$, $\boldsymbol{\beta}_{A_2}^T = \boldsymbol{\alpha}_B^T \otimes (\, 1, \mathbf{0}_{1 \times k}\,)$, we get

$$E\left[z^{N_{\mathcal{G},1} + \ldots + N_{\mathcal{G},n+1}} w^{N_{\mathcal{B},n+1}} \right] =$$

$$\{\boldsymbol{\alpha}_G^T \otimes (\, 1, \mathbf{0}_{1 \times k}\,)\}\, \boldsymbol{\psi}_{\mathcal{G};n+1}(\, z, \ldots, z\,;\, 1, \ldots, 1, w\,) +$$

$$\{\boldsymbol{\alpha}_B^T \otimes (\, 1, \mathbf{0}_{1 \times k}\,)\}\, \boldsymbol{\psi}_{\mathcal{B};n+1}(\, z, \ldots, z\,;\, 1, \ldots, 1, w\,),$$

from which it is seen by Proposition 6.1 with $w = 0$ that

$$\sum_{i=0}^{\infty} \Delta(k, i, n)\, z^i = \{\boldsymbol{\alpha}_G^T \otimes (\, 1, \mathbf{0}_{1 \times k}\,)\}\, \boldsymbol{\psi}_{\mathcal{G};n+1}(\, z, \ldots, z\,;\, 1, \ldots, 1, 0\,) +$$

$$\{\boldsymbol{\alpha}_B^T \otimes (\, 1, \mathbf{0}_{1 \times k}\,)\}\, \boldsymbol{\psi}_{\mathcal{B};n+1}(\, z, \ldots, z\,;\, 1, \ldots, 1, 0\,). \tag{6.16}$$

The terms $\boldsymbol{\psi}_{\mathcal{G};n+1}$ and $\boldsymbol{\psi}_{\mathcal{B};n+1}$ in (6.16) are given respectively by

$$\boldsymbol{\psi}_{\mathcal{G};n+1}(\, z, \ldots, z\,;\, 1, \ldots, 1, 0\,) =$$

$$\sum_{\ell=0}^{n} \left(\mathbf{\Phi}_{\mathscr{GB}}(z)\, \mathbf{\Phi}_{\mathscr{BG}}(1) \right)^{\ell} \mathbf{\Phi}_{\mathscr{G}\{\omega\}}(z) + \sum_{\ell=0}^{n-1} \left(\mathbf{\Phi}_{\mathscr{GB}}(z)\, \mathbf{\Phi}_{\mathscr{BG}}(1) \right)^{\ell} \mathbf{\Phi}_{\mathscr{GB}}(z)\, \mathbf{\Phi}_{\mathscr{B}\{\omega\}}(1) +$$

$$\left(\mathbf{\Phi}_{\mathscr{GB}}(z)\, \mathbf{\Phi}_{\mathscr{BG}}(1) \right)^{n} \mathbf{\Phi}_{\mathscr{GB}}(z)\, \mathbf{\Phi}_{\mathscr{B}\{\omega\}}(0) + \left(\mathbf{\Phi}_{\mathscr{GB}}(z)\, \mathbf{\Phi}_{\mathscr{BG}}(1) \right)^{n} \mathbf{\Phi}_{\mathscr{GB}}(z)\, \mathbf{\Phi}_{\mathscr{BG}}(0)\, \mathbf{1},$$

$$(6.17)$$

and

$$\mathbf{\Psi}_{\mathscr{B};n+1}(z, ..., z\,;\, 1, ..., 1, 0\,) =$$

$$\sum_{\ell=0}^{n-1} \left(\mathbf{\Phi}_{\mathscr{BG}}(1)\, \mathbf{\Phi}_{\mathscr{GB}}(z) \right)^{\ell} \mathbf{\Phi}_{\mathscr{B}\{\omega\}}(1) + \left(\mathbf{\Phi}_{\mathscr{BG}}(1)\, \mathbf{\Phi}_{\mathscr{GB}}(z) \right)^{n} \mathbf{\Phi}_{\mathscr{B}\{\omega\}}(0) +$$

$$\sum_{\ell=0}^{n-1} \left(\mathbf{\Phi}_{\mathscr{BG}}(1)\, (\mathbf{\Phi}_{\mathscr{GB}}(z) \right)^{\ell} \mathbf{\Phi}_{\mathscr{BG}}(1)\, \mathbf{\Phi}_{\mathscr{G}\{\omega\}}(z) + \left(\mathbf{\Phi}_{\mathscr{BG}}(1)\, \mathbf{\Phi}_{\mathscr{GB}}(z) \right)^{n} \mathbf{\Phi}_{\mathscr{BG}}(0)\, \mathbf{\Phi}_{\mathscr{G}\{\omega\}}(z) +$$

$$\left(\mathbf{\Phi}_{\mathscr{BG}}(1)\, \mathbf{\Phi}_{\mathscr{GB}}(z) \right)^{n} \mathbf{\Phi}_{\mathscr{BG}}(0)\, \mathbf{\Phi}_{\mathscr{GB}}(z)\, \mathbf{1}; \qquad\qquad (6.18)$$

see (2.63) and (2.65). An alternative form of (6.17) is obtained by using (2.63) as follows

$$\mathbf{\Psi}_{\mathscr{G};n+1}(z, ..., z\,;\, 1, ..., 1, 0\,) =$$

$$\sum_{\ell=0}^{n} \left(z\, \{\mathbf{I} - z\, \mathbf{Q}_{\mathscr{GG}}\}^{-1} \mathbf{Q}_{\mathscr{GB}}\, \{\mathbf{I} - \mathbf{Q}_{\mathscr{BB}}\}^{-1} \mathbf{Q}_{\mathscr{BG}} \right)^{\ell} z\, \{\mathbf{I} - z\, \mathbf{Q}_{\mathscr{GG}}\}^{-1} \mathbf{Q}_{\mathscr{G}\{\omega\}} +$$

$$\sum_{\ell=0}^{n-1} \left(z\, \{\mathbf{I} - z\, \mathbf{Q}_{\mathscr{GG}}\}^{-1} \mathbf{Q}_{\mathscr{GB}}\, \{\mathbf{I} - \mathbf{Q}_{\mathscr{BB}}\}^{-1} \mathbf{Q}_{\mathscr{BG}} \right)^{\ell} z\, \{\mathbf{I} - z\, \mathbf{Q}_{\mathscr{GG}}\}^{-1} \mathbf{Q}_{\mathscr{GB}} \times$$

$$\{\mathbf{I} - \mathbf{Q}_{\mathscr{BB}}\}^{-1} \mathbf{Q}_{\mathscr{B}\{\omega\}}. \qquad\qquad (6.19)$$

The corresponding equation for (6.18) will be subject to some further examination and, therefore, we rewrite it by reference to the individual terms as follows

$$\mathbf{\Psi}_{\mathscr{B};n+1}(z, ..., z\,;\, 1, ..., 1, 0\,) = \mathbf{h}_1 + \mathbf{h}_2, \qquad\qquad (6.20)$$

with

$$\mathbf{h}_1 = \sum_{\ell=0}^{n-1} \left(z \left\{ \mathbf{I} - \mathbf{Q}_{\mathscr{BB}} \right\}^{-1} \mathbf{Q}_{\mathscr{BG}} \left\{ \mathbf{I} - z\, \mathbf{Q}_{\mathscr{GG}} \right\}^{-1} \mathbf{Q}_{\mathscr{GB}} \right)^{\ell} \left\{ \mathbf{I} - \mathbf{Q}_{\mathscr{BB}} \right\}^{-1} \mathbf{Q}_{\mathscr{B}\{\omega\}}, \qquad (6.21)$$

$$\mathbf{h}_2 = \sum_{\ell=0}^{n-1} \left(z \left\{ \mathbf{I} - \mathbf{Q}_{\mathscr{BB}} \right\}^{-1} \mathbf{Q}_{\mathscr{BG}} \left\{ \mathbf{I} - z\, \mathbf{Q}_{\mathscr{GG}} \right\}^{-1} \mathbf{Q}_{\mathscr{GB}} \right)^{\ell} \times$$

$$\left\{ \mathbf{I} - \mathbf{Q}_{\mathscr{BB}} \right\}^{-1} \mathbf{Q}_{\mathscr{BG}}\, z \left\{ \mathbf{I} - z\, \mathbf{Q}_{\mathscr{GG}} \right\}^{-1} \mathbf{Q}_{\mathscr{G}\{\omega\}} . \qquad (6.22)$$

(6.16), (6.19), and (6.20) in conjunction with (6.10) completely describe the generating function of $\{ \Delta(k, i, n) : i \geq 0 \}$.

The Inversion of the generating function of $\{ \Delta(k, i, n) : i \geq 0 \}$. The expressions in (6.19) and (6.20) don't readily lend themselves to inversion. The following lemma shows, however, that (6.16) can be rewritten in a form which will allow the inverse to be established by simple inspection.

LEMMA 6.2. *For square matrices* $\mathbf{U}$ *and* $\mathbf{V}$ *of the same size and* $i \in \{ 1, 2, \ldots \}$, *define by (2.13) the matrix* $\mathbf{M}_i(\mathbf{U}, \mathbf{V})$ *by*

$$\mathbf{M}_i(\mathbf{U}, \mathbf{V}) = \Delta_i(\mathbf{U}, \ldots, \mathbf{U}; \mathbf{V}, \ldots, \mathbf{V}).$$

(Notice that $\mathbf{M}_1(\mathbf{U}, \mathbf{V}) = \mathbf{U}$ *by definition.) If* $\left\{ \mathbf{I} - z\, \mathbf{U} \right\}$ *is invertible, then the following holds for all column vectors* $\mathbf{a}$ *and* $\mathbf{b}$ *of appropriate length*

$$\sum_{\ell=0}^{i-1} \mathbf{a}^T \left(z \left\{ \mathbf{I} - z\, \mathbf{U} \right\}^{-1} \mathbf{V} \right)^{\ell} z \left\{ \mathbf{I} - z\, \mathbf{U} \right\}^{-1} \mathbf{b} =$$

$$z \left\{ \mathbf{a}^T \otimes (1, \mathbf{0}_{1 \times (i-1)}) \right\} \left\{ \mathbf{I} - z\, \mathbf{M}_i(\mathbf{U}, \mathbf{V}) \right\}^{-1} \left\{ \mathbf{b} \otimes \mathbf{1}_i \right\}. \qquad (6.23)$$

PROOF OF LEMMA 6.2. It is $\mathbf{I} - z\, \mathbf{M}_i(\mathbf{U}, \mathbf{V}) = \mathbf{M}_i(\mathbf{I} - z\, \mathbf{U}, - z\, \mathbf{V})$, and we know its inverse by Lemma 4.10. Thus,

$$z \left\{ \mathbf{a}^T \otimes (1, \mathbf{0}_{1 \times (i-1)}) \right\} \mathbf{M}_i(\mathbf{I} - z\, \mathbf{U}, - z\, \mathbf{V})^{-1} \left\{ \mathbf{b} \otimes \mathbf{1}_i \right\} =$$

$$\sum_{j=1}^{i} z\, \mathbf{a}^T \, \Xi_{1j}(\mathbf{I} - z\, \mathbf{U}, \ldots, \mathbf{I} - z\, \mathbf{U}; -z\, \mathbf{V}, \ldots, - z\, \mathbf{V})\, \mathbf{b} =$$

$$\sum_{j=1}^{i} z\, \mathbf{a}^T \left\{ \mathbf{I} - z\, \mathbf{U} \right\}^{-1} \left(z\, \mathbf{V} \left\{ \mathbf{I} - z\, \mathbf{U} \right\}^{-1} \right)^{j-1} \mathbf{b} =$$

$$\sum_{\ell=0}^{i-1} z\, \mathbf{a}^T \left(z \left\{ \mathbf{I} - z\, \mathbf{U} \right\}^{-1} \mathbf{V} \right)^{\ell} \left\{ \mathbf{I} - z\, \mathbf{U} \right\}^{-1} \mathbf{b},$$

which is the left hand side of (6.23). ∎

In the light of Lemma 6.2, and by (6.19) and (6.20), the two terms on the right-hand side of (6.16) will now be recast as follows. Using (6.19), the first term is seen to evaluate to

$$\{\alpha_G^T \otimes (1, 0_{1 \times k})\} \, \psi_{\mathscr{G};n+1}(z, ..., z \, ; 1, ..., 1, 0) \; =$$

$$z \, \Big\{ \{\alpha_G^T \otimes (1, 0_{1 \times k})\} \otimes (1, 0_{1 \times n}) \Big\} \times$$

$$\Big\{ I - z \, M_{n+1}(Q_{\mathscr{GG}}, Q_{\mathscr{GB}} \{I - Q_{\mathscr{BB}}\}^{-1} Q_{\mathscr{BG}}) \Big\}^{-1} \{ Q_{\mathscr{G}\{\omega\}} \otimes 1_{n+1}\} +$$

$$z \, \Big\{ \{\alpha_G^T \otimes (1, 0_{1 \times k})\} \otimes (1, 0_{1 \times (n-1)}) \Big\} \times$$

$$\Big\{ I - z \, M_n(Q_{\mathscr{GG}}, Q_{\mathscr{GB}} \{I - Q_{\mathscr{BB}}\}^{-1} Q_{\mathscr{BG}}) \Big\}^{-1} \times$$

$$\Big\{ (Q_{\mathscr{GB}} \{I - Q_{\mathscr{BB}}\}^{-1} Q_{\mathscr{B}\{\omega\}}) \otimes 1_n \Big\} \, I_{\{n \geq 1\}} \, . \tag{6.24}$$

To see that also the expressions in (6.20) (and thus the second term on the right hand side of (6.16)) can be handled by Lemma 6.2, we note that the terms h_1 and h_2 in (6.21) - (6.22) can be written as follows

$$h_1 = \Big\{ I - Q_{\mathscr{BB}} \Big\}^{-1} Q_{\mathscr{B}\{\omega\}} I_{\{n \geq 1\}} +$$

$$\sum_{\ell=0}^{n-2} \Big\{ I - Q_{\mathscr{BB}} \Big\}^{-1} Q_{\mathscr{BG}} \Big(z \, \{I - z \, Q_{\mathscr{GG}}\}^{-1} Q_{\mathscr{GB}} \{I - Q_{\mathscr{BB}}\}^{-1} Q_{\mathscr{BG}} \Big)^{\ell} \times$$

$$z \, \Big\{ I - z \, Q_{\mathscr{GG}} \Big\}^{-1} Q_{\mathscr{GB}} \{I - Q_{\mathscr{BB}}\}^{-1} Q_{\mathscr{B}\{\omega\}},$$

$$h_2 = \sum_{\ell=0}^{n-1} \Big\{ I - Q_{\mathscr{BB}} \Big\}^{-1} Q_{\mathscr{BG}} \Big(z \, \{I - z \, Q_{\mathscr{GG}}\}^{-1} Q_{\mathscr{GB}} \{I - Q_{\mathscr{BB}}\}^{-1} Q_{\mathscr{BG}} \Big)^{\ell} \times$$

$$z \, \Big\{ I - z \, Q_{\mathscr{GG}} \Big\}^{-1} Q_{\mathscr{G}\{\omega\}}.$$

Thus, by Lemma 6.2,

$$\{\alpha_B^T \otimes (1, 0_{1 \times k})\} \, \psi_{\mathscr{B};n+1}(z, ..., z \, ; 1, ..., 1, 0) = \{\alpha_B^T \otimes (1, 0_{1 \times k})\} \, (h_1 + h_2) =$$

$$\{\alpha_B^T \otimes (1, 0_{1 \times k})\} \, \{I - Q_{\mathscr{BB}}\}^{-1} Q_{\mathscr{B}\{\omega\}} I_{\{n \geq 1\}} +$$

$$z \, \Big\{ \{\{\alpha_B^T \otimes (1, 0_{1 \times k})\} \{I - Q_{\mathscr{BB}}\}^{-1} Q_{\mathscr{BG}}\} \otimes (1, 0_{1 \times (n-2)}) \Big\} \times$$

$$\left\{ \mathbf{I} - z\, \mathbf{M}_{n-1}\!\left(\mathbf{Q}_{\mathcal{GG}}, \mathbf{Q}_{\mathcal{GB}} \left\{ \mathbf{I} - \mathbf{Q}_{\mathcal{BB}} \right\}^{-1} \mathbf{Q}_{\mathcal{BG}} \right) \right\}^{-1} \times$$

$$\left\{ \left\{ \mathbf{Q}_{\mathcal{GB}} \left\{ \mathbf{I} - \mathbf{Q}_{\mathcal{BB}} \right\}^{-1} \mathbf{Q}_{\mathcal{B}\{\omega\}} \right\} \otimes \mathbf{1}_{n-1} \right\} \mathrm{I}_{\{\,n\geq 2\,\}} +$$

$$z \left\{ \left\{ \left\{ \boldsymbol{\alpha}_{\mathrm{B}}^{\mathrm{T}} \otimes (1, \mathbf{0}_{1\times k}) \right\} \left\{ \mathbf{I} - \mathbf{Q}_{\mathcal{BB}} \right\}^{-1} \mathbf{Q}_{\mathcal{BG}} \right\} \otimes (1, \mathbf{0}_{1\times(n-1)}) \right\} \times$$

$$\left\{ \mathbf{I} - z\, \mathbf{M}_{n}\!\left(\mathbf{Q}_{\mathcal{GG}}, \mathbf{Q}_{\mathcal{GB}} \left\{ \mathbf{I} - \mathbf{Q}_{\mathcal{BB}} \right\}^{-1} \mathbf{Q}_{\mathcal{BG}} \right) \right\}^{-1} \left\{ \mathbf{Q}_{\mathcal{G}\{\omega\}} \otimes \mathbf{1}_{n} \right\} \mathrm{I}_{\{\,n\geq 1\,\}}. \qquad (6.25)$$

The *inversion* of the generating function of $\{\ \Delta(k, i, n) : i \geq 0\ \}$ is now straigthforvard: it is easily deduced from (6.16) by power series expansions of (6.24) and (6.25) about the origin. This result is stated in the next theorem.

THEOREM 6.3. $\Delta(k, i, n)$ *is given by*

$$\Delta(k, i, n) =$$

$$\left\{ \boldsymbol{\alpha}_{\mathrm{B}}^{\mathrm{T}} \otimes (1, \mathbf{0}_{1\times k}) \right\} \left\{ \mathbf{I} - \mathbf{Q}_{\mathcal{BB}} \right\}^{-1} \mathbf{Q}_{\mathcal{B}\{\omega\}} \mathrm{I}_{\{\,n\geq 1\,\}} \mathrm{I}_{\{\,i=0\,\}} +$$

$$\left\{ \left\{ \boldsymbol{\alpha}_{\mathrm{G}}^{\mathrm{T}} \otimes (1, \mathbf{0}_{1\times k}) \right\} \otimes (1, \mathbf{0}_{1\times n}) \right\} \times$$

$$\left\{ \mathbf{M}_{n+1}\!\left(\mathbf{Q}_{\mathcal{GG}}, \mathbf{Q}_{\mathcal{GB}} \left\{ \mathbf{I} - \mathbf{Q}_{\mathcal{BB}} \right\}^{-1} \mathbf{Q}_{\mathcal{BG}} \right) \right\}^{(i-1)} \left\{ \mathbf{Q}_{\mathcal{G}\{\omega\}} \otimes \mathbf{1}_{n+1} \right\} \mathrm{I}_{\{\,i\geq 1\,\}} +$$

$$\left\{ \left\{ \boldsymbol{\alpha}_{\mathrm{G}}^{\mathrm{T}} \otimes (1, \mathbf{0}_{1\times k}) \right\} \otimes (1, \mathbf{0}_{1\times(n-1)}) \right\} \left\{ \mathbf{M}_{n}\!\left(\mathbf{Q}_{\mathcal{GG}}, \mathbf{Q}_{\mathcal{GB}} \left\{ \mathbf{I} - \mathbf{Q}_{\mathcal{BB}} \right\}^{-1} \mathbf{Q}_{\mathcal{BG}} \right) \right\}^{(i-1)} \times$$

$$\left\{ (\mathbf{Q}_{\mathcal{GB}} \left\{ \mathbf{I} - \mathbf{Q}_{\mathcal{BB}} \right\}^{-1} \mathbf{Q}_{\mathcal{B}\{\omega\}}) \otimes \mathbf{1}_{n} \right\} \mathrm{I}_{\{\,n\geq 1\,\}} \mathrm{I}_{\{\,i\geq 1\,\}} +$$

$$\left\{ \left\{ \left\{ \boldsymbol{\alpha}_{\mathrm{B}}^{\mathrm{T}} \otimes (1, \mathbf{0}_{1\times k}) \right\} \left\{ \mathbf{I} - \mathbf{Q}_{\mathcal{BB}} \right\}^{-1} \mathbf{Q}_{\mathcal{BG}} \right\} \otimes (1, \mathbf{0}_{1\times(n-2)}) \right\} \times$$

$$\left\{ \mathbf{M}_{n-1}\!\left(\mathbf{Q}_{\mathcal{GG}}, \mathbf{Q}_{\mathcal{GB}} \left\{ \mathbf{I} - \mathbf{Q}_{\mathcal{BB}} \right\}^{-1} \mathbf{Q}_{\mathcal{BG}} \right) \right\}^{(i-1)} \times$$

$$\left\{ \left\{ \mathbf{Q}_{\mathcal{GB}} \left\{ \mathbf{I} - \mathbf{Q}_{\mathcal{BB}} \right\}^{-1} \mathbf{Q}_{\mathcal{B}\{\omega\}} \right\} \otimes \mathbf{1}_{n-1} \right\} \mathrm{I}_{\{\,n\geq 2\,\}} \mathrm{I}_{\{\,i\geq 1\,\}} +$$

$$\left\{ \left\{ \left\{ \boldsymbol{\alpha}_{\mathrm{B}}^{\mathrm{T}} \otimes (1, \mathbf{0}_{1\times k}) \right\} \left\{ \mathbf{I} - \mathbf{Q}_{\mathcal{BB}} \right\}^{-1} \mathbf{Q}_{\mathcal{BG}} \right\} \otimes (1, \mathbf{0}_{1\times(n-1)}) \right\} \times$$

$$\left\{ \mathbf{M}_{n}\!\left(\mathbf{Q}_{\mathcal{GG}}, \mathbf{Q}_{\mathcal{GB}} \left\{ \mathbf{I} - \mathbf{Q}_{\mathcal{BB}} \right\}^{-1} \mathbf{Q}_{\mathcal{BG}} \right) \right\}^{(i-1)} \left\{ \mathbf{Q}_{\mathcal{G}\{\omega\}} \otimes \mathbf{1}_{n} \right\} \mathrm{I}_{\{\,n\geq 1\,\}} \mathrm{I}_{\{\,i\geq 1\,\}}. \quad (6.26)$$

$\blacksquare$

6.3 Application and computational experience

In this section, the scope of the practical applicability of Theorem 6.3 will be examined by the example of Model 5 from Section 1.2.1. We start with considerations pertaining to the numerical implementation of (6.5) and (6.26).

6.3.1 Computational implementation

The computation of $\Pr\{\ T_G(t_0) > t,\ M_B(t_0) \le n\ \}$ will be based on (6.5) and (6.26), the former of which is an infinite sum. Truncating it after the $k_0 th$ summand will result in an errror the modulus of which is bounded by

$$\sum_{k=k_0+1}^{\infty} \sum_{i=1}^{k+1} \frac{t_0^k \mu^k}{k!} \exp\{\ -t_0\ \mu\ \}\ \Delta(k, i, n) \le$$

$$\exp\{\ -t_0\ \mu\ \} \left\{\ \exp(\ t_0\ \mu\ ) - \sum_{k=0}^{k_0} \frac{t_0^k \mu^k}{k!}\ \right\} = \varepsilon, \tag{6.27}$$

say. Obviously, k_0 is to be selected such that ε in (6.27) does not exceed the desired error bound. k_0 in turn determines the size of the matrices involved in the evaluation of (6.26): the square matrix $\mathbf{M}_{n+1}$ in (6.26) is of size $(k+1)(n+1)|G|$, whereas the matrix $\mathbf{M}_n$ in the last term on the right-hand side of (6.26) is of size $(k+1)n|B|$. The size of the largest square matrix involved is thus

$$(k_0+1) \times \begin{cases} \max\{\ (n+1)\ |G|, n\ |B|\ \} & \text{if } \boldsymbol{\alpha}_G^T\ \mathbf{1} \ne 0 \text{ and } \boldsymbol{\alpha}_B^T\ \mathbf{1} \ne 0, \\ (n+1)\ |G| & \text{if } \boldsymbol{\alpha}_G^T\ \mathbf{1} \ne 0 \text{ and } \boldsymbol{\alpha}_B^T\ \mathbf{1} = 0, \\ n\ |B| & \text{if } \boldsymbol{\alpha}_G^T\ \mathbf{1} = 0 \text{ and } \boldsymbol{\alpha}_B^T\ \mathbf{1} \ne 0. \end{cases} \tag{6.28}$$

For any system proposed for the implementation of a given model, the number in (6.28) will determine whether or not this is feasible. The present application was implemented in the *full version* of MATLAB on the Apple Macintosh SE/30. Matrices of any size are possible with the full version of MATLAB; there is a limitation, however, due to the size of the computer's RAM (4 Mb in our case).

6.3.2 Application: two parallel units with a single repairman

The system comprises two parallel units with identical failure characteristics and a repairman according to Model 5 in Section 1.2.1. As in [GRO] pp 279-281, we assume a scaling which makes the mean time to failure of a single component unity, and thus $\lambda = 1$. Furthermore, we assume that $\mu = 4.0$, $n = 2$, $t_0 = 1.6$, and $t \in \{\ 0.5t_0, 0.75t_0, 0.9t_0\ \}$. The results obtained from a MATLAB implementation of (6.5) on the Macintosh SE/30 are shown in Table 6.1. Table 6.2 shows estimates of $\Pr\{\ T_G(t_0) > t,\ M_B(t_0) \le n\ \}$ obtained by simulation, also implemented in MATLAB. It is seen from Table 6.1 that increasing the cut-off point k_0 in the series representation (6.5) does not increase the achieved (relative) accuracy beyond a certain value. Furthermore, the CPU-time increases rapidly with k_0. The reason for not having been able to achieve the desired accuracy (i.e., the accuracy claimed by the error bound ε in (6.27)) by increasing k_0, is in the computational implementation rather than in the analytical results: (6.26) involves powers of large sparce matrices in course of the computation of which accuracy will have been lost. Reibman and Trivedi's paper [REI1] may serve as a starting point for further work on this topic.

Truncation point k_0	CPU-time [sec]	t	Pr{ $T_G(t_0) > t, M_B(t_0) \le n$ }	Error bound ε in (6.27)	Actual relative error
5	43	0.80	0.18733	8.0876×10^{-1}	80.9 %
		1.20	0.17742		81.3 %
		1.44	0.16601		81.2 %
10	531	0.80	0.80055	1.8411×10^{-1}	18.5 %
		1.20	0.74985		21.0 %
		1.44	0.66266		25.1 %
15	3,702	0.80	0.97048	8.2310×10^{-3}	1.2 %
		1.20	0.91071		4.1 %
		1.44	0.79507		10.1 %
20	16,892	0.80	0.97801	9.3968×10^{-5}	0.4 %
		1.20	0.91799		3.3 %
		1.44	0.80096		9.5 %
25	57,577	0.80	0.97809	3.5515×10^{-7}	0.4 %
		1.20	0.91807		3.3 %
		1.44	0.80103		9.5 %

Table 6.1 Results by (6.5) for n = 2, $t_0 = 1.6$, $\lambda = 1.0$, and $\mu = 4.0$

Sample size	CPU-time [sec]	t	Pr{ $T_G(t_0) > t, M_B(t_0) \le n$ }	Error bound with 0.95 confidence
10^3	40	0.80	0.98300	8.0123×10^{-3}
		1.20	0.93900	1.4834×10^{-2}
		1.44	0.88200	1.9995×10^{-2}
10^4	395	0.80	0.98160	2.6341×10^{-3}
		1.20	0.94840	4.3359×10^{-3}
		1.44	0.88450	6.2646×10^{-3}
10^5	3,935	0.80	0.98204	8.2314×10^{-4}
		1.20	0.94928	1.3600×10^{-3}
		1.44	0.88488	1.9782×10^{-3}

Table 6.2 Results by simulation for n = 2, $t_0 = 1.6$, $\lambda = 1.0$, and $\mu = 4.0$

Another inconvenience is the excessive computing time associated with the method. Thus, further work is needed to render the present result more applicable. One possible direction for this could be along the lines of De Souza e Silva and Gail [DES1] by trying to devise a recursive scheme for the computation of the quantities $\Delta(k, i, n)$.

Notwithstanding these important practical issues (they will have to be dealt with in future research), this chapter is primarily seen as an analytical contribution showing that the randomization technique can be employed to obtain a closed form expression for the compound dependability measure under consideration.

6.3.3 Implementation in MATLAB

The MATLAB implementations for both the analytical method and its validation (by simulation) are shown in Section 6.3.4. Various new MATLAB features are apparent. First of all, this is our first encounter with MATLAB *functions*, as opposed to *scripts*; the former being independent program units devised to evaluate an output value for a given input, roughly analogous to FORTRAN subroutines. Functions can return a scalar, like binom in (c), but also an array, like prob in (b). The variables used within a function are local. The function fact in (g) demonstrates that recursion is possible in MATLAB.

(It should be added, however, that, in my experience, any 'real' intended use of this feature has resulted in computing times which were unacceptable, or even in outright program failure.) The simulation program makes use of the system-provided random number generator rand.

6.3.4 MATLAB code

Main routine (a) and function files (b) - (h)

(a) Main routine

```
clear;
% Supply and show data...

lamt = input('Enter component failure rate in 1/year... ');
mut  = input('Enter component repair rate in 1/year ... ');

disp(fprintf('lamt = %12.5e\n',lamt));
disp(fprintf('mut = %12.5e\n',mut));

% Up states {0,1}, down state {2}.

L = [-2*lamt    2*lamt          0
        mut     -(mut+lamt)     lamt
         0       mut           -mut ];

E = ones(3) - eye(3);

disp('The rate parameter of the Poisson process W is ... ');
mu=max([L(1,:)*E(:,1) L(2,:)*E(:,2) L(3,:)*E(:,3)])
disp(...
'The transition probability matrix of the subordinated chain Z is ...');
P = (1/mu)*L + eye(3)

PGG = P(1:2,1:2);
PGB = P(1:2,3:3);
```

```
PBG = P(3:3,1:2);
PBB = P(3:3,3:3);

t0 = input('Enter t0 > 0.0 ...');
disp(fprintf('t0 = %12.5f\n',t0));
t1 = input('Enter t1 < t0 ...');
disp(fprintf('t1 = %12.5f\n',t1));
t2 = input('Enter t2 < t0 ...');
disp(fprintf('t2 = %12.5f\n',t2));
t3 = input('Enter t3 < t0 ...');
disp(fprintf('t3 = %12.5f\n',t3));
n = input('Enter n ≥ 0 ...');
disp(fprintf('n = %3.0f\n',n));

trpoint = ...
input('Truncation point k0 will be 5, 10, 15, ..., m*5; enter m... ');

for ii=1:trpoint
  k0 = ii*5;
  fix(clock)
  disp(fprintf('Truncation point k0 is %3.0f\n',k0));
  disp('Calculation started ...');
  fix(clock)
  [p1 p2 p3] = prob(PGG,PGB,PBG,PBB,mu,t0,t1,t2,t3,n,k0);
  eps = fehler(t0*mu,k0);
  disp('Calculation finished.');
  fix(clock)
  disp(fprintf('Pr[ TG(t0) > t1, MB(t0) ≤ n ] = %12.5e\n',p1));
  disp(fprintf('Pr[ TG(t0) > t2, MB(t0) ≤ n ] = %12.5e\n',p2));
  disp(fprintf('Pr[ TG(t0) > t3, MB(t0) ≤ n ] = %12.5e\n',p3));
  disp(fprintf('The absolute error is ≤ %12.5e\n',eps));
end
```

(b) Function `prob`

```
function [p1,p2,p3] = prob(PGG,PGB,PBG,PBB,mu,t0,t1,t2,t3,n,k0)
% Calculates the truncated version of (6.5) for three different
% values of t; namely t = t1, t2, t3

t = [t1 t2 t3];
dimt = max(size(t));
p = ones(1,dimt);

for k=1:k0
  temp1 = zeros(1,dimt);
  for i=1:(k+1)
    temp2 = zeros(1,dimt);
    for j=0:(i-1)
      bincoef = binom(k,j);
      for el=1:dimt
        temp2(1,el) = temp2(1,el) + ...
                   bincoef*((t(1,el)/t0)^j)*((1.0-(t(1,el)/t0))^(k-j));
      end
    end
    del = delta([1 0],PGG,PGB,PBG,PBB,k,i,n);
    temp1 = temp1 + del*temp2;
  end
  tempcoef = (((t0*mu)^k)/fact(k));
  p = p + tempcoef*temp1;
  p1 = p(1,1);
  p2 = p(1,2);
  p3 = p(1,3);
end
p1 = exp(- t0*mu)*p1;
```

```
p2 = exp(- t0*mu)*p2;
p3 = exp(- t0*mu)*p3;
```

(c) Function binom

```
function b=binom(n,k)
% Binomial coefficient
if (k>n)|(n<0)
    b = 0;
else
    b = fact(n)/(fact(n-k)*fact(k));
end
```

(d) Function delta

```
function d=delta(alphaG,PGG,PGB,PBG,PBB,k,i,n)
% alphaG is the initial probability (row-)vector of size 2
% PGG, PGB, PBG, PBB are the blocks of the transition prob. matrix P
% k is an integer ≥ 1
% i is an integer ≥ 0
% n is an integer ≥ 0
QGG = kron(E(k+1),PGG);
QBG = kron(E(k+1),PBG);
QGB = kron(E(k+1),PGB);
QBB = kron(E(k+1),PBB);

QGOM = ones(max(size(QGG)),1) - QGG*ones(max(size(QGG)),1) - ...
                               QGB*ones(max(size(QBB)),1);
QBOM = ones(max(size(QBB)),1) - QBB*ones(max(size(QBB)),1) - ...
                               QBG*ones(max(size(QGG)),1);

if n*i>0
    TEMP1 = (M(n,QGG,QGB*inv(eye(QBB)-QBB)*QBG))^(i-1);
    d1 = kron([1 zeros(1,n-1)],kron([1 zeros(1,k)],alphaG))* ...
        TEMP1*kron(ones(n,1),QGB*inv(eye(QBB)-QBB)*QBOM);
else
    d1 = 0;
end

if i>0
    TEMP2 = (M(n+1,QGG,QGB*inv(eye(QBB)-QBB)*QBG))^(i-1);
    d2= kron([1 zeros(1,n)],kron([1 zeros(1,k)],alphaG))*...
        TEMP2*kron(ones(n+1,1),QGOM);
else
    d2 = 0;
end

d = d1 + d2;
```

(e) Function E

```
function Ei=E(i)
Ei = zeros(i);
for j=1:i
    for el=1:i
        if el-j==1
            Ei(j,el) = 1;
        end
    end
```

```
end
```

(f) Function M

```
function Mi=M(i,U,V)
Mi = kron(eye(i),U) + kron(E(i),V);
```

(g) Function fact

```
function y=fact(x)
% factorial
[m,n]=size(x);
if (m+n)==2
   if x==0
      y=1;
   elseif x==1
      y=1;
   else
      y=x*fact(x-1);
   end
else
   disp('error');
end
```

(h) Function fehler

```
function f=fehler(x,k)
% Calculates exp(-x)[exp(x) - (1 + x/1! + x^2/2! + ... + x^k/k!)]
% x ≥ 0
% k = 1, 2, ...

temp = 1.0;
for i=1:k
    temp = temp + (x^i)/fact(i);
end

f = exp(-x)*(exp(x) - temp);
```

Simulation routine

```
clear;
% Supply and show data...

lamt = input('Enter component failure rate in 1/year... ');
mut  = input('Enter component repair rate in 1/year ...');

disp(fprintf('lamt = %12.5e\n',lamt));
disp(fprintf('mut = %12.5e\n',mut));

% Up states {0,1}, down state {2}.

L = [-2*lamt   2*lamt         0
       mut    -(mut+lamt)    lamt
        0        mut        -mut ];

t0 = input('Enter t0 > 0.0 ...');
```

```
disp(fprintf('t0 = %12.5f\n',t0));
t50 = t0*0.5;
t75 = t0*0.75;
t90 = t0*0.9;
n = input('Enter n ≥ 0 ...');
disp(fprintf('n = %3.0f\n',n));
init = input('Enter seed for random number generator... ');
magn = input('Sample sizes will be 10, 100, ..., 10**m; enter m');

for ii=1:magn
   ssize = 10^ii;
   fix(clock)
   disp(fprintf('Sample size is %12.0f\n',ssize));
   rand('seed',init);
   counter50 = 0;               % Number of simulation runs such that
                                % n>=downvisits and totaluptime>0.50*t0
   counter75 = 0;               % Number of simulation runs such that
                                % n>=downvisits and totaluptime>0.75*t0
   counter90 = 0;               % Number of simulation runs such that
                                % n>=downvisits and totaluptime>0.90*t0

   for i=1:ssize
      totaltime = 0.0;          % Total elapsed time of simulated history
      totaluptime = 0.0;        % Total time spent in G during [0,t0]
      state = 0;                % Initial state
      downvisits = 0;           % Holds the number visits to B during [0,t0]

      while 1
         if state==0
            rate = 2*lamt;
            holdingtime = - log(1.0 - rand)/rate;
            totaltime = totaltime + holdingtime;
            totaluptime = totaluptime + holdingtime;
            state = 1;
         end
         if totaltime>t0; break; end;

         if state==1
            rate = mut + lamt;
            holdingtime = - log(1.0 - rand)/rate;
            totaltime = totaltime + holdingtime;
            totaluptime = totaluptime + holdingtime;
            if rand<(mut/(mut+lamt))
               state = 0;
            else
               state = 2;
            end
         end
         if totaltime>t0; break; end;

         if state==2
            downvisits = downvisits + 1;
            rate = mut;
            holdingtime = - log(1.0 - rand)/rate;
            totaltime = totaltime + holdingtime;
            state = 1;
         end
         if totaltime>t0; break; end;
      end

      if ((n>=downvisits)&(totaluptime>t50))
         counter50 = counter50 + 1;
      end
      if ((n>=downvisits)&(totaluptime>t75))
         counter75 = counter75 + 1;
      end
```

```
          if ((n>=downvisits)&(totaluptime>t90))
              counter90 = counter90 + 1;
          end

      end

      prob50 = counter50/ssize;
      prob75 = counter75/ssize;
      prob90 = counter90/ssize;

      disp(...
      fprintf('Pr[ TG(t0) > %12.5f, MB(t0) ≤ n ] = %12.5e\n',t50,prob50));
      disp(...
      fprintf('Pr[ TG(t0) > %12.5f, MB(t0) ≤ n ] = %12.5e\n',t75,prob75));
      disp(...
      fprintf('Pr[ TG(t0) > %12.5f, MB(t0) ≤ n ] = %12.5e\n',t90,prob90));
end
fix(clock)
```

CHAPTER 7

A COMPOUND MEASURE OF DEPENDABILITY FOR CONTINUOUS-TIME ABSORBING MARKOV SYSTEMS

In this chapter, we deal with *absorbing* Markov reliability models whose state space S is partitioned as $S = G \cup B \cup \{\omega\}$, where G and B are transient sets and ω is an absorbing state. Let $Y = \{ Y_t : t \geq 0 \}$ be such a Markov process and let $T_G = TS_{G,\infty}$ and N_B stand respectively for the total time spent by Y in G, and the number of visits by Y to B, *until absorption*. We know T_G from Chapter 4; it can be interpreted as the total amount of work delivered by the system until system breakdown if G is the set of 'up' states and B is the set of *repairable* system 'down' states. Then, N_B is the number of repair periods until failure. With this interpretation then, it seems reasonable to specify a required *simultaneous* lower, and upper bound for T_G and N_B respectively. (Notice that N_B is an appropriate *cost measure* if every repair instance carries the same cost.) Individually, we know the distributions of N_B and T_G from Corollary 2.6 and Theorem 4.12 respectively. For the required probability, $\Pr\{ T_G > t, N_B \leq n \}$, however, the *joint* distribution of T_G and N_B needs examining. This will be accomplished in Section 7.1. The result is a closed-form expression in terms of exponentials of certain submatrices of the transition rate matrix of Y. Section 7.2 contains the proofs. In Section 7.3, we report on the MATLAB implementation of our dependability measure for Model 3 from Section 1.2.1. This chapter is based on material first reported in [CSE9].

7.1 The dependability measure

We start with some remarks on the notation. For any column-vector $\mathbf{u}$ and the compatible square matrices $\mathbf{\Phi}$ and $\mathbf{\Psi}$ for which $\mathbf{\Phi}^{-1}$ and $(\mathbf{\Phi} + \mathbf{\Psi})^{-1}$ exist, let the column-vector $v_n(\mathbf{\Phi}, \mathbf{\Psi}, \mathbf{u})$ be defined by

$$
v_n(\mathbf{\Phi}, \mathbf{\Psi}, \mathbf{u}) = \begin{bmatrix} (\mathbf{\Phi} + \mathbf{\Psi})^{-1} \{ \mathbf{I} - (-\mathbf{\Psi} \, \mathbf{\Phi}^{-1})^{n} \} \, \mathbf{u} \\ (\mathbf{\Phi} + \mathbf{\Psi})^{-1} \{ \mathbf{I} - (-\mathbf{\Psi} \, \mathbf{\Phi}^{-1})^{n-1} \} \, \mathbf{u} \\ \vdots \\ (\mathbf{\Phi} + \mathbf{\Psi})^{-1} \{ \mathbf{I} - (-\mathbf{\Psi} \, \mathbf{\Phi}^{-1})^{2} \} \, \mathbf{u} \\ (\mathbf{\Phi} + \mathbf{\Psi})^{-1} \{ \mathbf{I} - (-\mathbf{\Psi} \, \mathbf{\Phi}^{-1})^{1} \} \, \mathbf{u} \end{bmatrix}.
\tag{7.1}
$$

We shall use the matrix $\mathbf{M}_n$, $n \geq 1$, which is defined by

$$
\mathbf{M}_n(\mathbf{\Phi}, \mathbf{\Psi}) = \mathbf{\Delta}_n(\mathbf{\Phi}, ..., \mathbf{\Phi}; \mathbf{\Psi}, ..., \mathbf{\Psi}).
\tag{7.2}
$$

(Notice that $\mathbf{M}_n$ is already familiar from Theorem 4.9 and Lemma 6.2.) As usual, the transition rate matrix of Y is $\mathbf{\Lambda}$, and its initial probability vector is $\boldsymbol{\alpha} = (\boldsymbol{\alpha}_G^T, \boldsymbol{\alpha}_B^T, 0)^T$.

We are now in a position to state our main result which is a closed-form expression in terms of matrix exponentials for $\Pr\{ T_G > t, N_B \le n \}$. Its proof is given in Section 7.2.

THEOREM 7.1. *Under the above assumptions, we have for $t \ge 0$ and $n \in \{0, 1, \dots\}$ the following*

$$\Pr\{ T_G > t, N_B \le n \} =$$

$$+ (\boldsymbol{\alpha}_G^T, 0, \dots, 0) \exp\{ t \, \mathbf{M}_{n+1}(\mathbf{\Lambda}_{GG}, - \mathbf{\Lambda}_{GB} \mathbf{\Lambda}_{BB}^{-1} \mathbf{\Lambda}_{BG}) \} \times$$

$$v_{n+1}(\mathbf{\Lambda}_{GG}, - \mathbf{\Lambda}_{GB} \mathbf{\Lambda}_{BB}^{-1} \mathbf{\Lambda}_{BG}, \mathbf{\Lambda}_{GG} \mathbf{1} + \mathbf{\Lambda}_{GB} \mathbf{1}) -$$

$$- (\boldsymbol{\alpha}_G^T, 0, \dots, 0) \exp\{ t \, \mathbf{M}_n(\mathbf{\Lambda}_{GG}, - \mathbf{\Lambda}_{GB} \mathbf{\Lambda}_{BB}^{-1} \mathbf{\Lambda}_{BG}) \} \times$$

$$v_n(\mathbf{\Lambda}_{GG}, - \mathbf{\Lambda}_{GB} \mathbf{\Lambda}_{BB}^{-1} \mathbf{\Lambda}_{BG}, \mathbf{\Lambda}_{GB} \mathbf{\Lambda}_{BB}^{-1} (\mathbf{\Lambda}_{BG} \mathbf{1} + \mathbf{\Lambda}_{BB} \mathbf{1})) \, I_{\{ n \ge 1 \}} +$$

$$+ (\boldsymbol{\alpha}_B^T \mathbf{\Lambda}_{BB}^{-1} \mathbf{\Lambda}_{BG}, 0, \dots, 0) \exp\{ t \, \mathbf{M}_{n-1}(\mathbf{\Lambda}_{GG}, - \mathbf{\Lambda}_{GB} \mathbf{\Lambda}_{BB}^{-1} \mathbf{\Lambda}_{BG}) \} \times$$

$$v_{n-1}(\mathbf{\Lambda}_{GG}, - \mathbf{\Lambda}_{GB} \mathbf{\Lambda}_{BB}^{-1} \mathbf{\Lambda}_{BG}, \mathbf{\Lambda}_{GB} \mathbf{\Lambda}_{BB}^{-1} (\mathbf{\Lambda}_{BG} \mathbf{1} + \mathbf{\Lambda}_{BB} \mathbf{1})) \, I_{\{ n \ge 2 \}} -$$

$$- (\boldsymbol{\alpha}_B^T \mathbf{\Lambda}_{BB}^{-1} \mathbf{\Lambda}_{BG}, 0, \dots, 0) \exp\{ t \, \mathbf{M}_n(\mathbf{\Lambda}_{GG}, - \mathbf{\Lambda}_{GB} \mathbf{\Lambda}_{BB}^{-1} \mathbf{\Lambda}_{BG}) \} \times$$

$$v_n(\mathbf{\Lambda}_{GG}, - \mathbf{\Lambda}_{GB} \mathbf{\Lambda}_{BB}^{-1} \mathbf{\Lambda}_{BG}, \mathbf{\Lambda}_{GG} \mathbf{1} + \mathbf{\Lambda}_{GB} \mathbf{1}) \, I_{\{ n \ge 1 \}}. \tag{7.3}$$

NOTES.

(I)   (7.3) can be simplified by noting that for any $S' \subset S$,

$$\mathbf{\Lambda}_{S'G} \mathbf{1} + \mathbf{\Lambda}_{S'B} \mathbf{1} = - \mathbf{\Lambda}_{S'\{\omega\}} \mathbf{1}. \tag{7.4}$$

In applications, however, the process Y frequently arises from some non-absorbing Markov chain Z, say, by converting some of Z's states into absorbing ones and subsequently lumping them into ω. It is easily seen that then the $\{G, B\}$-blocks of the transition rate matrix of Y, i.e., $\mathbf{\Lambda}_{GG}, \mathbf{\Lambda}_{GB}, \mathbf{\Lambda}_{BG}$, and $\mathbf{\Lambda}_{BB}$, are identical to the corresponding submatrices of the transition rate matrix of Z, whereas this does *not* hold for $\mathbf{\Lambda}_{S'\{\omega\}}$. For this reason, it is expedient *not* to use (7.4) in (7.3).

(II)   The distribution function of N_B is easily deduced from (7.3) if Y is started in one of the states in G (i.e., $\boldsymbol{\alpha}_B = \mathbf{0}$) since in this case $\Pr\{ T_G > 0 \} = 1$. Then, it is

$$\Pr\{ N_B = 0 \} = \boldsymbol{\alpha}_G^T \mathbf{\Lambda}_{GG}^{-1} (\mathbf{\Lambda}_{GG} \mathbf{1} + \mathbf{\Lambda}_{GB} \mathbf{1}) = 1 + \boldsymbol{\alpha}_G^T \mathbf{\Lambda}_{GG}^{-1} \mathbf{\Lambda}_{GB} \mathbf{1}, \tag{7.5}$$

whereas for $n \ge 1$,

$$\Pr\{\, N_B \le n \,\} = 1 + \boldsymbol{\alpha_G}^T \boldsymbol{\Lambda_{GG}}^{-1} \boldsymbol{\Lambda_{GB}} \left(\boldsymbol{\Lambda_{BB}}^{-1} \boldsymbol{\Lambda_{BG}} \boldsymbol{\Lambda_{GG}}^{-1} \boldsymbol{\Lambda_{GB}} \right)^n \mathbf{1}. \tag{7.6}$$

From (7.6) it is easily seen that for $n \ge 1$,

$$\Pr\{\, N_B = n \,\} = - \boldsymbol{\alpha_G}^T \boldsymbol{\Lambda_{GG}}^{-1} \boldsymbol{\Lambda_{GB}} \left(\boldsymbol{\Lambda_{BB}}^{-1} \boldsymbol{\Lambda_{BG}} \boldsymbol{\Lambda_{GG}}^{-1} \boldsymbol{\Lambda_{GB}} \right)^n \times$$
$$\left(\mathbf{I} - \boldsymbol{\Lambda_{BB}}^{-1} \boldsymbol{\Lambda_{BG}} \boldsymbol{\Lambda_{GG}}^{-1} \boldsymbol{\Lambda_{GB}} \right) \mathbf{1}. \tag{7.7}$$

(7.5) and (7.7) are also easily deduced from Corollary 2.6 bearing in mind that $N_B \equiv M_B$ for the embedded (and hence discrete-parameter) Markov chain Z of Y. (The required expression for $\boldsymbol{\Lambda_{BB}}^{-1} \boldsymbol{\Lambda_{BG}} \boldsymbol{\Lambda_{GG}}^{-1} \boldsymbol{\Lambda_{GB}}$ in terms of the transition probability matrix of Z is given by (4.54); by this, the former is seen to be $\mathbf{H}(B,G)$ from (2.30).) The distribution function of N_B for *any* initial probability vector $\boldsymbol{\alpha}$ may be deduced likewise from (7.22) (see, Section 7.2) by letting there $t \to +\infty$; this will not be pursued here, however.

(III) The corresponding quantity for T_G (namely an expression for its distribution function) cannot be derived from (7.3) by taking $n \to +\infty$ since the size of the matrices involved increases with n. $\Pr\{\, T_G \le t \,\}$ is known from Theorem 4.12.

7.2 Proof of Theorem 7.1

7.2.1 Proof outline

The proof heavily relies on the Laplace transform of the joint distribution of the sojourn times of Y in G and B as given in Corollary 4.8; the notation from Chapter 4 applies. The Markov process Y alternates between G and B until absorption into ω. The following equivalences are obvious

$$N_B \le n \iff T_{B,n+1} = 0, \tag{7.8}$$

$$T_G = t, \, N_B \le n \iff T_{G,1} + \ldots + T_{G,n+1} = t, \, T_{B,n+1} = 0. \tag{7.9}$$

Moreover, using (7.8) and (7.9), the following iterrelationship in the Laplace transform domain is seen to hold

$$\int_0^{+\infty} \Pr\{\, T_G \le t, N_B \le n \,\} \exp\{ - t\, \sigma \,\} \, dt =$$

$$\Pr\{\, N_B \le n \,\} \int_0^{+\infty} \Pr\{\, T_G \le t \mid N_B \le n \,\} \exp\{ - t\, \sigma \,\} \, dt =$$

$$\Pr\{\, N_B \le n \,\} \frac{1}{\sigma} E[\exp\{ - \sigma\, T_G \,\} \mid N_B \le n] = \frac{1}{\sigma} E[\exp\{ - \sigma\, T_G \,\} \, I_{\{N_B \le n\}}] =$$

$$\frac{1}{\sigma} E\big[\exp\{ -\sigma \sum_{k=1}^{n+1} T_{G,k} \} \, I_{\{T_{B,n+1}=0\}}\big] =$$

$$\frac{1}{\sigma} \lim_{\rho \to +\infty} E\big[\exp\{ -\sigma \sum_{k=1}^{n+1} T_{G,k} - \rho\, T_{B,n+1} \}\big] . \tag{7.10}$$

(The last step in (7.10) holds by $\lim_{\rho \to +\infty} \exp\{ -\rho\, T_{B,n+1} \} = I_{\{T_{B,n+1}=0\}}$.) In what follows, the right hand side of (7.10) will be expressed in terms of the transition rate matrix $\mathbf{\Lambda}$ of Y. $\Pr\{ T_G \leq t, N_B \leq n \}$ is then obtained by Laplace transform inversion.

We note, in passing, that the reasoning leading to (7.10) is a Laplace transform analogue of the one used to establish (6.15). Also, (7.9) is the continuous-time counterpart of (6.14).

7.2.2 An auxiliary result

Before entering the detailed proof of Theorem 7.1, we shall restate three earlier results in order to facilitate an easy point reference.

PROPOSITION 7.2. *For any subset* $S' \subset S$ *with* $\omega \notin S'$, *it is* $\exp\{ t\, \mathbf{\Lambda}_{S'S'} \} \to \mathbf{0}$ *element-wise, as* $t \to +\infty$, *and thus for all* $\sigma \in \mathbb{C}$ *with* $\mathrm{Re}(\sigma) > 0$,

$$(\sigma\, \mathbf{I} - \mathbf{\Lambda}_{S'S'})^{-1} = \int_0^{+\infty} \exp\{ -\sigma\, t \} \exp\{ t\, \mathbf{\Lambda}_{S'S'} \} \, dt . \tag{7.11}$$

More generally, (7.11) holds if $\mathbf{M}_n\big(\mathbf{\Lambda}_{S_1 S_1}, -\mathbf{\Lambda}_{S_1 S_2}\, \mathbf{\Lambda}_{S_2 S_2}^{-1}\, \mathbf{\Lambda}_{S_2 S_1}\big)$ *(with* $n \geq 1$*) is substituted for* $\mathbf{\Lambda}_{S'S'}$ *with any disjoint* $S_1, S_2 \subset S$ *where* $\omega \notin S_1 \cup S_2$. *Also, for* $\mathrm{Re}(\sigma) \to +\infty$,

$$\big\{ \sigma\, \mathbf{I} - \mathbf{M}_n\big(\mathbf{\Lambda}_{S_1 S_1}, -\mathbf{\Lambda}_{S_1 S_2}\, \mathbf{\Lambda}_{S_2 S_2}^{-1}\, \mathbf{\Lambda}_{S_2 S_1}\big) \big\}^{-1} \to \mathbf{0} \tag{7.12}$$

element-wise. Finally, for $k \to +\infty$,

$$\big(\mathbf{\Lambda}_{S_1 S_1}^{-1}\, \mathbf{\Lambda}_{S_1 S_2}\, \mathbf{\Lambda}_{S_2 S_2}^{-1}\, \mathbf{\Lambda}_{S_2 S_1}\big)^k \to \mathbf{0} \tag{7.13}$$

element-wise, from which it follows that $\big(\mathbf{I} - \mathbf{\Lambda}_{S_1 S_1}^{-1}\mathbf{\Lambda}_{S_1 S_2}\mathbf{\Lambda}_{S_2 S_2}^{-1}\mathbf{\Lambda}_{S_2 S_1}\big)$ *is invertible.*

PROOF OF PROPOSITION 7.2. For (7.11) and (7.12), see Lemma 4.1, (4.26), and (4.50). For (7.13), use (4.54) and (4.55). ∎

7.2.3 Proof details

Corollary 4.8 will be applied with $A_1 = B$, $A_2 = G$, $\sigma_1 = \ldots = \sigma_{n+1} = \sigma$, $\tau_1 = \ldots = \tau_n = 0$, $\tau_{n+1} = \rho$, and $m = n \geq 0$. For $n = 0$ we have

$$E\Big[\exp\{-\sigma \sum_{k=1}^{n+1} T_{G,k} - \rho\, T_{B,n+1}\}\Big] = (T_{G,1}, T_{B,1})^*(\sigma, \rho) =$$

$$-\boldsymbol{\alpha}_G^T (\boldsymbol{\Lambda}_{GG} - \sigma\, \mathbf{I})^{-1}\, \boldsymbol{\Lambda}_{GB} (\boldsymbol{\Lambda}_{BB} - \rho\, \mathbf{I})^{-1}\, \boldsymbol{\Lambda}_{BB}\, \mathbf{1} +$$

$$+\boldsymbol{\alpha}_G^T (\boldsymbol{\Lambda}_{GG} - \sigma\, \mathbf{I})^{-1} (\boldsymbol{\Lambda}_{GG}\, \mathbf{1} + \boldsymbol{\Lambda}_{GB}\, \mathbf{1}) -$$

$$-\boldsymbol{\alpha}_B^T (\boldsymbol{\Lambda}_{BB} - \rho\, \mathbf{I})^{-1}\, \boldsymbol{\Lambda}_{BG} (\boldsymbol{\Lambda}_{GG} - \sigma\, \mathbf{I})^{-1}\, \boldsymbol{\Lambda}_{GG}\, \mathbf{1} +$$

$$+\boldsymbol{\alpha}_B^T (\boldsymbol{\Lambda}_{BB} - \rho\, \mathbf{I})^{-1} (\boldsymbol{\Lambda}_{BB}\, \mathbf{1} + \boldsymbol{\Lambda}_{BG}\, \mathbf{1}), \tag{7.14}$$

whereas for $n \geq 1$

$$E\Big[\exp\{-\sigma \sum_{k=1}^{n+1} T_{G,k} - \rho\, T_{B,n+1}\}\Big] =$$

$$(T_{G,1}, ..., T_{G,n+1}, T_{B,1}, ..., T_{B,n+1})^*(\sigma, ..., \sigma;\, 0, ..., 0, \rho) =$$

$$-\boldsymbol{\alpha}_G^T (\boldsymbol{\Lambda}_{GG} - \sigma\, \mathbf{I})^{-1}\, \boldsymbol{\Lambda}_{GB}\, \boldsymbol{\Lambda}_{BB}^{-1} \Big\{ \boldsymbol{\Lambda}_{BG} (\boldsymbol{\Lambda}_{GG} - \sigma\, \mathbf{I})^{-1}\, \boldsymbol{\Lambda}_{GB}\, \boldsymbol{\Lambda}_{BB}^{-1} \Big\}^{n-1} \times$$
$$\boldsymbol{\Lambda}_{BG} (\boldsymbol{\Lambda}_{GG} - \sigma\, \mathbf{I})^{-1}\, \boldsymbol{\Lambda}_{GB} (\boldsymbol{\Lambda}_{BB} - \rho\, \mathbf{I})^{-1}\, \boldsymbol{\Lambda}_{BB}\, \mathbf{1} +$$

$$+ \sum_{k=1}^{n+1} \boldsymbol{\alpha}_G^T (\boldsymbol{\Lambda}_{GG} - \sigma\, \mathbf{I})^{-1} \times$$
$$\Big\{ \boldsymbol{\Lambda}_{GB}\, \boldsymbol{\Lambda}_{BB}^{-1}\, \boldsymbol{\Lambda}_{BG} (\boldsymbol{\Lambda}_{GG} - \sigma\, \mathbf{I})^{-1} \Big\}^{k-1} (\boldsymbol{\Lambda}_{GG}\, \mathbf{1} + \boldsymbol{\Lambda}_{GB}\, \mathbf{1}) -$$

$$- \sum_{k=1}^{n} \boldsymbol{\alpha}_G^T (\boldsymbol{\Lambda}_{GG} - \sigma\, \mathbf{I})^{-1}\, \boldsymbol{\Lambda}_{GB}\, \boldsymbol{\Lambda}_{BB}^{-1} \times$$
$$\Big\{ \boldsymbol{\Lambda}_{BG} (\boldsymbol{\Lambda}_{GG} - \sigma\, \mathbf{I})^{-1}\, \boldsymbol{\Lambda}_{GB}\, \boldsymbol{\Lambda}_{BB}^{-1} \Big\}^{k-1} (\boldsymbol{\Lambda}_{BG}\, \mathbf{1} + \boldsymbol{\Lambda}_{BB}\, \mathbf{1}) -$$

$$- \boldsymbol{\alpha}_B^T\, \boldsymbol{\Lambda}_{BB}^{-1}\, \boldsymbol{\Lambda}_{BG} (\boldsymbol{\Lambda}_{GG} - \sigma\, \mathbf{I})^{-1} \Big\{ \boldsymbol{\Lambda}_{GB}\, \boldsymbol{\Lambda}_{BB}^{-1}\, \boldsymbol{\Lambda}_{BG} (\boldsymbol{\Lambda}_{GG} - \sigma\, \mathbf{I})^{-1} \Big\}^{n-1} \times$$
$$\boldsymbol{\Lambda}_{GB} (\boldsymbol{\Lambda}_{BB} - \rho\, \mathbf{I})^{-1}\, \boldsymbol{\Lambda}_{BG} (\boldsymbol{\Lambda}_{GG} - \sigma\, \mathbf{I})^{-1}\, \boldsymbol{\Lambda}_{GG}\, \mathbf{1} +$$

$$+ \sum_{k=1}^{n} \boldsymbol{\alpha}_B^T\, \boldsymbol{\Lambda}_{BB}^{-1} \Big\{ \boldsymbol{\Lambda}_{BG} (\boldsymbol{\Lambda}_{GG} - \sigma\, \mathbf{I})^{-1}\, \boldsymbol{\Lambda}_{GB}\, \boldsymbol{\Lambda}_{BB}^{-1} \Big\}^{k-1} (\boldsymbol{\Lambda}_{BB}\, \mathbf{1} + \boldsymbol{\Lambda}_{BG}\, \mathbf{1}) +$$

$$+ \alpha_B^T \Lambda_{BB}^{-1} \left\{ \Lambda_{BG} (\Lambda_{GG} - \sigma I)^{-1} \Lambda_{GB} \Lambda_{BB}^{-1} \right\}^{n-1} \times$$
$$\Lambda_{BG} (\Lambda_{GG} - \sigma I)^{-1} \Lambda_{GB} (\Lambda_{BB} - \rho I)^{-1} (\Lambda_{BB} 1 + \Lambda_{BG} 1) -$$

$$- \sum_{k=1}^{n} \alpha_B^T \Lambda_{BB}^{-1} \Lambda_{BG} (\Lambda_{GG} - \sigma I)^{-1} \times$$
$$\left\{ \Lambda_{GB} \Lambda_{BB}^{-1} \Lambda_{BG} (\Lambda_{GG} - \sigma I)^{-1} \right\}^{k-1} (\Lambda_{GB} 1 + \Lambda_{GG} 1). \tag{7.15}$$

From (7.12) it follows that both (7.14) and (7.15) can be written for $\mathrm{Re}(\rho) \to +\infty$ as

$$E\left[\exp\{ -\sigma \sum_{k=1}^{n+1} T_{G,k} - \rho\, T_{B,n+1} \} \right] = h_n^*(\sigma) + o(1), \tag{7.16}$$

where $h_n^*(\sigma)$ is defined by

$$h_n^*(\sigma) =$$

$$\sum_{k=1}^{n+1} \alpha_G^T (\Lambda_{GG} - \sigma I)^{-1} \times$$
$$\left\{ \Lambda_{GB} \Lambda_{BB}^{-1} \Lambda_{BG} (\Lambda_{GG} - \sigma I)^{-1} \right\}^{k-1} (\Lambda_{GG} 1 + \Lambda_{GB} 1) -$$

$$- \sum_{k=1}^{n} \alpha_G^T (\Lambda_{GG} - \sigma I)^{-1} \Lambda_{GB} \Lambda_{BB}^{-1} \times$$
$$\left\{ \Lambda_{BG} (\Lambda_{GG} - \sigma I)^{-1} \Lambda_{GB} \Lambda_{BB}^{-1} \right\}^{k-1} (\Lambda_{BG} 1 + \Lambda_{BB} 1) +$$

$$+ \sum_{k=1}^{n} \alpha_B^T \Lambda_{BB}^{-1} \times$$
$$\left\{ \Lambda_{BG} (\Lambda_{GG} - \sigma I)^{-1} \Lambda_{GB} \Lambda_{BB}^{-1} \right\}^{k-1} (\Lambda_{BG} 1 + \Lambda_{BB} 1) -$$

$$- \sum_{k=1}^{n} \alpha_B^T \Lambda_{BB}^{-1} \Lambda_{BG} (\Lambda_{GG} - \sigma I)^{-1} \times$$
$$\left\{ \Lambda_{GB} \Lambda_{BB}^{-1} \Lambda_{BG} (\Lambda_{GG} - \sigma I)^{-1} \right\}^{k-1} (\Lambda_{GG} 1 + \Lambda_{GB} 1). \tag{7.17}$$

By (7.10) and (7.16) it is now seen that

$$\int_0^{+\infty} \Pr\{ T_G \le t, N_B \le n \} \exp\{ - t \, \sigma \} \, dt = h_n^*(\sigma) / \sigma. \qquad (7.18)$$

The plan of the inversion of (7.18) is as follows. We first note that if a (generalised) function h_n can be found whose Laplace transform is h_n^* in (7.17), then the right hand side of (7.18) is the Laplace transform of the integral of h_n, i.e., with

$$h_n^*(\sigma) = \int_0^{+\infty} h_n(t) \exp\{ - t \, \sigma \} \, dt$$

we have

$$h_n^*(\sigma) / \sigma = \int_0^{+\infty} \int_0^t h_n(s) \, ds \, \exp\{ - t \, \sigma \} \, dt \, ,$$

from which it will follow that

$$\Pr\{ T_G \le t, N_B \le n \} = \int_0^t h_n(s) \, ds. \qquad (7.19)$$

For the required inversion of h_n^*, we shall need the following alternative form of (7.17) which is obtained from (7.17) by Lemma 4.10 as follows

$$h_n^*(\sigma) =$$

$$(\alpha_G^T, 0, ..., 0) \left\{ M_{n+1}(\Lambda_{GG}, - \Lambda_{GB} \Lambda_{BB}^{-1} \Lambda_{BG}) - \sigma I \right\}^{-1} \times$$
$$c_{n+1}(\Lambda_{GG} \mathbf{1} + \Lambda_{GB} \mathbf{1}) -$$

$$- (\alpha_G^T, 0, ..., 0) \left\{ M_n(\Lambda_{GG}, - \Lambda_{GB} \Lambda_{BB}^{-1} \Lambda_{BG}) - \sigma I \right\}^{-1} \times$$
$$c_n(\Lambda_{GB} \Lambda_{BB}^{-1} (\Lambda_{BG} \mathbf{1} + \Lambda_{BB} \mathbf{1})) I_{\{ n \ge 1 \}} +$$

$$+ \alpha_B^T \Lambda_{BB}^{-1} (\Lambda_{BG} \mathbf{1} + \Lambda_{BB} \mathbf{1}) I_{\{ n \ge 1 \}} +$$

$$+ (\alpha_B^T \Lambda_{BB}^{-1} \Lambda_{BG}, 0, ..., 0) \left\{ M_{n-1}(\Lambda_{GG}, - \Lambda_{GB} \Lambda_{BB}^{-1} \Lambda_{BG}) - \sigma I \right\}^{-1} \times$$
$$c_{n-1}(\Lambda_{GB} \Lambda_{BB}^{-1} (\Lambda_{BG} \mathbf{1} + \Lambda_{BB} \mathbf{1})) I_{\{ n \ge 2 \}} -$$

$$- (\alpha_B^T \Lambda_{BB}^{-1} \Lambda_{BG}, 0, ..., 0) \left\{ M_n(\Lambda_{GG}, - \Lambda_{GB} \Lambda_{BB}^{-1} \Lambda_{BG}) - \sigma I \right\}^{-1} \times$$
$$c_n(\Lambda_{GG} \mathbf{1} + \Lambda_{GB} \mathbf{1})) I_{\{ n \ge 1 \}}, \qquad (7.20)$$

where for any column-vector $\mathbf{v}$, $\mathbf{c}_n(\mathbf{v})$ is defined as the column-vector which is obtained by an n-fold replication of $\mathbf{v}$, i.e., $\mathbf{c}_n(\mathbf{v}) = (\mathbf{v}^T, ..., \mathbf{v}^T)^T$. On inversion of (7.20), we get by Proposition 7.2

$$h_n(t) =$$

$$- (\boldsymbol{\alpha}_G^T, 0, ..., 0) \exp\left\{ t\, M_{n+1}(\boldsymbol{\Lambda}_{GG}, - \boldsymbol{\Lambda}_{GB}\, \boldsymbol{\Lambda}_{BB}^{-1}\, \boldsymbol{\Lambda}_{BG}) \right\} \mathbf{c}_{n+1}(\boldsymbol{\Lambda}_{GG}\, \mathbf{1} + \boldsymbol{\Lambda}_{GB}\, \mathbf{1}) +$$

$$+ (\boldsymbol{\alpha}_G^T, 0, ..., 0) \exp\left\{ t\, M_n(\boldsymbol{\Lambda}_{GG}, - \boldsymbol{\Lambda}_{GB}\, \boldsymbol{\Lambda}_{BB}^{-1}\, \boldsymbol{\Lambda}_{BG}) \right\} \times$$
$$\mathbf{c}_n(\boldsymbol{\Lambda}_{GB}\, \boldsymbol{\Lambda}_{BB}^{-1}\, (\boldsymbol{\Lambda}_{BG}\, \mathbf{1} + \boldsymbol{\Lambda}_{BB}\, \mathbf{1}))\, I_{\{ n \geq 1 \}} +$$

$$+ \boldsymbol{\alpha}_B^T\, \boldsymbol{\Lambda}_{BB}^{-1}\, (\boldsymbol{\Lambda}_{BG}\, \mathbf{1} + \boldsymbol{\Lambda}_{BB}\, \mathbf{1})\, \delta(t)\, I_{\{ n \geq 1 \}} +$$

$$- (\boldsymbol{\alpha}_B^T\, \boldsymbol{\Lambda}_{BB}^{-1}\, \boldsymbol{\Lambda}_{BG}, 0, ..., 0) \exp\left\{ t\, M_{n-1}(\boldsymbol{\Lambda}_{GG}, - \boldsymbol{\Lambda}_{GB}\, \boldsymbol{\Lambda}_{BB}^{-1}\, \boldsymbol{\Lambda}_{BG}) \right\} \times$$
$$\mathbf{c}_{n-1}(\boldsymbol{\Lambda}_{GB}\, \boldsymbol{\Lambda}_{BB}^{-1}\, (\boldsymbol{\Lambda}_{BG}\, \mathbf{1} + \boldsymbol{\Lambda}_{BB}\, \mathbf{1}))\, I_{\{ n \geq 2 \}} +$$

$$+ (\boldsymbol{\alpha}_B^T\, \boldsymbol{\Lambda}_{BB}^{-1}\, \boldsymbol{\Lambda}_{BG}, 0, ..., 0) \exp\left\{ t\, M_n(\boldsymbol{\Lambda}_{GG}, - \boldsymbol{\Lambda}_{GB}\, \boldsymbol{\Lambda}_{BB}^{-1}\, \boldsymbol{\Lambda}_{BG}) \right\} \times$$
$$\mathbf{c}_n(\boldsymbol{\Lambda}_{GG}\, \mathbf{1} + \boldsymbol{\Lambda}_{GB}\, \mathbf{1})\, I_{\{ n \geq 1 \}}, \tag{7.21}$$

where $\delta(t)$ stands for the Dirac operator; see, e.g., [GUE]. Integrating (7.21), we get by (7.19)

$$\Pr\{ T_G \leq t, N_B \leq n \} =$$

$$(\boldsymbol{\alpha}_G^T, 0, ..., 0) \left\{ I - \exp\left\{ t\, M_{n+1}(\boldsymbol{\Lambda}_{GG}, - \boldsymbol{\Lambda}_{GB}\, \boldsymbol{\Lambda}_{BB}^{-1}\, \boldsymbol{\Lambda}_{BG}) \right\} \right\} \times$$
$$\left(M_{n+1}(\boldsymbol{\Lambda}_{GG}, - \boldsymbol{\Lambda}_{GB}\, \boldsymbol{\Lambda}_{BB}^{-1}\, \boldsymbol{\Lambda}_{BG}) \right)^{-1} \mathbf{c}_{n+1}(\boldsymbol{\Lambda}_{GG}\, \mathbf{1} + \boldsymbol{\Lambda}_{GB}\, \mathbf{1}) -$$

$$- (\boldsymbol{\alpha}_G^T, 0, ..., 0) \left\{ I - \exp\left\{ t\, M_n(\boldsymbol{\Lambda}_{GG}, - \boldsymbol{\Lambda}_{GB}\, \boldsymbol{\Lambda}_{BB}^{-1}\, \boldsymbol{\Lambda}_{BG}) \right\} \right\} \times$$
$$\left(M_n(\boldsymbol{\Lambda}_{GG}, - \boldsymbol{\Lambda}_{GB}\, \boldsymbol{\Lambda}_{BB}^{-1}\, \boldsymbol{\Lambda}_{BG}) \right)^{-1} \mathbf{c}_n(\boldsymbol{\Lambda}_{GB}\, \boldsymbol{\Lambda}_{BB}^{-1}\, (\boldsymbol{\Lambda}_{BG}\, \mathbf{1} + \boldsymbol{\Lambda}_{BB}\, \mathbf{1}))\, I_{\{ n \geq 1 \}} +$$

$$+ \boldsymbol{\alpha}_B^T\, \boldsymbol{\Lambda}_{BB}^{-1}\, (\boldsymbol{\Lambda}_{BG}\, \mathbf{1} + \boldsymbol{\Lambda}_{BB}\, \mathbf{1})\, I_{\{ n \geq 1 \}} +$$

$$+ (\boldsymbol{\alpha}_B^T\, \boldsymbol{\Lambda}_{BB}^{-1}\, \boldsymbol{\Lambda}_{BG}, 0, ..., 0) \left\{ I - \exp\left\{ t\, M_{n-1}(\boldsymbol{\Lambda}_{GG}, - \boldsymbol{\Lambda}_{GB}\, \boldsymbol{\Lambda}_{BB}^{-1}\, \boldsymbol{\Lambda}_{BG}) \right\} \right\} \times$$
$$M_{n-1}(\boldsymbol{\Lambda}_{GG}, - \boldsymbol{\Lambda}_{GB}\, \boldsymbol{\Lambda}_{BB}^{-1}\, \boldsymbol{\Lambda}_{BG})\, \mathbf{c}_{n-1}(\boldsymbol{\Lambda}_{GB}\, \boldsymbol{\Lambda}_{BB}^{-1}\, (\boldsymbol{\Lambda}_{BG}\, \mathbf{1} + \boldsymbol{\Lambda}_{BB}\, \mathbf{1}))\, I_{\{ n \geq 2 \}} -$$

$$- (\boldsymbol{\alpha}_B^T\, \boldsymbol{\Lambda}_{BB}^{-1}\, \boldsymbol{\Lambda}_{BG}, 0, ..., 0) \left\{ I - \exp\left\{ t\, M_n(\boldsymbol{\Lambda}_{GG}, - \boldsymbol{\Lambda}_{GB}\, \boldsymbol{\Lambda}_{BB}^{-1}\, \boldsymbol{\Lambda}_{BG}) \right\} \right\} \times$$
$$M_n(\boldsymbol{\Lambda}_{GG}, - \boldsymbol{\Lambda}_{GB}\, \boldsymbol{\Lambda}_{BB}^{-1}\, \boldsymbol{\Lambda}_{BG})\, \mathbf{c}_n(\boldsymbol{\Lambda}_{GG}\, \mathbf{1} + \boldsymbol{\Lambda}_{GB}\, \mathbf{1})\, I_{\{ n \geq 1 \}}. \tag{7.22}$$

(7.3) now follows from (7.22) with Proposition 7.2 by putting in (7.1) $\Phi = \Lambda_{GG}$, and $\Psi = -\Lambda_{GB}\,\Lambda_{BB}^{-1}\,\Lambda_{BG}$. Notice in particular that the invertibility of $(\Phi + \Psi)$ follows from (7.13) since Φ is invertible. ∎

7.3 Application: the Markov model of the three-unit power trans-mission system revisited

The above results will now be applied to Model 3 from Section 1.2.1. Some aspects of this model have already been examined in Section 4.4.1. The system's transition rate matrix is given in (4.67). The quantity under consideration, $\Pr\{\,T_G > t,\ N_B \leq n\,\}$, is the probability that by the time of its first major breakdown, the system will have been in operation for at least t time units (years) *with the proviso* that the number of smaller breakdowns observed until then does not exceed n. The values assumed for the rate parameters are shown in Table 7.1. The system is started at $t = 0$ with all its components working, i.e., the initial probability vector α is given by $(1, 0, 0, 0, 0)^T$. Figure 7.1 shows the values of $\Pr\{\,T_G > t,\ N_B \leq n \mid Y_0 = 1\,\}$ for $n = 0, ..., 8$, and $\Pr\{\,T_G > t \mid Y_0 = 1\,\}$ for t in the range of up to 100 years. (The calculation of the latter is based on (4.60).) It is seen that each of the curves decays exponentially as $t \to +\infty$, and that for fixed t, they tend (as one would expect) to $\Pr\{\,T_G > t \mid Y_0 = 1\,\}$ as $n \to +\infty$.

The system used to produce the data for Figure 7.1 was the 'educational version' of MATLAB on the Apple Macintosh SE/30. Each of the curves in Figure 7.1 is based on 34 equidistant points on the time axis. Their respective computing times for $n = 0, ..., 8$ were (in secs) 9 , 26 , 64 , 134 , 247 , 413 , 641, 940, 1322, whereas the points for $n = +\infty$ took only 9 secs to compute. (The size of the matrices $\mathbf{M}_n$ in (7.3) increases with n which explains the observed increase in CPU-time.) The code will not be shown here since no new features of MATLAB are involved.

λ_1	λ_2	λ_T	$\lambda_c{}'$	μ_1	μ_2	μ_T
0.25	0.25	0.005	0.01	1095	1095	0

Table 7.1 Assumed parameter values in [1/years]

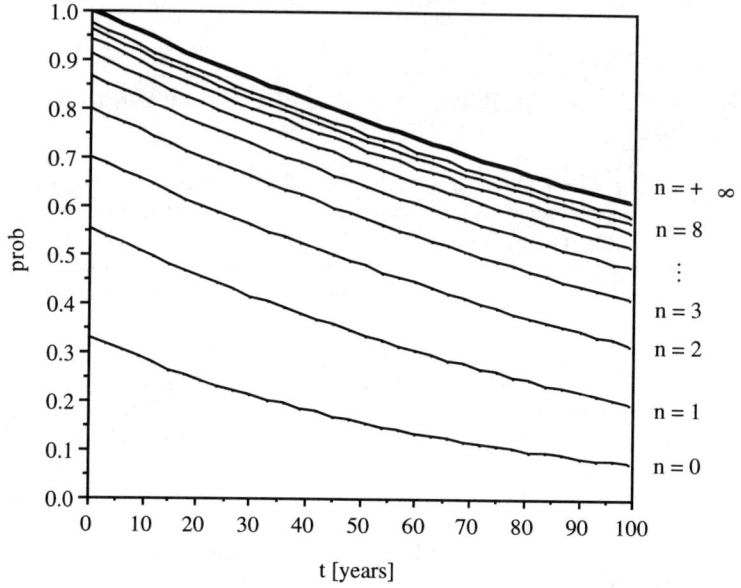

Figure 7.1 Pr { $T_G > t$, $N_B \leq n \mid Y_0 = 1$ } for $n = 0, 1, ..., 8, +\infty$

CHAPTER 8

SOJOURN TIMES FOR FINITE SEMI-MARKOV PROCESSES

Semi-Markov modelling is an important tool in the reliability field as has been indicated by the examples in Section 1.2.2. Moreover, semi-Markov processes are also interesting from a mathematical viewpoint since they are obtained by relaxing one of the key assumptions of Markov theory: here, the distribution of the (conditional) holding time in any state need not be memoryless but it is arbitrary. Notice the deliberate use of the word *memoryless* here in order to account for both the discrete-, and the continuous-time case. As is well-known, in these two cases the holding time distributions are geometric and exponential respectively. In what follows, $Y = \{ Y_t : t \geq 0 \}$ is a semi-Markov process with finite state space, S. The transitions of Y between the states in S are governed by the embedded Markov chain $X = \{ X_n : n = 0, 1, ... \}$ whose transition probability matrix is denoted by **R**. The cumulative distribution functions of the holding times of Y in s are denoted by $F_{s,s'}$, $s' \neq s$. These are the *conditional* distribution functions of the holding time in s given that the next state to be visited is s'. $F_{s,s'}$ is a proper distribution function for non-absorbing s only. The semi-Markov process Y will be assumed either irreducible or absorbing. We partition S into $n+1$ (≥ 3) non-empty subsets if Y is absorbing, and then $S = A_1 \cup ... \cup A_n \cup A_{n+1}$ with A_{n+1} denoting the set of all absorbing states; otherwise, the partitioning is $S = A_1 \cup ... \cup A_n$, $n \geq 2$. (In the latter case, A_{n+1} is *defined* to be the empty set in order to be able to cover both cases by the same set of formulae.) Following the notation introduced in Chapter 4, for $i \in \{ 1, ..., n \}$, $T_{A_i,j}$ stands for the length of the *j*th sojourn of Y in A_i if A_i is visited by Y at least j times. We put $T_{A_i,j} = 0$ otherwise.

The present chapter is devoted to the examination of the sojourn time vector

$$\mathbf{T} = \{ T_{A_i,j} : i = 1, ..., n; j = 1, ..., m \}.$$

In [RUB1], Rubino and Sericola derived an expression for the cumulative distribution function of the, say, *m*th sojourn of Y in a subset of $S \setminus A_{n+1}$ for an absorbing Y in terms of the cumulative distribution function of Y's first sojourn in the same subset. Also in [RUB1], this latter quantity was shown to satisfy a certain integral equation. These developments will be deduced here from one of our general results (Theorem 8.2). In this chapter we also establish a method for deriving the distribution of sojourn times (in the Laplace transform domain) if $n \geq 3$ for processes with a specific transition structure.

The key tool is a recurrence relation for the Laplace transform of **T** and it will be presented in Section 8.1. The Laplace transform of **T** itself will be deduced in Section 8.2 from which then most known results on sojourn times in Markov and semi-Markov processes easily follow; a few examples of this are discussed in Section 8.2 too. Furthermore, Section 8.2 also contains an indication of how the key theorem (which will be formulated for semi-Markov processes whose state space is partitioned into *any finite number* of classes) can be used to arrive at the Laplace transform of the vector of sojourn times if S is partitioned into three non-absorbing classes. Section 8.3 is devoted to the proof of the key tool.

All the results discussed here easily carry over to semi-Markov reward processes. The realisations of the holding times are then interpreted as the (random) rewards generated by the process when visiting the individual states. The sequence of visits is defined by the underlying (discrete-time) embedded Markov chain.

Let us single out a few of the more recent references from the wealth of literature on issues related in some sense to the subject matter of the present, and subsequent chapters. Rossiter [ROSS] arrives at an expression in the Laplace transform domain for the time spent in, say, the 'up' state by an alternating renewal process during a finite time interval. Masuda and Sumita [MAS4] have considered multivariate reward processes on semi-Markov processes, again in the Laplace transform domain, with a view towards asymptotic analysis. Other papers dealing with related analytical (as opposed to numerical) issues are [MAS2], [SUM4], and [RUB5]. The following papers are devoted more to (semi-)Markov processes with a special structure for applications and/or to issues concerning the computation of various dependability characteristics for (special) systems modelled by semi-Markov processes: [KUB], [MAS1], [MAS3], [KUL1], [KUL2], [REI1], and [SUM3]. Finally, [TRI2] is a review paper devoted to the numerical evaluation of dependability characteristics of (computer) systems modelled by (semi-)Markov reward processes.

The core material in this chapter is based on work first reported in [CSE1].

8.1 A recurrence relation for the Laplace transform of the vector of sojourn times

The result here will be formulated for a semi-Markov process Y which is absorbing and whose state space S is partitioned into $n+1$ (≥ 3) disjoint classes as described above. Assuming Y to be absorbing does not entail any restrictions since, as has already been demonstrated earlier (see, Corollary 2.3 and Corollary 4.3), the corresponding formulae for the irreducible case are obtained by substituting zero for the transition probabilities to A_{n+1} for the underlying embedded Markov chain.

We start with some remarks on the notation. $\boldsymbol{\alpha} = (\boldsymbol{\alpha}_{A_1}^T, ..., \boldsymbol{\alpha}_{A_n}^T, \mathbf{0})^T$ is the initial probability vector. Laplace transform techniques will be the main tool in this and all the subsequent chapters. Thus, we shall need the matrix $\mathbf{Q}^*(\rho)$ whose entries are defined for $\rho \in \mathbb{C}^+ = \{ \rho \in \mathbb{C} : \mathrm{Re}(\rho) \geq 0 \}$, $s \in S \setminus A_{n+1}$, and $s' \in S$ by

$$
q_{ss'}^*(\rho) = \begin{cases} r_{ss'} \displaystyle\int_{[0, +\infty)} \exp\{ -\rho\, t \}\, F_{s,s'}\{dt\} & \text{if } s \neq s', \\ 0 & \text{if } s = s'. \end{cases} \tag{8.1}
$$

(The integral in (8.1) is what is usually termed the Laplace-Stieltjes transform of $F_{s,s'}$. An alternative interpretation is that it is the Laplace transform of the probability *measure* whose distribution function is $F_{s,s'}$. We prefer the latter interpretation for it is rooted in the more general measure-theoretic framework.) Let $m \geq 1$ be a fixed integer and define the $m \times m$ matrix $\mathbf{U}$ by

$$
u_{ij} = \begin{cases} 1 & \text{for } j - i = 1, \\ 0 & \text{otherwise.} \end{cases}
$$

For $s \in S \setminus A_{n+1}$ and $\boldsymbol{\tau}_i = (\tau_{1,i}, ..., \tau_{m,i})^T \in \mathbb{C}^{+m}$, $i = 1, ..., n$, put

$$\psi_s(\boldsymbol{\tau}_1, ..., \boldsymbol{\tau}_n) = E\left[\exp\left\{ -\sum_{i=1}^{n} (T_{A_i,1}, ..., T_{A_i,m}) \boldsymbol{\tau}_i \right\} \mid Y_0 = s \right].$$

Define the column-vectors $\boldsymbol{\psi}_{A_k}(\boldsymbol{\tau}_1, ..., \boldsymbol{\tau}_n)$ by

$$\boldsymbol{\psi}_{A_k}(\boldsymbol{\tau}_1, ..., \boldsymbol{\tau}_n) = \{ \psi_s(\boldsymbol{\tau}_1, ..., \boldsymbol{\tau}_n) : s \in A_k \}, k = 1, ..., n.$$

In Theorem 8.1, a recurrence relation is established for these vectors of Laplace transforms. It is as follows.

THEOREM 8.1. *For all* $i \in \{ 1, ..., n \}$, *the matrix* $\mathbf{I} - \mathbf{Q}^*_{A_iA_i}(\rho)$ *is invertible and we have*

$$\boldsymbol{\psi}_{A_i}(\boldsymbol{\tau}_1, ..., \boldsymbol{\tau}_n) =$$

$$\sum_{\substack{k=1 \\ k \neq i}}^{n} \mathbf{K}^*_{A_iA_k}(\mathbf{1}^T(\mathbf{I}-\mathbf{U})\boldsymbol{\tau}_i) \, \boldsymbol{\psi}_{A_k}(\boldsymbol{\tau}_1, ..., \boldsymbol{\tau}_{i-1}, \mathbf{U}\boldsymbol{\tau}_i, \boldsymbol{\tau}_{i+1}, ..., \boldsymbol{\tau}_n) +$$

$$\mathbf{K}^*_{A_iA_{n+1}}(\mathbf{1}^T(\mathbf{I}-\mathbf{U})\boldsymbol{\tau}_i) \, \mathbf{1}, \tag{8.2}$$

where the matrices $\mathbf{K}^*_{A_iA_k}(\rho)$, $k \in \{1, ..., n+1\} \setminus \{ i \}$, *are defined for* $\rho \in \mathbb{C}^+$ *by*

$$\mathbf{K}^*_{A_iA_k}(\rho) = \left(\mathbf{I} - \mathbf{Q}^*_{A_iA_i}(\rho)\right)^{-1} \mathbf{Q}^*_{A_iA_k}(\rho). \tag{8.3}$$

The proof of Theorem 8.1 is the subject of Section 8.3.

NOTE. The last term on the right hand side of (8.2) is zero if Y is irreducible since $\mathbf{Q}^*_{A_iA_{n+1}}(\rho) \equiv \mathbf{0}$. (This follows from $\mathbf{R}_{(S \setminus A_{n+1})A_{n+1}} = \mathbf{0}$.)

8.2 Laplace transforms of vectors of sojourn times

8.2.1 S is partitioned into three subsets (n = 2)

This is the case examined hitherto (in the literature) for special cases only, e.g., under the Markov assumption (see, Section 4.4 and [CSE4]) or for the semi-Markov case but only for the marginal distributions of the sojourn time vector (see, [RUB1]). The following holds true in general.

THEOREM 8.2. *Let* Y *be an irreducible (or absorbing) semi-Markov process with* $S = A_1 \cup A_2$ *(or* $S = A_1 \cup A_2 \cup A_3$*). Then, the Laplace transform of* $\{T\backslash s\backslash do3(A_i\backslash j) : i = 1, 2; j = 1, ..., m \}$, $\boldsymbol{\psi}(\boldsymbol{\tau}_1, \boldsymbol{\tau}_2)$, *is given by*

$$\psi(\tau_1, \tau_2) =$$

$$\alpha_{A_1}^T \left\{ \prod_{k=0}^{m-1} \left\{ K^*_{A_1A_2}(1^T(I-U)U^k\tau_1) \, K^*_{A_2A_1}(1^T(I-U)U^k\tau_2) \right\} \right\} 1 +$$

$$\alpha_{A_2}^T \left\{ \prod_{k=0}^{m-1} \left\{ K^*_{A_2A_1}(1^T(I-U)U^k\tau_2) \, K^*_{A_1A_2}(1^T(I-U)U^k\tau_1) \right\} \right\} 1 +$$

$$\alpha_{A_1}^T \sum_{\ell=0}^{m-1} \left\{ \prod_{k=0}^{\ell-1} \left\{ K^*_{A_1A_2}(1^T(I-U)U^k\tau_1) \, K^*_{A_2A_1}(1^T(I-U)U^k\tau_2) \right\} \right\} \times$$

$$\left\{ K^*_{A_1A_2}(1^T(I-U)U^\ell\tau_1) \, K^*_{A_2A_3}(1^T(I-U)U^\ell\tau_2) + K^*_{A_1A_3}(1^T(I-U)U^\ell\tau_1) \right\} 1 +$$

$$\alpha_{A_2}^T \sum_{\ell=0}^{m-1} \left\{ \prod_{k=0}^{\ell-1} \left\{ K^*_{A_2A_1}(1^T(I-U)U^k\tau_2) \, K^*_{A_1A_2}(1^T(I-U)U^k\tau_1) \right\} \right\} \times$$

$$\left\{ K^*_{A_2A_1}(1^T(I-U)U^\ell\tau_2) \, K^*_{A_1A_3}(1^T(I-U)U^\ell\tau_1) + K^*_{A_2A_3}(1^T(I-U)U^\ell\tau_2) \right\} 1,$$

$$(8.4)$$

where the matrices $K^*_{A_iA_k}$ $(i \neq k)$ *are defined by* (8.3).

PROOF OF THEOREM 8.2. Applying (8.2) twice, gives

$$\psi_{A_1}(\tau_1, \tau_2) =$$

$$K^*_{A_1A_2}(1^T(I-U)\tau_1) \, K^*_{A_2A_1}(1^T(I-U)\tau_2) \, \psi_{A_1}(U\tau_1, U\tau_2) +$$

$$\left\{ K^*_{A_1A_2}(1^T(I-U)\tau_1) \, K^*_{A_2A_3}(1^T(I-U)\tau_2) + K^*_{A_1A_3}(1^T(I-U)\tau_1) \right\} 1. \qquad (8.5)$$

From (8.5), we have for $j = 1, 2, \ldots$ by induction that

$$\psi_{A_1}(\tau_1, \tau_2) =$$

$$\prod_{k=0}^{j-1} \left\{ K^*_{A_1A_2}(1^T(I-U)U^k\tau_1) K^*_{A_2A_1}(1^T(I-U)U^k\tau_2) \right\} \psi_{A_1}(U^j\tau_1, U^j\tau_2) +$$

$$\sum_{\ell=0}^{j-1} \left\{ \prod_{k=0}^{\ell-1} \left\{ K^*_{A_1A_2}(1^T(I-U)U^k\tau_1) \, K^*_{A_2A_1}(1^T(I-U)U^k\tau_2) \right\} \right\} \times$$

$$\left\{ K^*_{A_1A_2}(1^T(I-U)U^\ell\tau_1) \, K^*_{A_2A_3}(1^T(I-U)U^\ell\tau_2) + \right.$$

$$+ \mathbf{K}^*_{A_1 A_3}(\mathbf{1}^T(\mathbf{I}\text{-}\mathbf{U})\mathbf{U}^\ell \boldsymbol{\tau}_1) \Big\} \; \mathbf{1}. \tag{8.6}$$

Using $\mathbf{U}^m = \mathbf{0}$ and $\boldsymbol{\psi}_{A_1}(0,0) = \mathbf{1}$, (8.4) follows from (8.6) with $j = m$ since $\psi(\boldsymbol{\tau}_1, \boldsymbol{\tau}_2) = \boldsymbol{\alpha}_{A_1}{}^T \boldsymbol{\psi}_{A_1}(\boldsymbol{\tau}_1, \boldsymbol{\tau}_2) + \boldsymbol{\alpha}_{A_2}{}^T \boldsymbol{\psi}_{A_2}(\boldsymbol{\tau}_1, \boldsymbol{\tau}_2)$. ∎

Notice, for later use, that the arguments of the matrices $\mathbf{K}^*$ on the right hand side of (8.4) are given by

$$\mathbf{1}^T (\mathbf{I} - \mathbf{U}) \, \mathbf{U}^k \, \boldsymbol{\tau}_i = \boldsymbol{\tau}_{k+1,i} \, , \, k = 0, 1, ..., m - 1 \, . \tag{8.7}$$

Two special cases will now be examined. As a first application of Theorem 8.2, we show that it can be used to rederive Corollary 4.8. Let us first examine what $\mathbf{Q}^*$ from (8.1) evaluates to if Y is a continuous-time Markov process. In that case, the conditional holding time distribution in $s \in S \setminus A_{n+1}$ is exponential and it depends only on s, i.e.,

$$F_{s,s'}(t) = 1 - \exp\{ - c_s \, t \}, \, t \geq 0, \tag{8.8}$$

for some $c_s \in (0, +\infty)$. It is by (8.1) and (8.8)

$$q_{ss'}^*(\rho) = \begin{cases} r_{s\,s'} \dfrac{c_s}{c_s + \rho} & \text{if } s \neq s', \\ 0 & \text{if } s = s'. \end{cases} \tag{8.9}$$

Let $\mathbf{C}$ be the diagonal matrix with $\mathbf{C} = \text{diag}\{ c_s : s \in S \setminus A_{n+1} \}$. Then (8.9) in matrix form becomes

$$\mathbf{Q}^*(\rho) = \mathbf{C} \, (\mathbf{C} + \rho \, \mathbf{I})^{-1} \, \mathbf{R}_{S \setminus A_{n+1} S},$$

which in terms of the submatrices of $\mathbf{C}$ and $\mathbf{R}$ reads as

$$\mathbf{Q}_{A_i A_i}^*(\rho) = \mathbf{C}_{A_i A_i} \, (\mathbf{C}_{A_i A_i} + \rho \, \mathbf{I})^{-1} \, \mathbf{R}_{A_i A_i}, \tag{8.10}$$

$$\mathbf{Q}_{A_i A_j}^*(\rho) = \mathbf{C}_{A_i A_i} \, (\mathbf{C}_{A_i A_i} + \rho \, \mathbf{I})^{-1} \, \mathbf{R}_{A_i A_j}, \tag{8.11}$$

$i \in \{1, 2\}, j \in \{1, 2, 3\} \setminus \{ i \}$. On the other hand, the entries of the transition rate matrix $\boldsymbol{\Lambda}$ of Y and those of $\mathbf{R}$, are interrelated for $s \in S \setminus A_{n+1}$ by

$$\lambda_{ss'} = \begin{cases} r_{s\,s'} \, c_s & \text{if } s \neq s', \\ - c_s & \text{if } s = s', \end{cases}$$

which, because of $r_{ss} = 0$, in matrix form is written as

$$\boldsymbol{\Lambda}_{S \setminus A_{n+1} S} = - \, \mathbf{C}_{S \setminus A_{n+1} S \setminus A_{n+1}} \left(\mathbf{I} - \mathbf{R}_{S \setminus A_{n+1} S} \right). \tag{8.12}$$

Again for $i \in \{1, 2\}, j \in \{1, 2, 3\} \setminus \{ i \}$, (8.12) implies for the submatrices of $\mathbf{\Lambda}$ the following

$$\mathbf{\Lambda}_{A_iA_i} = - C_{A_iA_i} (\mathbf{I} - \mathbf{R}_{A_iA_i}), \tag{8.13}$$

$$\mathbf{\Lambda}_{A_iA_j} = C_{A_iA_i} \mathbf{R}_{A_iA_j}. \tag{8.14}$$

(Notice that (8.13) and (8.14) have already been encountered in the proof of Lemma 4.11.) From (8.3), (8.10), (8.11), (8.13), and (8.14) it is easily seen that

$$\mathbf{K}^*_{A_iA_j}(\rho) = - (\mathbf{\Lambda}_{A_iA_i} - \rho \mathbf{I})^{-1} \mathbf{\Lambda}_{A_iA_j}, \quad i \in \{1, 2\}, j \in \{1, 2, 3\} \setminus \{ i \}. \tag{8.15}$$

For the sake of simplicity, assume that Y is an *irreducible* Markov process. By (8.7) and (8.15), (8.4) then becomes

$$\psi(\tau_1, \tau_2) =$$

$$\alpha_{A_1}{}^T \prod_{k=0}^{m-1} \left\{ (\mathbf{\Lambda}_{A_1A_1} - \tau_{k+1,1} \mathbf{I})^{-1} \mathbf{\Lambda}_{A_1A_2} (\mathbf{\Lambda}_{A_2A_2} - \tau_{k+1,2} \mathbf{I})^{-1} \mathbf{\Lambda}_{A_2A_1} \right\} \mathbf{1} +$$

$$\alpha_{A_2}{}^T \prod_{k=0}^{m-1} \left\{ (\mathbf{\Lambda}_{A_2A_2} - \tau_{k+1,2} \mathbf{I})^{-1} \mathbf{\Lambda}_{A_2A_1} (\mathbf{\Lambda}_{A_1A_1} - \tau_{k+1,1} \mathbf{I})^{-1} \mathbf{\Lambda}_{A_1A_2} \right\} \mathbf{1},$$

a result which is easily seen to agree with (4.42).

As a second application of Theorem 8.2, the distribution of the mth sojourn time in A_1 will be represented under the general semi-Markov assumption in terms of the first sojourn time in A_1. This result can be found in Rubino and Sericola [RUB1] and it is as follows.

COROLLARY 8.3. *Define the vector of Laplace transforms* $\mathbf{\phi}_{A_1}(\rho) = \{ \phi_a(\rho) : a \in A_1 \}$ *by*

$$\phi_a(\rho) = E[\exp\{ - \rho \, T_{A_1,1} \} \mid Y_0 = a], \quad a \in A_1.$$

Then, under the assumptions of Theorem 8.1, the Laplace transform of the mth *sojourn in* A_1 *is given for* $m = 1, 2, \ldots$ *by*

$$E[\exp\{ - \rho \, T_{A_1,m} \}] = 1 - \{ \alpha_{A_1}{}^T + \alpha_{A_2}{}^T \mathbf{G} \} \, \mathbf{H}^{m-1} (1 - \mathbf{\phi}_{A_1}(\rho)), \tag{8.16}$$

where $\mathbf{G}$ *and* $\mathbf{H}$ *are respectively defined by*

$$\mathbf{G} = (\mathbf{I} - \mathbf{R}_{A_2A_2})^{-1} \mathbf{R}_{A_2A_1},$$

$$\mathbf{H} = (\mathbf{I} - \mathbf{R}_{A_1A_1})^{-1} \mathbf{R}_{A_1A_2} (\mathbf{I} - \mathbf{R}_{A_2A_2})^{-1} \mathbf{R}_{A_2A_1}. \tag{8.17}$$

NOTES.

(I)   Notice that we know the matrix $\mathbf{H}$ in (8.17) from discrete-parameter theory: it is identical to that defined by (2.30). From (4.55), we know in particular that

$$\mathbf{H}^k \to \mathbf{0}, \text{ as } k \to +\infty. \tag{8.18}$$

(II)   In [RUB1], the above result was presented in terms of distribution functions rather than Laplace transforms; in fact, there was no use made of Laplace transforms there. (8.16) implies the result for distribution functions by first dividing both sides of (8.16) by $\rho > 0$ (which then gives the corresponding equation for the *Laplace transforms of the distribution functions*). This implies the desired form:

$$\Pr\{ T_{A_1,m} \leq t \} = 1 - $$

$$- \{ \boldsymbol{\alpha}_{A_1}{}^T + \boldsymbol{\alpha}_{A_2}{}^T \mathbf{G} \} \, \mathbf{H}^{m-1} \left[\mathbf{1} - \{ \Pr\{ T_{A_1,1} \leq t \mid Y_0 = a \} : a \in A_1 \} \right], t \geq 0.$$

This result shows that once the distribution functions $\Pr\{ T_{A_1,1} \leq t \mid Y_0 = a \}$, $a \in A_1$, are available (as closed form expressions or in a numerical form) the knowledge of the transition probability matrix of the embedded Markov chain of Y suffices to evaluate the distribution function of the m*th* sojourn time in A_1.

PROOF OF COROLLARY 8.3.   For $\rho \in \mathbb{C}^+$ put $\boldsymbol{\tau}_1 = (0, 0, ..., 0, \rho)^T$, $\boldsymbol{\tau}_2 = \mathbf{0} \in \mathbb{C}^{+m}$, and notice that

$$\mathbf{1}^T (\mathbf{I} - \mathbf{U}) \, \mathbf{U}^k \, \boldsymbol{\tau}_1 = \begin{cases} 0 & \text{for } k = 0, 1, ..., m - 2, \\ \rho & \text{for } k = m - 1, \end{cases}$$

and, of course,

$$\mathbf{1}^T (\mathbf{I} - \mathbf{U}) \, \mathbf{U}^k \, \boldsymbol{\tau}_2 = 0 \text{ for } k = 0, ..., m - 1.$$

From Theorem 8.2 it follows that

$$E\left[\exp\{ - \rho \, T_{A_1,m} \}\right] = \psi(\boldsymbol{\tau}_1, \boldsymbol{\tau}_2) = $$

$$\boldsymbol{\alpha}_{A_1}{}^T \left\{ \mathbf{K}^*{}_{A_1A_2}(0) \, \mathbf{K}^*{}_{A_2A_1}(0) \right\}^{m-1} \mathbf{K}^*{}_{A_1A_2}(\rho) \, \mathbf{K}^*{}_{A_2A_1}(0) \, \mathbf{1} + $$

$$\boldsymbol{\alpha}_{A_2}{}^T \left\{ \mathbf{K}^*{}_{A_2A_1}(0) \, \mathbf{K}^*{}_{A_1A_2}(0) \right\}^{m-1} \mathbf{K}^*{}_{A_2A_1}(0) \, \mathbf{K}^*{}_{A_1A_2}(\rho) \, \mathbf{1} + $$

$$\boldsymbol{\alpha}_{A_1}{}^T \sum_{\ell=0}^{m-2} \left\{ \mathbf{K}^*{}_{A_1A_2}(0) \, \mathbf{K}^*{}_{A_2A_1}(0) \right\}^{\ell} \left\{ \mathbf{K}^*{}_{A_1A_2}(0) \, \mathbf{K}^*{}_{A_2A_3}(0) + \mathbf{K}^*{}_{A_1A_3}(0) \right\} \mathbf{1} + $$

$$\boldsymbol{\alpha}_{A_1}{}^T \left\{ \mathbf{K}^*{}_{A_1A_2}(0) \, \mathbf{K}^*{}_{A_2A_1}(0) \right\}^{m-1} \left\{ \mathbf{K}^*{}_{A_1A_2}(\rho) \, \mathbf{K}^*{}_{A_2A_3}(0) + \mathbf{K}^*{}_{A_1A_3}(\rho) \right\} \mathbf{1} + $$

$$\boldsymbol{\alpha}_{A_2}{}^T \sum_{\ell=0}^{m-2} \left\{ \mathbf{K}^*_{A_2A_1}(0)\, \mathbf{K}^*_{A_1A_2}(0) \right\}^{\ell} \left\{ \mathbf{K}^*_{A_2A_1}(0)\, \mathbf{K}^*_{A_1A_3}(0) + \mathbf{K}^*_{A_2A_3}(0) \right\} \mathbf{1} +$$

$$\boldsymbol{\alpha}_{A_2}{}^T \left\{ \mathbf{K}^*_{A_2A_1}(0)\, \mathbf{K}^*_{A_1A_2}(0) \right\}^{m-1} \left\{ \mathbf{K}^*_{A_2A_1}(0)\, \mathbf{K}^*_{A_1A_3}(\rho) + \mathbf{K}^*_{A_2A_3}(0) \right\} \mathbf{1}.$$

$$(8.19)$$

Using now

$$\mathbf{K}^*_{A_2A_1}(0) = \mathbf{G}, \quad \mathbf{K}^*_{A_1A_2}(0)\, \mathbf{K}^*_{A_2A_1}(0) = \mathbf{H},$$

and

$$\mathbf{K}^*_{A_iA_j}(0)\, \mathbf{1} + \mathbf{K}^*_{A_iA_3}(0)\, \mathbf{1} = \mathbf{1}, \ i,j \in \{1,2\}, \ i \neq j, \qquad (8.20)$$

(with $i = 2$, $j = 1$) we get from (8.19)

$$E\big[\exp\{-\rho\, T_{A_1,m}\}\big] =$$

$$\left\{ \boldsymbol{\alpha}_{A_1}{}^T + \boldsymbol{\alpha}_{A_2}{}^T \mathbf{G} \right\} \mathbf{H}^{m-1} \left\{ \mathbf{K}^*_{A_1A_2}(\rho)\, \mathbf{1} + \mathbf{K}^*_{A_1A_3}(\rho)\, \mathbf{1} \right\} + c, \qquad (8.21)$$

with some constant c *not* depending on ρ. From (8.21) (with $\rho = 0$) and (8.20) (with $i = 1$, $j = 2$) it follows that

$$c = 1 - \left\{ \boldsymbol{\alpha}_{A_1}{}^T + \boldsymbol{\alpha}_{A_2}{}^T \mathbf{G} \right\} \mathbf{H}^{m-1}\, \mathbf{1}.$$

It is therefore

$$E\big[\exp\{-\rho\, T_{A_1,m}\}\big] =$$

$$1 - \left\{ \boldsymbol{\alpha}_{A_1}{}^T + \boldsymbol{\alpha}_{A_2}{}^T \mathbf{G} \right\} \mathbf{H}^{m-1} \left[\mathbf{1} - \left\{ \mathbf{K}^*_{A_1A_2}(\rho)\, \mathbf{1} + \mathbf{K}^*_{A_1A_3}(\rho)\, \mathbf{1} \right\} \right]. \qquad (8.22)$$

Putting in (8.22) $m = 1$ and $\boldsymbol{\alpha}_{A_2} = \mathbf{0}$ gives

$$\boldsymbol{\phi}_{A_1}(\rho) = 1 - \left[\mathbf{1} - \left\{ \mathbf{K}^*_{A_1A_2}(\rho)\, \mathbf{1} + \mathbf{K}^*_{A_1A_3}(\rho)\, \mathbf{1} \right\} \right] = \mathbf{K}^*_{A_1A_2}(\rho)\, \mathbf{1} + \mathbf{K}^*_{A_1A_3}(\rho)\, \mathbf{1}.$$

$$(8.23)$$

(8.16) now follows from (8.22) and (8.23). ∎

Let us now address the question of how (8.23) can be used to obtain the system of integral equations alluded to earlier. Using (8.3), we deduce from (8.23) that

$$\boldsymbol{\phi}_{A_1}(\rho) - \mathbf{Q}^*{}_{A_1A_1}(\rho)\,\boldsymbol{\phi}_{A_1}(\rho) = \mathbf{Q}^*{}_{A_1A_2}(\rho)\,\mathbf{1} + \mathbf{Q}^*{}_{A_1A_3}(\rho)\,\mathbf{1},$$

which, written element-wise, is

$$\phi_{a_1}(\rho) - \sum_{a \in A_1} q_{a_1a}{}^*(\rho)\,\phi_a(\rho) = \sum_{a \in A_2 \cup A_3} q_{a_1a}{}^*(\rho),\ a_1 \in A_1. \tag{8.24}$$

(8.24) is, more explicitly,

$$\mathrm{E}[\exp\{-\rho\,T_{A_1,1}\} \mid Y_0 = a_1] -$$

$$- \sum_{\substack{a \in A_1 \\ a \neq a_1}} r_{a_1a} \int_{[0,\,+\infty)} \exp\{-\rho\,t\}\,F_{a_1,a}\{dt\}\,\mathrm{E}[\exp\{-\rho\,T_{A_1,1}\} \mid Y_0 = a] =$$

$$\sum_{a \in A_2 \cup A_3} r_{a_1a} \int_{[0,\,+\infty)} \exp\{-\rho\,t\}\,F_{a_1,a}\{dt\},\ a_1 \in A_1. \tag{8.25}$$

Using now the well-known fact that

$$\eta_1{}^*(\tau)\,\eta_2{}^*(\tau) = \tau \int_{[0,\,+\infty)} \int_{[0,\,t]} \Pr\{\eta_2 \leq t - x\}\,\Pr\{\eta_1 \in dx\}\,\exp\{-t\,\tau\}\,dt$$

for any two non-negative random variables η_1 and η_2 with respective Laplace transforms $\eta_1{}^*$ and $\eta_2{}^*$, (8.25) corresponds in the time domain to

$$\Pr\{T_{A_1,1} \leq t \mid Y_0 = a_1\} - \sum_{\substack{a \in A_1 \\ a \neq a_1}} r_{a_1a} \int_{[0,\,t]} \Pr\{T_{A_1,1} \leq t - x \mid Y_0 = a\}\,F_{a_1,a}\{dx\} =$$

$$\sum_{a \in A_2 \cup A_3} r_{a_1a}\,F_{a_1,a}(t),\ a_1 \in A_1. \tag{8.26}$$

Now, *if* the distribution corresponding to $F_{a_1,a}$ *has Lebesgue density*, $f_{a_1,a}$ say, and if $g_a(t)$ is defined by

$$g_a(t) = \Pr\{T_{A_1,1} \leq t \mid Y_0 = a\},\ a \in A_1,$$

then, (8.26) can be rewritten by a change of variable in the integral as

$$g_{a_1}(t) = \sum_{\substack{a \in A_1 \\ a \neq a_1}} \int_{[0,\,t]} g_a(x)\, r_{a_1 a}\, f_{a_1,a}(t - x)\, dx + \sum_{a \in A_2 \cup A_3} r_{a_1 a}\, F_{a_1,a}(t),\ a_1 \in A_1. \qquad (8.27)$$

(8.27) is a system of linear Volterra equations of the second kind (convolution equations). Defining the kernel $\mathbf{K}$ by

$$k_{a_1 a}(t, x) = k_{a_1 a}(t - x) = \begin{cases} r_{a_1 a}\, f_{a_1,a}(t - x) & \text{for } a_1,\, a \in A_1,\ a \neq a_1, \\ \\ 0 & \text{for } a_1 = a \in A_1, \end{cases}$$

and putting

$$\mathbf{h}(t) = \{ \sum_{a \in A_2 \cup A_3} r_{a_1 a}\, F_{a_1,a}(t) : a_1 \in A_1 \},$$

(8.27) is seen to be equivalent to

$$\mathbf{g}(t) = \int_{[0,\,t]} \mathbf{K}(t - x)\, \mathbf{g}(x)\, dx + \mathbf{h}(t), \qquad (8.28)$$

with the unknown function in (8.28) being $\mathbf{g}(t) = \{ g_a(t) : a \in A_1 \}$. Equations of the type (8.28) are well-researched, see, for example Linz's book [LIN]. In [LIN], Chapter 12, PASCAL code is provided for the numerical solution of equations of the type (8.28).

8.2.2 S is partitioned into four subsets (n = 3)

Theorem 8.1 is a suitable tool for determining the Laplace transform of sojourn times with a reasonable effort for $n \geq 3$ in special cases only. We shall restrict ourselves to the discussion of the case $n = 3$ which already gives an insight into the difficulties arising in general for $n \geq 3$.

Theorem 8.1, applied consecutively three times, gives of what could be thought of as an analogue of (8.5) for $n = 3$:

$$\psi_{A_1}(\tau_1, \tau_2, \tau_3) =$$

$$\mathbf{K}^*_{A_1 A_2}(\mathbf{1}^T(\mathbf{I}-\mathbf{U})\tau_1) \times$$

$$\left\{ \mathbf{K}^*_{A_2 A_1}(\mathbf{1}^T(\mathbf{I}-\mathbf{U})\tau_2)\, \psi_{A_1}(\mathbf{U}\tau_1, \mathbf{U}\tau_2, \tau_3) + \right.$$

$$\mathbf{K}^*_{A_2 A_3}(\mathbf{1}^T(\mathbf{I}-\mathbf{U})\tau_2)\, \{ \mathbf{K}^*_{A_3 A_1}(\mathbf{1}^T(\mathbf{I}-\mathbf{U})\tau_3)\, \psi_{A_1}(\mathbf{U}\tau_1, \mathbf{U}\tau_2, \mathbf{U}\tau_3) +$$

$$K^*_{A_3A_2}(\mathbf{1}^T(I-U)\tau_3)\,\psi_{A_2}(U\tau_1, U\tau_2, U\tau_3) +$$

$$K^*_{A_3A_4}(\mathbf{1}^T(I-U)\tau_3)\,\mathbf{1}\} +$$

$$K^*_{A_2A_4}(\mathbf{1}^T(I-U)\tau_2)\,\mathbf{1}\Big\} +$$

$$K^*_{A_1A_3}(\mathbf{1}^T(I-U)\tau_1)\times$$

$$\Big\{ K^*_{A_3A_1}(\mathbf{1}^T(I-U)\tau_3)\,\psi_{A_1}(U\tau_1, \tau_2, U\tau_3) +$$

$$K^*_{A_3A_2}(\mathbf{1}^T(I-U)\tau_3)\,\{K^*_{A_2A_1}(\mathbf{1}^T(I-U)\tau_2)\,\psi_{A_1}(U\tau_1, U\tau_2, U\tau_3) +$$

$$K^*_{A_2A_3}(\mathbf{1}^T(I-U)\tau_2)\,\psi_{A_3}(U\tau_1, U\tau_2, U\tau_3) +$$

$$K^*_{A_2A_4}(\mathbf{1}^T(I-U)\tau_2)\,\mathbf{1}\} +$$

$$K^*_{A_3A_4}(\mathbf{1}^T(I-U)\tau_3)\,\mathbf{1}\Big\} +$$

$$K^*_{A_1A_4}(\mathbf{1}^T(I-U)\tau_1)\,\mathbf{1}. \tag{8.29}$$

Unfortunately, unlike in the case $n = 2$, the right hand side of (8.29) involves terms which cannot be reduced to an already known quantity by a repeated application of (8.29); this will namely still contain the terms

$$\psi_{A_1}(U^{\ell}\,\tau_1, U^{\ell}\,\tau_2, \tau_3) \text{ and } \psi_{A_1}(U^{\ell}\,\tau_1, \tau_2, U^{\ell}\,\tau_3)$$

after applying it ℓ times. Furthermore, additional difficulty arises from the ψ_{A_2}-, and ψ_{A_3}- terms on the right hand side of (8.29).

The latter problem is easily avoided though by considering only those semi-Markov processes for which no transitions can occur between A_2 and A_3, i.e.,

$$R_{A_2A_3} = 0, \ \ R_{A_3A_2} = 0. \tag{8.30}$$

This class of processes is still interesting enough as far as reliability applications are concerned: A_1 will then stand for the system's working states; A_2 and A_3 are two distinct sets of states from which error recovery is possible and which do not communicate with each other; A_4 stands for ultimate system breakdown. The initial probability vector will be concentrated on A_1 if at time $t = 0$ the system is operational - in this case it suffices to evaluate ψ_{A_1} to characterize fully the joint distribution of the system's sojourn times. (A software-hardware system along somewhat similar lines was described and analysed by Sumita and Masuda in [SUM2].)

We shall describe a procedure by means of which $\psi_{A_1}(\tau_1, \tau_2, \tau_3)$ can be evaluated *for any given m* if (8.30) holds. It should be noted, however, that it is a recursive scheme and no closed form expression is available for ψ_{A_1} in any other case than n = 2. Assuming thus (8.30), (8.29) can be rewritten as

$$\psi_{A_1}(\tau_1, \tau_2, \tau_3) =$$

$$K^*_{A_1A_2}(1^T(I-U)\tau_1)\, K^*_{A_2A_1}(1^T(I-U)\tau_2)\, \psi_{A_1}(U\tau_1, U\tau_2, \tau_3) +$$

$$K^*_{A_1A_3}(1^T(I-U)\tau_1)\, K^*_{A_3A_1}(1^T(I-U)\tau_3)\, \psi_{A_1}(U\tau_1, \tau_2, U\tau_3) +$$

$$\left\{ K^*_{A_1A_2}(1^T(I-U)\tau_1)\, K^*_{A_2A_4}(1^T(I-U)\tau_2) + \right.$$

$$\left. K^*_{A_1A_3}(1^T(I-U)\tau_1)\, K^*_{A_3A_4}(1^T(I-U)\tau_3) \right\}\, 1 + K^*_{A_1A_4}(1^T(I-U)\tau_1)\, 1. \qquad (8.31)$$

Due to $U^m = 0$, a repeated application of (8.31) will eventually result in an equation expressing $\psi_{A_1}(\tau_1, \tau_2, \tau_3)$ in terms of $\psi_{A_1}(0, \tau_2, 0)$ and $\psi_{A_1}(0, 0, \tau_3)$. But, the latter two quantities are easily obtainable from (8.31). To obtain $\psi_{A_1}(0, 0, \tau_3)$, for example, put in (8.31) $\tau_1 = 0$ and $\tau_2 = 0$:

$$\psi_{A_1}(0,0,\tau_3) = K^*_{A_1A_2}(0)\, K^*_{A_2A_1}(0)\, \psi_{A_1}(0,0,\tau_3) +$$

$$K^*_{A_1A_3}(0)\, K^*_{A_3A_1}(1^T(I-U)\tau_3)\, \psi_{A_1}(0,0, U\tau_3) +$$

$$\left\{ K^*_{A_1A_2}(0)\, K^*_{A_2A_4}(0) + K^*_{A_1A_3}(0)\, K^*_{A_3A_4}(1^T(I-U)\tau_3) \right\}\, 1 + K^*_{A_1A_4}(0)\, 1.$$

$$(8.32)$$

(8.32) allows $\psi_{A_1}(0, 0, \tau_3)$ to be expressed in terms of $\psi_{A_1}(0, 0, U\tau_3)$ as follows

$$\psi_{A_1}(0, 0, \tau_3) =$$

$$(I - H)^{-1} \left\{ K^*_{A_1A_3}(0)\, K^*_{A_3A_1}(1^T(I-U)\tau_3)\, \psi_{A_1}(0, 0, U\tau_3) + \right.$$

$$\{ K^*_{A_1A_2}(0)\, K^*_{A_2A_4}(0) + K^*_{A_1A_3}(0)\, K^*_{A_3A_4}(1^T(I-U)\tau_3) \}\, 1 +$$

$$\left. K^*_{A_1A_4}(0)\, 1 \right\}, \qquad (8.33)$$

where H is defined by (8.17). ($I - H$ is invertible because of (8.18).) Repeated application of (8.33) gives of course $\psi_{A_1}(0, 0, \tau_3)$ by virtue of $U^m = 0$.

8.3 Proof of Theorem 8.1

The proof is based on a repeated application of what could be termed a 'generalised renewal argument'; we have encountered this technique already in Section 2.1.3. In classical renewal theory, (see, e.g., [BHA2], or [KAR]), an event is sought at which the process under consideration is regenerated, i.e., at which a probabilistic replica of the process is started again. There is no such unique event in our case; but, it is possible to identify a sequence of events interrelated by a system of equations in the Laplace transform domain. This system of equations will turn out to be (8.2) and (8.3).

The renewal technique itself has, of course, been available in the semi-Markov context, too, for a long time; see, for example, [NOL], Chapter 3. The crucial (and perhaps novel) feature in our employment of this device lies in the nested structure of our reasoning: in the course of the main proof, which itself is based on the renewal idea, an auxiliary result (Lemma 8.4) will be employed whose proof is also based on the renewal idea. This is, informally speaking, due to the fact that the structure of an individual sojourn time (which can be taken to be the first one) has to be explored *before* a collection of sojourn times can be considered. The key to both steps is the renewal technique.

Without loss of generality, we may restrict ourselves to $i = 1$ in (8.2). We start with an auxiliary result for which some definitions are needed. Define for $a \in A_1$, $a' \in S \setminus A_1$ and $\rho \geq 0$ measures $\mu_{a,a';\rho}$ on $[0, \infty)$ by

$$\mu_{a,a';\rho} (D) = \Pr\{ \{ \rho T_{A_1,1} \in D \} \cap E(a') \mid Y_0 = a \},$$

where the event $E(a')$ is defined by

$$E(a') = \{ Y \text{ visits } a' \text{ immediately after leaving } A_1 \}.$$

Put for $a \in A_1$ and $a' \in S \setminus A_1$

$$\kappa_{a,a'} = \mu_{a,a';1}. \tag{8.34}$$

We shall need the following lemma.

LEMMA 8.4. *For any* $k \in \{2, 3, ..., n+1\}$, *the Laplace transforms* $\{ \kappa_{a,a'}{}^* : a \in A_1, a' \in A_k \}$ *of the measures defined by* (8.34) *are given in matrix form by* (8.3), *i.e., written element-wise,*

$$\kappa_{a,a'}{}^* (\rho) = \kappa^*_{a,a'} (\rho), \rho \in \mathbb{C}^+, a \in A_1, a' \in A_k. \tag{8.35}$$

Furthermore, the Laplace transform of $\mu_{a,a';\rho}$ *is given by*

$$\mu_{a,a';\rho}{}^* (\tau) = \kappa_{a,a'}{}^* (\rho \tau), \tau \in \mathbb{C}^+. \tag{8.36}$$

PROOF OF LEMMA 8.4. It suffices to show (8.35) and (5.36) for $\rho, \tau \in [0, +\infty)$. Assume $\sigma > 0$ to be arbitrary but fixed. An analogue of the renewal argument gives for $t \geq 0$

$$\mu_{a,a';\sigma} ([0, t]) =$$

$$\sum_{i=1}^{n+1} \sum_{a'' \in A_i} \Pr\{\{ \sigma \, T_{A_1,1} \le t \} \cap E(a') \mid X_0 = a, X_1 = a'' \} \Pr\{ X_1 = a'' \mid X_0 = a \} =$$

$$\sum_{a'' \in A_1} \Pr\{\{ \sigma \, T_{A_1,1} \le t \} \cap E(a') \mid X_0 = a, X_1 = a'' \} \Pr\{ X_1 = a'' \mid X_0 = a \} +$$

$$\Pr\{\{ \sigma \, T_{A_1,1} \le t \} \cap E(a') \mid X_0 = a, X_1 = a' \} \Pr\{ X_1 = a' \mid X_0 = a \} =$$

$$\sum_{a'' \in A_1} r_{aa''} \int_{[0, +\infty)} \Pr\{\{ T_{A_1,1} \le \frac{t}{\sigma} - v \} \cap E(a') \mid X_0 = a'' \} \, F_{a,a''}\{dv\} + r_{aa'} \, F_{a,a'}(t / \sigma).$$

The above can be written in terms of $\kappa_{a,a'}$ (defined in (8.34))

$$\kappa_{a,a'} ([0, t / \sigma]) = \sum_{a'' \in A_1} r_{aa''} \int_{[0, +\infty)} \kappa_{a'',a'} ([0, \frac{t}{\sigma} - v]) \, F_{a,a''}\{dv\} + r_{aa'} \, F_{a,a'}(t / \sigma), \, t \ge 0.$$

$$(8.37)$$

Taking Laplace transforms in (8.37), we get for $\rho \ge 0$

$$\kappa_{a,a'}^* (\rho) = \sum_{a'' \in A_1} q_{aa''}^*(\rho) \, \kappa_{a'',a'}^* (\rho) + q_{aa'}^*(\rho),$$

which in matrix form is written as

$$\mathbf{K}_{A_1 A_k}^* (\rho) = \mathbf{Q}_{A_1 A_1}^*(\rho) \, \mathbf{K}_{A_1 A_k}^* (\rho) + \mathbf{Q}_{A_1 A_k}^*(\rho). \tag{8.38}$$

(8.38) implies

$$\mathbf{K}_{A_1 A_k}^* (\rho) = \left(\mathbf{I} - \mathbf{Q}_{A_1 A_1}^*(\rho) \right)^{-1} \mathbf{Q}_{A_1 A_k}^*(\rho),$$

and thus (8.35), since $\mathbf{I} - \mathbf{Q}_{A_1 A_1}^*(\rho)$ is invertible. (This follows from the fact that, for fixed $\rho > 0$, $\mathbf{Q}_{A_1 A_1}^*(\rho)$ can be thought of as the (A_1, A_1)-submatrix of the transition probability matrix of a suitably defined *absorbing* Markov chain, from which we have the invertibility of $\mathbf{I} - \mathbf{Q}_{A_1 A_1}^*(\rho)$ by Theorem 2.1.(a).) To see (8.36), notice that for $\rho > 0$

$$\mu_{a,a';\rho} ([0, t]) = \kappa_{a,a'} ([0, t / \rho]),$$

from which (8.36) follows. For $\rho = 0$, (8.36) follows from

$$\mu_{a,a';0} ([0, t]) \equiv \Pr\{ E(a') \mid Y_0 = a \} = \kappa_{a,a'} ([0, +\infty)) = \kappa_{a,a'}^* (0). \quad \blacksquare$$

PROOF OF THEOREM 8.1. The reasoning applied in the proof of Lemma 8.4 to show that $\mathbf{I} - \mathbf{Q}^*_{A_1 A_1}(\rho)$ is invertible for $\rho > 0$, can be applied to show the same for $\rho \in \mathbb{C}^+$ since, element-wise, $|\mathbf{Q}^*_{A_1 A_1}(\rho)| \leq \mathbf{Q}^*_{A_1 A_1}(|\rho|)$. We are going to examine the distribution of the random variable W defined for a fixed set of vectors $\tau_1, ..., \tau_n \in [0, \infty)^m$ by

$$W(\tau_1, ..., \tau_n) = \sum_{i=1}^{n} (T_{A_i,1}, ..., T_{A_i,m}) \tau_i.$$

It is for $a \in A_1$,

$$\Pr\{ W(\tau_1, ..., \tau_n) \leq t \mid Y_0 = a \} =$$

$$\sum_{k=2}^{n+1} \sum_{a' \in A_k} \Pr\{ \{ W(\tau_1, ..., \tau_n) \leq t \} \cap E(a') \mid Y_0 = a \}. \tag{8.39}$$

If $k \in \{2, ..., n\}$ and $a' \in A_k$ then,

$$\Pr\{ \{ W(\tau_1, ..., \tau_n) \leq t \} \cap E(a') \mid Y_0 = a \} =$$

$$\Pr\{\{ \tau_{1,1} T_{A_1,1} + \sum_{j=2}^{m} \tau_{1j} T_{A_1 j} + \sum_{i=2}^{n} (T_{A_i,1}, ..., T_{A_i,m}) \tau_i \leq t \} \cap E(a') \mid Y_0 = a \} =$$

$$\int_{[0, t]} \Pr\{ \sum_{j=1}^{m-1} \tau_{1,j+1} T_{A_1 j} + \sum_{i=2}^{n} (T_{A_i,1}, ..., T_{A_i,m}) \tau_i \leq t - v \mid Y_0 = a' \} \mu_{a,a';\tau_{1,1}} \{dv\} =$$

$$\int_{[0, t]} \Pr\{ W(U\tau_1, \tau_2, ..., \tau_n) \leq t - v \mid Y_0 = a' \} \mu_{a,a';\tau_{1,1}} \{dv\}. \tag{8.40}$$

If $a' \in A_{n+1}$, we have

$$\Pr\{ \{ W(\tau_1, ..., \tau_n) \leq t \} \cap E(a') \mid Y_0 = a \} =$$

$$\Pr\{ \{ \tau_{1,1} T_{A_1,1} \leq t \} \cap E(a') \mid Y_0 = a \} = \mu_{a,a';\tau_{1,1}} ([0, t]). \tag{8.41}$$

Substituting (8.40) and (8.41) into (8.39) gives

$$\Pr\{ W(\tau_1, ..., \tau_n) \leq t \mid Y_0 = a \} =$$

$$\sum_{k=2} \sum_{a' \in A_k} \int_{[0, t]} \Pr\{ W(U\tau_1, \tau_2, ..., \tau_n) \le t - v \mid Y_0 = a' \} \, \mu_{a,a';\tau_{1,1}} \{dv\} +$$

$$\sum_{a' \in A_{n+1}} \mu_{a,a';\tau_{1,1}} ([0, t]). \tag{8.42}$$

Taking Laplace transforms in (8.42), we get for $\rho \ge 0$

$$E\left[\exp\{ - W(\tau_1, ..., \tau_n) \rho \} \mid Y_0 = a \right] =$$

$$\sum_{k=2}^{n} \sum_{a' \in A_k} E\left[\exp\{ - W(U\tau_1, \tau_2, ..., \tau_n) \rho \} \mid Y_0 = a' \right] \mu_{a,a';\tau_{1,1}}{}^* (\rho) +$$

$$\sum_{a' \in A_{n+1}} \mu_{a,a';\tau_{1,1}}{}^* (\rho). \tag{8.43}$$

Putting $\rho = 1$ in (8.43), we see by (8.35), (8.36) and

$$\tau_{1,1} = \mathbf{1}^T (\mathbf{I} - \mathbf{U}) \, \tau_1,$$

that

$$\psi_a (\tau_1, ..., \tau_n) = \sum_{k=2}^{n} \sum_{a' \in A_k} \psi_{a'} (U\tau_1, \tau_2, ..., \tau_n) \, \kappa^*_{a,a'} (\mathbf{1}^T (\mathbf{I} - \mathbf{U}) \, \tau_1) +$$

$$\sum_{a' \in A_{n+1}} \kappa^*_{a,a'} (\mathbf{1}^T (\mathbf{I} - \mathbf{U}) \, \tau_1). \tag{8.44}$$

The matrix form of (8.44) is (8.2).   ∎

CHAPTER 9

THE NUMBER OF VISITS TO A SUBSET OF THE STATE SPACE BY AN IRREDUCIBLE SEMI-MARKOV PROCESS DURING A FINITE TIME INTERVAL: MOMENT RESULTS

9.1 Preliminaries on the moments of $M_{A_1}(t)$

The present and the next chapter are devoted to the examination of the variable $M_{A_1}(t)$ from Chapter 5, now under the semi-Markov assumption. The notation and the assumptions of Chapter 8 apply. We assume here that Y is irreducible and that the state space is partitioned into two subsets, i.e., $S = A_1 \cup A_2$. Furthermore, we require that the holding time cumulative distribution functions satisfy $F_{s,s'}(0) < 1$ for all s, s' $\in$ S, s $\neq$ s'. As we want to concentrate here on the moments of $M_{A_1}(t)$, the following elementary lemma will prove useful.

LEMMA 9.1. *Let M be a random variable taking values in* { 0, 1, 2, ... }. *If the kth moment of M is finite, then it obeys the recurrence relation*

$$E[M^k] = \sum_{\ell=0}^{k-1} \binom{k}{\ell} \sum_{i=1}^{\infty} i^\ell \Pr\{ M \geq i \} - \sum_{\ell=1}^{k-1} \binom{k}{\ell} E[M^\ell], \; k = 1, 2, \dots . \quad (9.1)$$

PROOF OF LEMMA 9.1. Let the functions $f_0, f_1, \dots$ be defined by

$$f_k(m) = \sum_{i=1}^{m} i^k, \; m = 0, 1, \dots .$$

Then, obviously,

$$f_k(m+1) = f_k(m) + \sum_{\ell=0}^{k} \binom{k}{\ell} m^\ell. \quad (9.2)$$

On the other hand, it is easily seen that

$$f_k(m+1) = 1 + \sum_{\ell=0}^{k} \binom{k}{\ell} f_\ell(m). \quad (9.3)$$

Equating (9.2) and (9.3), we get

$$m^k = 1 + \sum_{\ell=0}^{k-1} \binom{k}{\ell} (f_\ell(m) - m^\ell). \tag{9.4}$$

From (9.4), we have

$$E[M^k] = 1 + \sum_{\ell=0}^{k-1} \binom{k}{\ell} (E[f_\ell(M)] - E[M^\ell]). \tag{9.5}$$

(9.1) now follows from (9.5) by noting that

$$E[f_\ell(M)] = \sum_{i=1}^{\infty} i^\ell \Pr\{ M \geq i \}. \qquad\qquad \blacksquare$$

Assume now that the kth moment of $M_{A_1}(t)$ is finite. Let for any interval $[0, t)$ with $t > 0$, and $\ell = 0, ..., k$, the interval functions $U_\ell([0, t))$ be defined by

$$U_\ell([0, t)) = \sum_{i=1}^{\infty} i^\ell \Pr\{ M_{A_1}(t) \geq i \}. \tag{9.6}$$

Then, by (9.1), we get the following recurrence relation for the moments of $M_{A_1}(t)$ in terms of $U_0([0, t))$, $U_1([0, t))$, ...

$$E[(M_{A_1}(t))^k] = \sum_{\ell=0}^{k-1} \binom{k}{\ell} U_\ell([0, t)) - \sum_{\ell=1}^{k-1} \binom{k}{\ell} E[(M_{A_1}(t))^\ell]. \tag{9.7}$$

From (9.7), say, the first three moments of $M_{A_1}(t)$ read as follows

$$E[M_{A_1}(t)] = U_0([0, t)), \tag{9.8}$$

$$E[(M_{A_1}(t))^2] = 2 U_1([0, t)) - U_0([0, t)), \tag{9.9}$$

$$E[(M_{A_1}(t))^3] = 3 U_2([0, t)) - 3 U_1([0, t)) + U_0([0, t)). \tag{9.10}$$

It is seen from (9.7) - (9.10) that in order to examine the behaviour of the ℓth moment of $M_{A_1}(t)$ for $t \to +\infty$, it suffices to consider $U_0([0, t))$, ..., $U_{\ell-1}([0, t))$ as $t \to +\infty$. The latter is of course best performed in the Laplace transform domain by using a Tauberian argument [BHA2]. We shall make use of the fact that by (9.6) a *finite measure* is defined on (the Borel subsets of) $[0, +\infty)$; this measure, too, will be denoted by U_ℓ.

The main goal of the present chapter is the derivation of a closed form expression for the Laplace transform of the measures U_ℓ. This result will be presented in Section 9.2. The proof, again by a renewal argument, is given Section 9.3. In Section 9.4, the theory is applied to the reliability Models 6 and 7 from Section 1.2.2. For the more complicated of these two models (Model 7), a Tauberian argument on the Laplace transform of U will be used in conjunction with (9.8) and (9.9) to deduce the behaviour of the first two moments of $M_{A_1}(t)$ for $t \to +\infty$.

The present chapter is an elaboration of [CSE2]. An interesting paper of related interest is [MAS2] in which the authors derived a closed form expression for the Laplace transform generating fuction of $M_{A_1}(t)$, i.e., for the double transform

$$\sum_{m=0}^{\infty} z^m \, \mathbf{h}^*(\tau \, ; m) \ (\, \tau \in \mathbb{C}^+, z \in \mathbb{C}, |z| \leq 1),$$

where the Laplace transforms $\mathbf{h}^*(\tau \, ; m)$ are defined by

$$\mathbf{h}^*(\tau \, ; m) = \int_{[0, +\infty)} \exp\{-\tau \, t\} \, \mathbf{h}(t \, ; m) \, dt, \qquad (9.11)$$

$$\mathbf{h}(t \, ; m) = \{\, \Pr\{\, M_{A_1}(t) = m \mid Y_0 = s \,\} : s \in S \,\}, m = 0, 1, \ldots . \qquad (9.12)$$

This is, of course, not immediately comparable with what we set out to achieve. In addition, however, they derive an expression for the Laplace transform of the moments of $M_{A_1}(t)$: this result is highly relevant in that it offers a related alternative to the approach proposed here for the asymptotic analysis of the moments of $M_{A_1}(t)$. It should be added that the investigations in [MAS2] are different from a methodological point of view: there, the analysis is based on a trivariate process whose first component is Y.

Let us note in passing that the family of functions defined in (9.12) will be examined in the next chapter where a method for their numerical evaluation is presented which is based on a system of equations involving the Laplace transforms in (9.11).

9.2 Main result: the Laplace transform of the measures U_ℓ

The measures U_ℓ, defined by (9.6), can be expressed in terms of sojourn times. We denote, as in Chapter 4, the sum of the first m sojourns of Y in A_i by $TS_{A_i,m} = T_{A_i,1} + \ldots + T_{A_i,m}$. ($T_{A_i,0} \equiv 0$ by definition.) Then, it is

$$\{\, M_{A_1}(t) \geq m \,\} = \{\, Y_0 \in A_2, TS_{A_2,m} + TS_{A_1,m-1} < t \,\} \cup$$

$$\{\, Y_0 \in A_1, TS_{A_2,m-1} + TS_{A_1,m-1} < t \,\}.$$

This gives the representation

$$U_\ell([0, t)) =$$

$$\sum_{a_2 \in A_2} V_{a_2,\ell}([0, t)) \, \Pr\{\, Y_0 = a_2 \,\} + \sum_{a_1 \in A_1} W_{a_1,\ell}([0, t)) \, \Pr\{\, Y_0 = a_1 \,\}, \quad (9.13)$$

where the measures $V_{a_2,\ell}$ and $W_{a_1,\ell}$ are respectively defined for any measurable $D \subseteq [0, +\infty)$

by

$$V_{a_2,\ell}(D) = \sum_{m=1}^{\infty} m^\ell \, Pr\{ \, TS_{A_2,m} + TS_{A_1,m-1} \in D \mid Y_0 = a_2 \, \}, \tag{9.14}$$

$$W_{a_1,\ell}(D) = \sum_{m=1}^{\infty} m^\ell \, Pr\{ \, TS_{A_2,m-1} + TS_{A_1,m-1} \in D \mid Y_0 = a_1 \, \}. \tag{9.15}$$

We note for later reference that (9.15) can be rewritten as

$$W_{a_1,\ell}(D) = \varepsilon_0(D) + \sum_{m=2}^{\infty} m^\ell \, Pr\{ \, TS_{A_2,m-1} + TS_{A_1,m-1} \in D \mid Y_0 = a_1 \, \}, \tag{9.16}$$

where ε_0 stands for the probability measure assigning unity to the origin. The Laplace transforms of $V_{a_2,\ell}$ and $W_{a_1,\ell}$ are defined for $\tau \in \mathbb{C}^+$, as usual, by

$$V_{a_2,\ell}^{*}(\tau) = \int_{[0,\,+\infty)} \exp\{ -\tau t \} \, V_{a_2,\ell}\{dt\},$$

$$W_{a_1,\ell}^{*}(\tau) = \int_{[0,\,+\infty)} \exp\{ -\tau t \} \, W_{a_1,\ell}\{dt\}.$$

The Laplace transform of the measure U_ℓ is then obtained from (9.13) as

$$U_\ell^{*}(\tau) = \int_{[0,\,+\infty)} \exp\{ -\tau t \} \, U_\ell\{dt\} = \boldsymbol{\alpha}_{A_2}^{T} \, \mathbf{V}_\ell^{*}(\tau) + \boldsymbol{\alpha}_{A_1}^{T} \, \mathbf{W}_\ell^{*}(\tau), \tag{9.17}$$

where $\mathbf{V}_\ell^{*}(\tau)$ and $\mathbf{W}_\ell^{*}(\tau)$ stand for the column vectors $\{ \, V_{a_2,\ell}^{*}(\tau) : a_2 \in A_2 \, \}$ and $\{ \, W_{a_1,\ell}^{*}(\tau) : a_1 \in A_1 \, \}$ respectively. In what follows, $i, j \in \{ \, 1, 2 \, \}$ will be related by $i \neq j$. The next theorem is the main result.

THEOREM 9.2. *For* $\tau \in \mathbb{C}^+$, *let* $\mathbf{K}^{*}_{A_iA_j}(\tau)$ *be defined by* (8.3). *Then, the matrix* $\mathbf{M}(A_i, A_j)(\tau)$, *defined by*

$$\mathbf{M}(A_i, A_j)(\tau) = \left\{ \mathbf{I} - \mathbf{K}^{*}_{A_iA_j}(\tau) \, \mathbf{K}^{*}_{A_jA_i}(\tau) \right\}^{-1},$$

exists and the following representations hold:

$$\mathbf{V}_0^{*}(\tau) = \mathbf{M}(A_2, A_1)(\tau) \, \mathbf{K}^{*}_{A_2A_1}(\tau) \, \mathbf{1}, \tag{9.18}$$

$$\mathbf{W}_0^{*}(\tau) = \mathbf{M}(A_1, A_2)(\tau) \, \mathbf{1}, \tag{9.19}$$

and for $\ell = 1, 2, \ldots$,

$$V_\ell^* (\tau) = V_0^* (\tau) + \left(M(A_2, A_1)(\tau) - I \right) \sum_{k=0}^{\ell-1} \binom{\ell}{k} V_k^* (\tau), \qquad (9.20)$$

$$W_\ell^* (\tau) = W_0^* (\tau) + \left(M(A_1, A_2)(\tau) - I \right) \sum_{k=0}^{\ell-1} \binom{\ell}{k} W_k^* (\tau). \qquad (9.21)$$

The proof of Theorem 9.2 is the subject of the next section.

9.3 Proof of Theorem 9.2

The method of proof bears some resemblance to the one employed to prove Theorem 8.1: the renewal technique is used again twice in succession and one of the measures used in Section 8.3 will reappear. For $a_i \in A_i$, define the event $E_i(a_i)$ by

$E_i(a_i) = \{$ (re-)entry by Y into A_i is via $a_i \}$,

i.e., more formally,

$$E_i(a_i) = \{ Y_{T_{A_j,1}+0} = a_i, Y_0 \in A_j \} \cup \{ Y_{T_{A_1,1}+T_{A_2,1}+0} = a_i, Y_0 \in A_i \}. \qquad (9.22)$$

Let us now define the family of finite measures $\{ \kappa_{s,s'} : s, s' \in S \}$ on $[0, +\infty)$ as follows. For any measurable $D \subseteq [0, +\infty)$, put

$$\kappa_{a_j,a_i} (D) = \Pr\{ \{ T_{A_j,1} \in D \} \cap E_i(a_i) \mid Y_0 = a_j \}, \qquad (9.23)$$

$$\kappa_{a_i',a_i} (D) = \Pr\{ \{ T_{A_1,1} + T_{A_2,1} \in D \} \cap E_i(a_i) \mid Y_0 = a_i' \}, \qquad (9.24)$$

where $a_i, a_i' \in A_i$ and $a_j \in A_j = S \setminus A_i$. Define the Laplace transforms of these, as usual, by

$$\kappa_{s,s'}^* (\tau) = \int_{[0, +\infty)} \exp\{ -\tau t \} \kappa_{s,s'} \{dt\}, \tau \in \mathbb{C}^+.$$

We know the measures in (9.23) from Chapter 8, equation (8.34). We also know from Lemma 8.4 (see (8.35)) that the Laplace transforms of the measures in (9.23) are given by

$$\{ \kappa_{a_j,a_i}^* (\tau) : a_j \in A_j, a_i \in A_i \} = \left(I - Q^*_{A_jA_j}(\tau) \right)^{-1} Q^*_{A_jA_i}(\tau). \qquad (9.25)$$

The next lemma deals with the Laplace transforms of the measures in (9.24).

LEMMA 9.3. *The Laplace transforms of the measures in (9.24) are given by*

$$\{ \kappa_{a_i',a_i}{}^* (\tau) : a_i', a_i \in A_i \} =$$

$$\left(\mathbf{I} - \mathbf{Q}^*{}_{A_iA_i}(\tau)\right)^{-1} \mathbf{Q}^*{}_{A_iA_j}(\tau) \left(\mathbf{I} - \mathbf{Q}^*{}_{A_jA_j}(\tau)\right)^{-1} \mathbf{Q}^*{}_{A_jA_i}(\tau). \qquad (9.26)$$

PROOF OF LEMMA 9.3. We use the renewal argument to see that

$$\kappa_{a_i',a_i} ([0, t]) =$$

$$\sum_{\substack{a_i" \in A_i \\ a_i" \neq a_i'}} r_{a_i'a_i"} \, \Pr\{\{ TS_{A_1,1} + TS_{A_2,1} \leq t \} \cap E_i(a_i) \mid X_0 = a_i', X_1 = a_i" \} +$$

$$\sum_{a_j \in A_j} r_{a_i'a_j} \, \Pr\{\{ TS_{A_1,1} + TS_{A_2,1} \leq t \} \cap E_i(a_i) \mid X_0 = a_i', X_1 = a_j \} =$$

$$\sum_{\substack{a_i" \in A_i \\ a_i" \neq a_i'}} r_{a_i'a_i"} \int_{[0, +\infty)} \Pr\{\{ TS_{A_1,1} + TS_{A_2,1} \leq t - s \} \cap E_i(a_i) \mid X_0 = a_i" \} \, F_{a_i',a_i"} \{ds\} +$$

$$\sum_{a_j \in A_j} r_{a_i'a_j} \int_{[0, +\infty)} \Pr\{\{ TS_{A_j,1} \leq t - s \} \cap E_i(a_i) \mid X_0 = a_j \} \, F_{a_i',a_j} \{ds\} =$$

$$\sum_{\substack{a_i" \in A_i \\ a_i" \neq a_i'}} r_{a_i'a_i"} \int_{[0, +\infty)} \kappa_{a_i",a_i} ([0, t - s]) \, F_{a_i',a_i"} \{ds\} +$$

$$\sum_{a_j \in A_j} r_{a_i'a_j} \int_{[0, +\infty)} \kappa_{a_j,a_i} ([0, t - s]) \, F_{a_i',a_j} \{ds\}. \qquad (9.27)$$

(9.27) in the Laplace transform domain reads as follows

$$\{ \kappa_{a_i',a_i}{}^* (\tau) : a_i', a_i \in A_i \} = \mathbf{Q}^*{}_{A_iA_i}(\tau) \{ \kappa_{a_i',a_i}{}^* (\tau) : a_i', a_i \in A_i \} +$$

$$\mathbf{Q}^*{}_{A_iA_j}(\tau) \{ \kappa_{a_j,a_i}{}^* (\tau) : a_j \in A_j, a_i \in A_i \}.$$

This shows that

$$\{ \kappa_{a_i',a_i}{}^* (\tau) : a_i', a_i \in A_i \} =$$

$$\left(\mathbf{I} - \mathbf{Q}^*{}_{A_iA_i}(\tau)\right)^{-1} \mathbf{Q}^*{}_{A_iA_j}(\tau) \{ \kappa_{a_j,a_i}{}^* (\tau) : a_j \in A_j, a_i \in A_i \}. \qquad (9.28)$$

(9.25) and (9.28) give (9.26). ∎

We note, in passing, that (9.25) and (9.26) can be considered to be evaluating the matrix assembled as

$$
\left[
\begin{array}{c|c}
\mathbf{K}_{A_1A_1}{}^{*}(\tau) & \mathbf{K}_{A_1A_2}{}^{*}(\tau) \\
\hline
\mathbf{K}_{A_2A_1}{}^{*}(\tau) & \mathbf{K}_{A_2A_2}{}^{*}(\tau)
\end{array}
\right].
\tag{9.29}
$$

(9.25) shows that the off-diagonal blocks in (9.29) are equal to the matrices $\mathbf{K}^{*}_{A_iA_j}(\tau)$ which we first met in Chapter 8. Thus, the diagonal blocks in (9.29) are by Lemma 9.3

$$
\mathbf{K}_{A_iA_i}{}^{*}(\tau) = \mathbf{K}_{A_iA_j}{}^{*}(\tau)\,\mathbf{K}_{A_jA_i}{}^{*}(\tau) = \mathbf{K}^{*}_{A_iA_j}(\tau)\,\mathbf{K}^{*}_{A_jA_i}(\tau).
\tag{9.30}
$$

Henceforth, for brevity, the blocks in (9.29) will be used in preference to the element-wise notation.

We are now in a position to prove Theorem 9.2.

Proof of Theorem 9.2. The matrix $\mathbf{M}(A_i, A_j)(\tau)$ exists because, element-wise,

$$
\left| \left(\mathbf{I} - \mathbf{Q}^{*}_{A_iA_i}(\tau)\right)^{-1} \mathbf{Q}^{*}_{A_iA_j}(\tau) \left(\mathbf{I} - \mathbf{Q}^{*}_{A_jA_j}(\tau)\right)^{-1} \mathbf{Q}^{*}_{A_jA_i}(\tau) \right| \leq \mathbf{H},
$$

where $\mathbf{H} = \mathbf{H}(A_i, A_j)$ is defined in terms of the transition probability matrix $\mathbf{R}$ of the embedded Markov chain of Y according to (8.17). The invertibility of $\left(\mathbf{I} - \mathbf{M}(A_i, A_j)(\tau)\right)$ now readily follows by (8.18).

To establish (9.18) and (9.20), we note that a renewal argument shows for $t > 0$ and $m = 2$, 3, ... that

$$
\Pr\{ TS_{A_2,m} + TS_{A_1,m-1} \leq t \mid Y_0 = a_2 \} =
$$

$$
\sum_{a_2' \in A_2} \Pr\left\{ \left\{ T_{A_2,1} + T_{A_1,1} + \sum_{i=2}^{m} T_{A_2,i} + \sum_{i=2}^{m-1} T_{A_1,i} \leq t \right\} \cap E_2(a_2') \,\Big|\, Y_0 = a_2 \right\} =
$$

$$
\sum_{a_2' \in A_2} \int_{[0,\,t]} \Pr\{ TS_{A_2,m-1} + TS_{A_1,m-2} \leq t - s \mid Y_0 = a_2' \}\, \kappa_{a_2,a_2'}\{ds\}.
\tag{9.31}
$$

For $m = 1$, we have (with no renewal argument involved)

$$
\Pr\{ TS_{A_2,1} \leq t \mid Y_0 = a_2 \} =
$$

$$
\sum_{a_1 \in A_1} \Pr\{ \{ TS_{A_2,m} \leq t \} \cap E(a_1) \mid Y_0 = a_2 \} = \sum_{a_1 \in A_1} \kappa_{a_2,a_1}([0,\,t]).
\tag{9.32}
$$

Substituting (9.31) and (9.32) into (9.14) with $\ell = 0$ and $D = [0, t]$, we get

$$V_{a_2,0}([0, t]) = \sum_{a_1 \in A_1} \kappa_{a_2,a_1}([0, t]) +$$

$$\sum_{a_2' \in A_2} \sum_{m=2}^{\infty} \int_{[0, t]} \Pr\{ TS_{A_2,m-1} + TS_{A_1,m-2} \leq t - s \mid Y_0 = a_2' \} \kappa_{a_2,a_2'} \{ds\},$$

i.e., in terms of Laplace transforms,

$$V_{a_2,0}^*(\tau) = \sum_{a_1 \in A_1} \kappa_{a_2,a_1}^*(\tau) + \sum_{a_2' \in A_2} V_{a_2',0}^*(\tau) \kappa_{a_2,a_2'}^*(\tau),$$

or, in matrix form,

$$\mathbf{V}_0^*(\tau) = \mathbf{K}_{A_2 A_1}^*(\tau)\, \mathbf{1} + \mathbf{K}_{A_2 A_2}^*(\tau)\, \mathbf{V}_0^*(\tau). \tag{9.33}$$

(9.25) and (9.33) now imply (9.18) by (9.30). Equation (9.20) is shown by noting that from (9.14) for $\ell \geq 1$ and $D = [0, t]$, we have by (9.31)

$$V_{a_2,\ell}([0, t]) =$$

$$V_{a_2,0}([0, t]) + \sum_{k=1}^{\ell} \binom{\ell}{k} \sum_{m=2}^{\infty} (m - 1)^k \Pr\{ TS_{A_2,m} + TS_{A_1,m-1} \leq t \mid Y_0 = a_2 \} =$$

$$V_{a_2,0}([0, t]) + \sum_{k=1}^{\ell} \binom{\ell}{k} \sum_{a_2' \in A_2} \int_{[0, t]} \sum_{m=2}^{\infty} (m - 1)^k \times$$

$$\Pr\{ TS_{A_2,m-1} + TS_{A_1,m-2} \leq t - s \mid Y_0 = a_2' \} \kappa_{a_2,a_2'} \{ds\},$$

which in terms of Laplace transforms becomes

$$V_{a_2,\ell}^*(\tau) = V_{a_2,0}^*(\tau) + \sum_{k=1}^{\ell} \binom{\ell}{k} \sum_{a_2' \in A_2} V_{a_2',k}^*(\tau) \kappa_{a_2,a_2'}^*(\tau),$$

or, in matrix form,

$$\mathbf{V}_\ell^*(\tau) = \mathbf{V}_0^*(\tau) + \sum_{k=1}^{\ell} \binom{\ell}{k} \mathbf{K}_{A_2 A_2}^*(\tau)\, \mathbf{V}_k^*(\tau). \tag{9.34}$$

(9.34) now implies (9.20) since by (9.30),

$$\left(\mathbf{I} - \mathbf{K}_{A_iA_i}{}^*(\tau)\right)^{-1} = \mathbf{M}(A_i, A_j)(\tau),$$ (9.35)

and thus

$$\left(\mathbf{I} - \mathbf{K}_{A_iA_i}{}^*(\tau)\right)^{-1}\mathbf{K}_{A_iA_i}{}^*(\tau) = \left(\mathbf{I} - \mathbf{K}_{A_iA_i}{}^*(\tau)\right)^{-1} - \mathbf{I} = \mathbf{M}(A_i, A_j)(\tau) - \mathbf{I}.$$ (9.36)

The proofs of (9.19) and (9.21) are rather similar to that of (9.18) and (9.20). Instead of (9.31), we now apply

$$\Pr\{ TS_{A_2,m} + TS_{A_1,m} \le t \mid Y_0 = a_1 \} =$$

$$\sum_{a_1' \in A_1} \int_{[0, t]} \Pr\{ TS_{A_2,m-1} + TS_{A_1,m-1} \le t - s \mid Y_0 = a_1' \} \kappa_{a_1,a_1'}\{ds\}.$$ (9.37)

(Again, a renewal argument is used to obtain (9.37).) Substituting (9.37) into (9.16) (with $\ell = 0$) gives for $t > 0$, as before,

$$W_{a_1,0}([0, t]) = 1 + \sum_{a_1' \in A_1} \int_{[0, t]} W_{a_1',0}([0, t - s]) \kappa_{a_1,a_1'}\{ds\},$$

which in terms of Laplace transforms and vector notation can be written as

$$\mathbf{W}_0{}^*(\tau) = 1 + \mathbf{K}_{A_1A_1}{}^*(\tau)\mathbf{W}_0{}^*(\tau).$$

This implies (9.19) by (9.35). For $\ell \ge 1$, we have, now by (9.15) and (9.37), that

$$W_{a_1,\ell}([0, t]) = W_{a_1,0}([0, t]) + \sum_{k=1}^{\ell} \binom{\ell}{k} \sum_{a_1' \in A_1} \int_{[0, t]} W_{a_1',k}([0, t - s]) \kappa_{a_1,a_1'}\{ds\},$$

or, in terms of Laplace transforms and matrices,

$$\mathbf{W}_\ell{}^*(\tau) = \mathbf{W}_0{}^*(\tau) + \sum_{k=1}^{\ell} \binom{\ell}{k} \mathbf{K}_{A_1A_1}{}^*\mathbf{W}_k{}^*(\tau).$$

This proves (9.21) by (9.35) and (9.36). ∎

9.4 Reliability applications

9.4.1 The alternating renewal process

Our first application of Theorem 9.2 is to the alternating renewal process, introduced here in Section 1.2.2 as Model 6. Put $A_1 = G = \{ 1 \}$ and $A_2 = B = \{ 2 \}$. We are interested in $M_{A_1}(t)$, the number of working periods during $[0, t]$. It is

$$
\mathbf{Q}^*(\tau) = \begin{array}{c} \\ 1 \\ 2 \end{array}\begin{array}{cc} 1 & 2 \\ \left[\begin{array}{cc} 0 & \phi(\tau) \\ \mathcal{Y}(\tau) & 0 \end{array} \right]. \end{array}
$$

From Theorem 9.2, we have the (now scalar) quantities

$$
M(A_1, A_2)(\tau) = M(A_2, A_1)(\tau) = \frac{1}{1 - \phi(\tau)\,\mathcal{Y}(\tau)},
$$

$$
V_0^*(\tau) = \frac{\mathcal{Y}(\tau)}{1 - \phi(\tau)\,\mathcal{Y}(\tau)}, \quad W_0^*(\tau) = \frac{1}{1 - \phi(\tau)\,\mathcal{Y}(\tau)},
$$

and therefore,

$$
U_0^*(\tau) = \frac{\alpha_2\,\mathcal{Y}(\tau) + \alpha_1}{1 - \phi(\tau)\,\mathcal{Y}(\tau)}.
$$

If $\alpha_2 = 1$, $U_0^*(\tau) / \tau$ is the Laplace transform of the renewal function for the renewal process which is defined by transitions of Y from $\{ 2 \}$ to $\{ 1 \}$. (Notice that each such transition marks the beginning of a new sojourn of Y in $A_1 = G = \{ 1 \}$.) See, equation (8.4.2) in [BHA2].

9.4.2 Two units in parallel with an arbitrary change out time distribution

Our second application of Theorem 9.2 is to Model 7 in Section 1.2.2. Here, the relevant subsets of S are $A_1 = G = \{ 1, 2 \}$ and $A_2 = B = \{ 3, 4 \}$. Again, the number of working periods during $[0, t]$, $M_{A_1}(t)$, is of interest. Assume that we start with both units working, i.e.,

$$
\alpha_{A_1} = (1, 0)^T, \quad \alpha_{A_2} = \mathbf{0}. \tag{9.38}
$$

Denote by ξ^* the Laplace transform of the variable ξ, i.e., of the time needed to identify the faulty unit (the 'change out time'). Then, it is

$$Q^*(\tau) = \begin{array}{c} \\ 1 \\ 2 \\ 3 \\ 4 \end{array} \begin{bmatrix} \overset{1}{0} & \overset{2}{0} & \overset{3}{\dfrac{2\lambda}{2\lambda+\tau}} & \overset{4}{0} \\[2ex] \dfrac{\mu}{\mu+\lambda+\tau} & 0 & 0 & \dfrac{\lambda}{\mu+\lambda+\tau} \\[2ex] 0 & \xi^*(\tau) & 0 & 0 \\[2ex] 0 & \dfrac{2\mu}{2\mu+\tau} & 0 & 0 \end{bmatrix}.$$

It is easily seen that

$$K_{A_1 A_1}^*(\tau) = \begin{array}{c} \\ 1 \\ 2 \end{array} \begin{bmatrix} \overset{1}{0} & \overset{2}{\dfrac{\xi^*(\tau)2\lambda}{2\lambda+\tau}} \\[2ex] 0 & g(\tau) \end{bmatrix},$$

with $g(\tau)$ defined by

$$g(\tau) = \frac{2\lambda\mu}{\mu+\lambda+\tau} \left\{ \frac{\xi^*(\tau)}{2\lambda+\tau} + \frac{1}{2\mu+\tau} \right\}.$$

It is thus

$$M(A_1, A_2)(\tau) = \frac{1}{1 - g(\tau)} \begin{bmatrix} 1-g(\tau) & \dfrac{\xi^*(\tau)2\lambda}{2\lambda+\tau} \\[2ex] 0 & 1 \end{bmatrix}. \tag{9.39}$$

From (9.17) and (9.38) it follows that for $\ell = 0, 1, \ldots$

$$U_\ell^*(\tau) = \alpha_{A_1}{}^T W_\ell^*(\tau) = W_{1,\ell}^*(\tau). \tag{9.40}$$

Substituting (9.39) into (9.19) and (9.21), we see that

$$W_{1,0}^*(\tau) = 1 + \frac{2\xi^*(\tau)\lambda}{(2\lambda+\tau)(1-g(\tau))}, \tag{9.41}$$

$$W_{1,1}{}^*(\tau) = (1, 0)\left(M(A_1, A_2)(\tau)\right)^2 \mathbf{1} = 1 + \frac{2\xi^*(\tau)\lambda}{(2\lambda+\tau)(1-g(\tau))} + \frac{2\xi^*(\tau)\lambda}{(2\lambda+\tau)(1-g(\tau))^2}.$$

$$(9.42)$$

In actual applications, the Laplace transform of ξ and thus the formulae in (9.41) and (9.42) will be available either as closed form expressions or as subroutines. Thus, by (9.40), the Laplace transform of the distribution function of U_ℓ will also be available (since it is the function $U_\ell{}^*(\tau)/\tau$, $\tau \in \mathbb{C}^+\backslash\{0\}$). This means that a *numerical* computation of $U_\ell([0, t])$ is possible for any fixed $t > 0$ by numerical inversion of Laplace transforms. The methods for the numerical inversion of Laplace transforms are numerous; one of them will be applied in next chapter to obtain the probability mass function of $M_{A_1}(t)$.

To explore the moments of $M_{A_1}(t)$ as $t \rightarrow +\infty$, we note that by $g(0) = 1$, (9.41) and (9.42) imply for $\tau \rightarrow 0$ the following

$$\tau\,W_{1,0}{}^*(\tau) \rightarrow -\frac{1}{g'(0)}, \quad \tau^2\,W_{1,1}{}^*(\tau) \rightarrow -\frac{1}{(g'(0))^2},$$

where

$$g'(0) = -\frac{\mu + \lambda}{2\,\mu\,\lambda} - \frac{\mu}{\mu + \lambda}\,E(\xi).$$

By a Tauberian argument (see, e.g., Theorem B.2.2 in [BHA2]) it seen that for $t \rightarrow +\infty$,

$$\frac{W_{1,0}([0, t])}{t} \rightarrow -\frac{1}{g'(0)}, \quad \frac{W_{1,1}([0, t])}{t^2} \rightarrow -\frac{1}{2(g'(0))^2}.$$

Thus, by (9.8) and (9.9), we get for $t \rightarrow +\infty$,

$$\frac{E[M_{A_1}(t)]}{t} \rightarrow -\frac{1}{g'(0)}, \qquad\qquad (9.43)$$

$$\frac{E[(M_{A_1}(t))^2]}{t^2} \rightarrow -\frac{1}{(g'(0))^2}.$$

((9.43) is an analogue of what is known as the Elementary Renewal Theorem; see, for example, [KAR].)

CHAPTER 10

THE NUMBER OF VISITS TO A SUBSET OF THE STATE SPACE BY AN IRREDUCIBLE SEMI-MARKOV PROCESS DURING A FINITE TIME INTERVAL: THE PROBABILITY MASS FUNCTION

This chapter is concerned with a method for the numerical evaluation of the probability mass function of $M_{A_1}(t)$ under the semi-Markov assumption. The notation from Chapters 8 and 9 is retained here. The method considered here is based on explicit expressions for the Laplace transforms of the family of vector-valued functions defined in (9.12). In Section 10.1, which is devoted to the theory, the 'generalised renewal argument', already familiar in the semi-Markov context from the previous two chapters, is used to arrive at a set of recursive integral equations for the family of functions in (9.12). These equations are then solved in the Laplace transform domain. In Section 10.2, the method which will be used later for the numerical inversion of the Laplace transforms and its NAG implementation are discussed. In Section 10.3, two reliability examples are considered. The first one is the Markov model of a two-unit system in a fluctuating environment, formulated as Model 1 in Section 1.2.1. This example allows the proposed method to be assessed in the light of the results by the closed form expressions for the Markov case from Section 5.2.1. The second example is the semi-Markov model of a two-unit system of transformers, known from Section 1.2.2 as Model 7. The results for this case are validated by simulation. Section 10.4 is devoted to implementation issues. The language of implementation of the present method was FORTRAN 77 on the VAX mainframe using the commercially available, and in the U.K. very popular, numerical subroutine library NAG [NUM]. Some of the most important features of this library will be summarized. The alternative to using a software library is, of course, writing one's own procedures. To corroborate MATLAB's power and to discuss the notion and the use of what is termed a 'function function', we also provide a MATLAB implementation of the Laplace transform inversion algorithm used in this chapter; the MATLAB code will turn out to be very concise.

The core material of the present chapter is based on work first presented in [CSE7]. Most of the material in Sections 10.2 and 10.4 has not been presented before.

10.1 The Laplace transform of the probability mass function of $M_{A_1}(t)$

10.1.1 A recurrence relation in the Laplace transform domain

Here, we shall derive a recurrence relation for the Laplace transforms of $\mathbf{h}(t ; m)$, defined by (9.11) and (9.12). The recurrence relation will be written in terms of the two arrays of vectors

$$\{ \mathbf{h}^*_{A_i}(\tau ; m) : m = 0, 1, \ldots \}, \quad i = 1, 2.$$

The renewal argument along the lines of the previous two chapters is seen to yield (10.1) - (10.3) below. They are in terms of the measures defined by (9.22) and (9.23) and in terms of the functions $h_s(.; m)$, where $m = 0, 1, \ldots$ and $s \in S$. The device which allows (10.2) and

(10.3) to be established is *conditioning* with respect to the sojourn time variable $T_{A_i,1}$ if Y starts in some $s \in A_i$. The key equations are as follows.

m = 0:

$$h_s(t\,;0) = \begin{cases} 0 & \text{if } s \in A_1, \\ \Pr\{\,T_{A_2,1} > t \mid Y_0 = s\,\} & \text{if } s \in A_2, \end{cases} \tag{10.1}$$

m = 1:

$$h_s(t\,;1) = \begin{cases} \Pr\{\,T_{A_1,1} > t \mid Y_0 = s\,\} + \\[2mm] \quad \sum_{a_2 \in A_2} \int_{[0,\,t]} h_{a_2}(t - t_1; 0)\, \kappa_{s,a_2}\{dt_1\} & \text{if } s \in A_1, \\[4mm] \quad \sum_{a_1 \in A_1} \int_{[0,\,t]} h_{a_1}(t - t_2; 1)\, \kappa_{s,a_1}\{dt_2\} & \text{if } s \in A_2, \end{cases} \tag{10.2}$$

m ≥ 2:

$$h_s(t\,;m) = \begin{cases} \sum_{a_2 \in A_2} \int_{[0,\,t]} h_{a_2}(t - t_1; m\text{-}1)\, \kappa_{s,a_2}\{dt_1\} & \text{if } s \in A_1, \\[4mm] \sum_{a_1 \in A_1} \int_{[0,\,t]} h_{a_1}(t - t_2; m)\, \kappa_{s,a_1}\{dt_2\} & \text{if } s \in A_2. \end{cases} \tag{10.3}$$

We note in passing that (10.1) - (10.3) is a *system* of integral equations which bears some resemblance to the integral equation used by Jack [JAC1] and Christer and Jack [CHR] in their papers on finite time horizon replacement modelling. The solution technique adopted in [JAC1] is numerical integration. That approach is not feasible here, however, since, as is easily seen from (10.1) - (10.3), for any given m, direct integration would require the computation of a sum of integrals in m dimensions; this is bound to become quickly unfeasible for even very moderate values of m.

The technique adopted here is solving (10.1) - (10.3) for $\{\,\mathbf{h}^*_{A_i}(\tau\,;m) : m = 0, 1, \ldots\,\}$, i = 1, 2 in the Laplace transform domain and subsequently inverting the Laplace transforms numerically. As a first step, (10.1) - (10.3) will be written in terms of the Laplace transforms of the measures defined in (9.23). For this, it will be used that the Laplace transforms of the probability terms in (10.1) and (10.2) are given by

$$\int_0^{+\infty} \exp\{-\tau t\} \, \Pr\{\, T_{A_i,1} > t \mid Y_0 = s \,\} \, dt = (1 - T_{A_i,1}{}^*(\tau\,;s))/\tau,$$

where $T_{A_i,1}{}^*(\tau\,;s) = E[\exp\{-\tau\, T_{A_i,1}\} \mid Y_0 = s]$ is the Laplace transform of the length of the first sojourn of Y in A_i given that Y starts in $s \in A_i$ at $t = 0$. With this notation then, (10.1) - (10.3) are written in the Laplace transform domain as follows.

m = 0:

$$h_s{}^*(\tau\,;0) = \begin{cases} 0 & \text{if } s \in A_1, \\[2ex] (1 - T_{A_2,1}{}^*(\tau\,;s))/\tau & \text{if } s \in A_2, \end{cases} \tag{10.4}$$

m = 1:

$$h_s{}^*(\tau\,;1) = \begin{cases} (1 - T_{A_1,1}{}^*(\tau\,;s))/\tau + \\[2ex] \displaystyle\sum_{a_2 \in A_2} h_{a_2}{}^*(\tau\,;0)\, \kappa_{s,a_2}{}^*(\tau) & \text{if } s \in A_1, \\[3ex] \displaystyle\sum_{a_1 \in A_1} h_{a_1}{}^*(\tau\,;1)\, \kappa_{s,a_1}{}^*(\tau) & \text{if } s \in A_2, \end{cases} \tag{10.5}$$

m ≥ 2:

$$h_s{}^*(\tau\,;m) = \begin{cases} \displaystyle\sum_{a_2 \in A_2} h_{a_2}{}^*(\tau\,;m\text{-}1)\, \kappa_{s,a_2}{}^*(\tau) & \text{if } s \in A_1, \\[3ex] \displaystyle\sum_{a_1 \in A_1} h_{a_1}{}^*(\tau\,;m)\, \kappa_{s,a_1}{}^*(\tau) & \text{if } s \in A_2. \end{cases} \tag{10.6}$$

By (8.23) (which applies with $\mathbf{K}^*_{A_iA_3}(\tau) \equiv \mathbf{0}$ because of the irreducibility of Y), and by (8.35), we now know that

$$\{\, T_{A_i,1}{}^*(\tau\,;s) : s \in A_i \,\} = \boldsymbol{\phi}_{A_i}(\tau) = \mathbf{K}^*_{A_iA_j}(\tau)\,\mathbf{1} = \left(\mathbf{I} - \mathbf{Q}^*_{A_iA_i}(\tau)\right)^{-1} \mathbf{Q}^*_{A_iA_j}(\tau). \tag{10.7}$$

By (9.25) and (10.7), we get the following matrix versions of (10.4) - (10.6) m = 0:

$$\tau\, \mathbf{h}^*_{A_i}(\tau\,;0) = \begin{cases} \mathbf{0} & \text{if } i = 1, \\ \mathbf{1} - (\mathbf{I} - \mathbf{Q}^*_{A_2 A_2}(\tau))^{-1}\, \mathbf{Q}^*_{A_2 A_1}(\tau)\, \mathbf{1} & \text{if } i = 2, \end{cases} \quad (10.8)$$

m = 1:

$$\tau\, \mathbf{h}^*_{A_i}(\tau\,;1) = \begin{cases} \mathbf{1} - (\mathbf{I} - \mathbf{Q}^*_{A_1 A_1}(\tau))^{-1}\, \mathbf{Q}^*_{A_1 A_2}(\tau)\, (\mathbf{1} - \tau\, \mathbf{h}^*_{A_2}(\tau\,;0)) & \text{if } i = 1, \\ (\mathbf{I} - \mathbf{Q}^*_{A_2 A_2}(\tau))^{-1}\, \mathbf{Q}^*_{A_2 A_1}(\tau)\, \tau\, \mathbf{h}^*_{A_1}(\tau\,;1) & \text{if } i = 2, \end{cases}$$

$$(10.9)$$

m ≥ 2:

$$\tau\, \mathbf{h}^*_{A_i}(\tau\,;m) = \begin{cases} (\mathbf{I} - \mathbf{Q}^*_{A_1 A_1}(\tau))^{-1}\, \mathbf{Q}^*_{A_1 A_2}(\tau)\, \tau\, \mathbf{h}^*_{A_2}(\tau\,;m\text{-}1) & \text{if } i = 1, \\ (\mathbf{I} - \mathbf{Q}^*_{A_2 A_2}(\tau))^{-1}\, \mathbf{Q}^*_{A_2 A_1}(\tau)\, \tau\, \mathbf{h}^*_{A_1}(\tau\,;m) & \text{if } i = 2. \end{cases}$$

$$(10.10)$$

10.1.2 The direct computation of Laplace transforms

The family of vectors $\{\ \tau\, \mathbf{h}^*_{A_i}(\tau\,;m) : i = 1, 2; m \geq 0 \}$ is uniquely determined by the recurrence relation (10.8) - (10.10). Explicit expressions for $\tau\, \mathbf{h}^*_{A_i}(\tau\,;m)$ will now be obtained in terms of the matrices $\mathbf{K}^*_{A_i A_j}(\tau) = (\mathbf{I} - \mathbf{Q}^*_{A_i A_i}(\tau))^{-1}\, \mathbf{Q}^*_{A_i A_j}(\tau)$ which we know from (8.3). For m = 0 and m = 1, it is seen from (10.8) and (10.9) that

$$\tau\, \mathbf{h}^*_{A_1}(\tau\,;0) = \mathbf{0}, \quad (10.11)$$

$$\tau\, \mathbf{h}^*_{A_2}(\tau\,;0) = \mathbf{1} - \mathbf{K}^*_{A_2 A_1}(\tau)\, \mathbf{1}, \quad (10.12)$$

$$\tau\, \mathbf{h}^*_{A_1}(\tau\,;1) = (\mathbf{I} - \mathbf{K}^*_{A_1 A_2}(\tau)\, \mathbf{K}^*_{A_2 A_1}(\tau))\, \mathbf{1}, \quad (10.13)$$

$$\tau\, \mathbf{h}^*_{A_2}(\tau\,;1) = \mathbf{K}^*_{A_2 A_1}(\tau)\, (\mathbf{I} - \mathbf{K}^*_{A_1 A_2}(\tau)\, \mathbf{K}^*_{A_2 A_1}(\tau))\, \mathbf{1}. \quad (10.14)$$

For m ≥ 2, we get from (10.10) by induction

$$\tau\, \mathbf{h}^*_{A_1}(\tau\,;m) = (\mathbf{K}^*_{A_1 A_2}(\tau)\, \mathbf{K}^*_{A_2 A_1}(\tau))^{m-2}\, \mathbf{K}^*_{A_1 A_2}(\tau)\, \tau\, \mathbf{h}^*_{A_2}(\tau\,;1), \quad (10.15)$$

$$\tau\, \mathbf{h}^*_{A_2}(\tau\,;m) = (\mathbf{K}^*_{A_2 A_1}(\tau)\, \mathbf{K}^*_{A_1 A_2}(\tau))^{m-1}\, \tau\, \mathbf{h}^*_{A_2}(\tau\,;1). \quad (10.16)$$

(10.13) - (10.16) show that for m ≥ 1,

$$\tau\, h^*_{A_1}(\tau\,;m) = (K^*_{A_1A_2}(\tau)\, K^*_{A_2A_1}(\tau))^{m-1}\, (I - K^*_{A_1A_2}(\tau)\, K^*_{A_2A_1}(\tau))\, 1, \quad (10.17)$$

$$\tau\, h^*_{A_2}(\tau\,;m) =$$

$$(K^*_{A_2A_1}(\tau)\, K^*_{A_1A_2}(\tau))^{m-1}\, K^*_{A_2A_1}(\tau)\, (I - K^*_{A_1A_2}(\tau)\, K^*_{A_2A_1}(\tau))\, 1. \quad (10.18)$$

The system comprising (10.11), (10.12), (10.17), and (10.18) is the desired set of explicit equations.

10. 2 Numerical inversion of Laplace transforms using Laguerre polynomials and fast Fourier transform

The examples in Section 10.3 will show that the expressions obtained in Section 10.1 for the Laplace transforms of the occupation frequencies in A_1 can be usefully applied in practice when combined with a numerical Laplace transform inversion scheme. In Section 10.2.1, we give a summary of the numerical Laplace transform inversion scheme used in this paper. It is based on a method due to Weeks [WEE] in conjunction with the Fast Fourier Transform as reported by Cooley, Lewis and Welch in [COO]. Weeks' method in its original form has also found its way into the NAG FORTRAN subroutine library [NUM] via the more recent works by Garbow, Giunta and Murli [GAR1] and Garbow, Giunta, Lyness and Murli [GAR2]. In our account of the inversion method, we are following the notation and the terminology from [COO]; this will help to highlight in Section 10.2.1 a minor error in [COO]. In Section 10.2.2, the inversion algorithm is re-expressed in the notation of the NAG manual [NUM] and the references cited therein.

10.2.1 A summary of the numerical Laplace transform inversion scheme

The Laguerre polynomials $L_0, L_1, \ldots$ are defined by the Rodrigues Formula

$$L_n(t) = \frac{1}{n!}\, \exp\{\,t\,\}\, \frac{d^n}{dt^n}\,(\,t^n \exp\{\,-t\,\})\,\Big(= \sum_{j=0}^{n} \binom{n}{j}(-1)^j \frac{t^j}{j!}\,\Big)\,,\; n = 0,\,1,\,\ldots\,. \quad (10.19)$$

The Laguerre functions $\ell_n(t) = \exp\{\,-t/2\,\}\, L_n(t)$, $n = 0, 1, \ldots$, are an orthonormal basis of the set of square integrable measurable functions $\mathcal{L}_2[0, +\infty)$ with respect to the scalar product

$$< f, g > = \int_{[0,\,+\infty)} f(t)\, g(t)\, dt.$$

The Laguerre functions thus can be used for a series representation of functions f in $\mathcal{L}_2[0, +\infty)$ by

$$f(t) = \sum_{n=0}^{\infty} \int_{[0,\,+\infty)} f(s)\, \ell_n(s)\, ds\; \ell_n(t),$$

184

which in terms of Laguerre polynomials is written as

$$f(t) = \exp\{ -t/2 \} \sum_{n=0}^{\infty} \int_{[0, +\infty)} f(s) L_n(s) \exp\{ -s/2 \} ds \, L_n(t). \tag{10.20}$$

An alternative representation of f is of course

$$f(t) = \exp\{ -t \} \sum_{n=0}^{\infty} C_n L_n(t), \tag{10.21}$$

where the coefficients $\{ C_n : n \geq 0 \}$ now correspond to those of $f(t) \exp\{ -t/2 \}$ in (10.20). Noting that the Laplace transform of L_n is

$$L_n^*(\tau) = \int_0^{+\infty} \exp\{ -t \, \tau \} L_n(t) \, dt = \frac{(\tau - 1)^n}{\tau^{n+1}}, \tag{10.22}$$

(10.21) reads in the Laplace transform domain as

$$f^*(\tau) = \sum_{n=0}^{\infty} C_n \frac{\tau^n}{(\tau+1)^{n+1}} . \tag{10.23}$$

For a given Laplace transform f^*, the coefficients C_n can be inferred from (10.23) by repeated differentiation at $\tau = 0$. However, this does not yield a practical method for computing the C_n if f^* is available in a numerical form only. Weeks' method is devised to cater for this case as follows. Given a function g and two as yet unspecified but fixed parameters $c \geq 0$ and $T > 0$, the expansion in (10.21) is considered for f which is defined in terms of g by

$$f(x) = g(\frac{x}{T}) \exp\{ -x \, (\frac{c}{T} + \frac{1}{2}) \}.$$

(10.21) is then rewritten as

$$g(t) = \exp\{ (c - \frac{T}{2}) t \} \sum_{n=0}^{\infty} C_n L_n(t \, T). \tag{10.24}$$

(10.24) in the Laplace transform domain reads as

$$g^*(\tau) = \sum_{n=0}^{\infty} C_n L_n^* \left(\frac{\tau - (c - \frac{T}{2})}{T} \right) \frac{1}{T} ,$$

which then by (10.22) becomes

$$g^*(\tau) = \sum_{n=0}^{\infty} C_n \frac{\left((\tau - c) - \frac{T}{2} \right)^n}{\left((\tau - c) + \frac{T}{2} \right)^{n+1}} . \qquad (10.25)$$

NOTE. In [COO], (10.25) is arrived at via (10.24) in the above manner, too, but the argument of L_n in what corresponds to (10.24) (equation (90) of [COO]) is erroneously stated as t / T. The same printing error can be spotted in equation (102) of [COO]. This does not affect the validity of the final formulae in [COO], however.

In what follows, it will be shown how the coefficients C_n can be recovered from the knowledge of $g^*(\tau)$ along the line $\tau \in \{ c + i \omega \in \mathbb{C} : \omega \in \mathbb{R} \}$ by using (10.25). First, (10.25) is rewritten for $\tau = c + i \omega$, $\omega \in \mathbb{R}$, as

$$\left(\frac{T}{2} + i \omega \right) g^*(c + i \omega) = \sum_{n=0}^{\infty} C_n \left(\frac{i \omega - \frac{T}{2}}{i \omega + \frac{T}{2}} \right)^n . \qquad (10.26)$$

Since $(i \omega - \frac{T}{2}) / (i \omega + \frac{T}{2}) \in \mathbb{C}$ has modulus 1 and is $\neq 1$, it can be uniquely represented as

$$(i \omega - \frac{T}{2}) / (i \omega + \frac{T}{2}) = \exp\{ i \theta \}, \theta \in (0, 2\pi). \qquad (10.27)$$

Reading in (10.27) ω as a function of θ, we get

$$\omega(\theta) = \frac{T}{2} \frac{\exp\{ i \theta \} + 1}{\exp\{ i \theta \} - 1} = \frac{T}{2} \cot(\theta / 2). \qquad (10.28)$$

Thus the function $h : (0, 2\pi) \to \mathbb{C}$ defined by

$$h(\theta) = \frac{T}{2} (1 + i \cot(\theta / 2)) g^*(c + i \frac{T}{2} \cot(\theta / 2)), \qquad (10.29)$$

can be written by (10.26) and (10.28) as

$$h(\theta) = \sum_{n=0}^{\infty} C_n \exp\{ i n \theta \}. \qquad (10.30)$$

The scheme of Weeks [WEE] comprises the following steps:

1. Given the Laplace transform g^* (perhaps as a subroutine), select the parameters $c \geq 0$ and $T > 0$, and define h by (10.29);

2. Select $N \in \{ 2, 3, \dots \}$ and truncate in (10.30) the Fourier series after its N*th* summand and work out approximate values for $C_0, \dots, C_{N-1}$ based on

$$h(\theta) \approx \sum_{n=0}^{N-1} C_n \exp\{\, i\, n\, \theta\, \};$$

3. With these approximate coefficients, $\tilde{C}_0, \ldots, \tilde{C}_{N-1}$, say, obtain an approximate value of $g(t)$ by (10.24) as

$$g(t) \approx \exp\{\,(\,c - \tfrac{T}{2}\,)\, t\,\} \sum_{n=0}^{N-1} \tilde{C}_n\, L_n(\, t\, T\,). \qquad (10.31)$$

The computation of the Laguerre polynomials for (10.31) is best based on the recurrence relation

$$L_0(t) = 1, \qquad (10.32)$$

$$L_1(t) = 1 - t, \qquad (10.33)$$

$$L_{n+1}(t) = \frac{1}{n+1}\, ((2n + 1 - t)\, L_n(t) - n\, L_{n-1}(t)),\ n \geq 1, \qquad (10.34)$$

which is more efficient than (10.19). (The recurrence relation (10.32) - (10.34) is the usual method to generate the Laguerre polynomials for numerical computation; see, for example, [KUB], [MAS1], [SUM1], and [SUM2].) Step 2 of the above scheme can be performed by applying the fast Fourier transform to the values

$$\{\, h(j\, 2\pi\, /\, N) : j = 0, 1, \ldots, N - 1\, \};$$

see [COO].

There is a vaguely related methodology to solve systems of convolution equations; it is what has become known as the Laguerre Transform Method. [KUB], [MAS1], and [SUM1] - [SUM3] is but a small selection of papers on this topic. This method also uses the Laguerre series representation of the function f in the time domain but it makes use of a subsequent mapping of the coefficients $\{\, <f, \ell_n> : n = 0, 1, \ldots\, \}$ onto a new sequence of coefficients, the so-called 'Laguerre sharp coefficients', such that the combined mapping becomes an isomorphism with respect to the operation 'convolution'. The Laplace transform inversion method afforded by that technique requires the sharp coefficients of f to be expressed in terms of those of the other, known, functions. The method therefore requires more analytical work, in contrast to the one just described in which the provision of a program subroutine for f^* suffices.

10.2.2 The inversion scheme in the NAG implementation

The NAG FORTRAN library subroutines C06LBF and C06LCF are an implementation of Weeks' account [WEE] of the inversion method just explained. To be able to talk about these subroutines via the notation and in the terminology of the NAG manual [NUM], we are going to re-express here the key elements of the scheme.

The symbols σ, $\{\, a_n : n = 0, 1, \ldots\, \}$, and b in [GAR1] and [GAR2] stand respectively for our c, $\{ C_n : n = 0, 1, ... \}$, and T. (10.24) is restated with these variables as

$$g(t) = \exp\{ \sigma\, t \} \sum_{n=0}^{\infty} a_n \exp\{ -b\,t / 2 \} L_n(bt). \tag{10.35}$$

In [GAR1] and [GAR2], a function ϕ is *defined* in terms of $\{ a_n : n = 0, 1, ... \}$ as

$$\phi(z) = \sum_{n=0}^{\infty} a_n z^n, \tag{10.36}$$

which then in terms of the Laplace transform of g can be written as

$$\phi(z) = \frac{b}{1-z} g^*\!\left(\frac{b}{1-z} + \sigma - \frac{b}{2} \right), \tag{10.37}$$

since by (10.25),

$$\frac{b}{1-z} g^*\!\left(\frac{b}{1-z} + \sigma - \frac{b}{2} \right) = \frac{T}{1-z} g^*\!\left(\frac{T}{1-z} + c - \frac{T}{2} \right) =$$

$$\frac{T}{1-z} \sum_{n=0}^{\infty} C_n \frac{\left(\left(\frac{T}{1-z} + c - \frac{T}{2} - c \right) - \frac{T}{2} \right)^n}{\left(\left(\frac{T}{1-z} + c - \frac{T}{2} - c \right) + \frac{T}{2} \right)^{n+1}} = \sum_{n=0}^{\infty} C_n z^n = \phi(z).$$

Given g^* (and thus by (10.37) ϕ), Weeks' method evaluates approximations for the first ℓ of ϕ's Taylor coefficients in (10.36), $a_0, ..., a_{\ell-1}$, which then are substituted into the truncated version of (10.35) to obtain an approximation of g. The parameters b (> 0) and σ in (10.35) and (10.37) are auxiliary in nature and can be chosen, within certain limitations, arbitrarily. The degree of accuracy achieved in the course of the computation is, however, dependent on the choice of these parameters. For details on these and the procedure's other two parameters, σ_0, ϵ_{tol}, the reader is referred to the papers [GAR1], [GAR2] and the NAG manual [NUM].

10.3 Reliability applications

10.3.1 The Markov model of the two-unit power transmission system revisited

The probability mass function of $M_{A_1}(t)$ for the Markov model in Section 5.2.1 has been recomputed by the present method. Notice that, since Y is now Markovian, $K^*_{A_iA_j}(\tau)$ can be expressed in terms of Λ by (8.15). Our Laplace transform inversion scheme from Section 10.2 has been implemented on a VAX 8650 VMS machine in FORTRAN 77 using the NAG subroutines C06LBF and C06LCF; the FORTRAN code is shown, and will be discussed, in Section 10.4.2. The numerical results are shown in Table 10.1, the parameter values for the NAG subroutines C06LBF and C06LCF are shown in Table 10.2. Both methods, the direct one (based on Theorem 5.1) and the present one, deliver nearly identical reslts, as can be seen by comparing the Tables 5.1 and 10.1. It must be borne in mind, however, that the appropriate values for the NAG parameters σ_0 (= SIGMA0), σ (= SIGMA), b (= B), and ε_{tol} (= EPSTOL) had to be established by preliminary tests. This drawback of the present procedure is compensated for, however, by an almost instantaneous calculation of each of the entries in Table 10.1 on the VAX 8650. By contrast, the calculation of each of the entries in the last row in Table 5.1 took 2 CPU min on the Macintosh.

t [yr]	m = 1	m = 2	m = 3	m = 4	m = 5
1.0	$9.979\ 677_1$	$2.027\ 623_3$	$4.674\ 874_6$	$9.959\ 458_9$	$2.013\ 779_{11}$
	$(5.237\ 3_8)$	$(1.345\ 7_7)$	$(1.776\ 5_{10})$	$(8.426\ 1_{14})$	$(2.908\ 6_{15})$
2.5	$9.949\ 246_1$	$5.055\ 997_3$	$1.937\ 155_5$	$6.303\ 408_8$	$1.845\ 145_{10}$
	$(5.237\ 3_8)$	$(1.345\ 7_7)$	$(1.776\ 5_{10})$	$(8.426\ 1_{14})$	$(2.908\ 6_{15})$
5.0	$9.898\ 714_1$	$1.006\ 410_2$	$6.414\ 074_5$	$3.215\ 179_7$	$1.380\ 910_9$
	$(5.237\ 3_8)$	$(1.345\ 7_7)$	$(1.776\ 5_{10})$	$(8.426\ 1_{14})$	$(2.908\ 6_{15})$
10.0	$9.798\ 868_1$	$1.998\ 287_2$	$2.284\ 391_4$	$1.928\ 222_6$	$1.332\ 423_8$
	$(5.237\ 3_8)$	$(1.345\ 7_7)$	$(1.776\ 5_{10})$	$(8.426\ 1_{14})$	$(2.908\ 6_{15})$
20.0	$9.657\ 100_1$	$3.034\ 680_2$	$8.063\ 777_4$	$1.229\ 200_5$	$1.425\ 699_7$
	$(7.833\ 3_8)$	$(7.729\ 9_8)$	$(5.833\ 9_{11})$	$(7.806\ 8_{11})$	$(2.205\ 2_{13})$
50.0	$9.032\ 007_1$	$9.183\ 655_2$	$4.787\ 544_3$	$1.895\ 455_4$	$4.658\ 305_6$
	$(1.118\ 0_8)$	$(1.283\ 9_8)$	$(8.060\ 7_{11})$	$(1.664\ 8_{10})$	$(2.142\ 1_{11})$
100.0	$8.158\ 166_1$	$1.658\ 483_1$	$1.708\ 253_2$	$1.187\ 296_3$	$6.264\ 906_5$
	$(1.188\ 0_8)$	$(1.283\ 9_8)$	$(8.060\ 7_{11})$	$(5.483\ 5_{11})$	$(2.142\ 1_{11})$

Table 10.1 $\Pr\{\ M_{A_1}(t) = m \mid Y_0 = 1\}$ by numerical Laplace transform inversion for the Markov example. (The values returned for ERRVEC(1) by the NAG system are noted in brackets.)
(Negative powers of 10 are indicated by a subscript, e.g., 9.432_1 stands for 9.432×10^{-1}.)

t [yr]	σ_0	σ	b	ε_{tol}	ℓ
1.0	0.10	0.80	1.75	0.0001	8
2.5	0.10	0.80	1.75	0.0001	8
5.0	0.10	0.80	1.75	0.0001	8
10.0	0.10	0.80	1.75	0.0001	8
20.0	0.01	0.71	1.75	0.000001	8
50.0	0.01	0.10	0.225	0.0001	8
100.0	0.01	0.10	0.225	0.0001	8

Table 10.2 NAG parameter values for the Markov example

The NAG parameters EPSTOL and ERRVEC(1) are respectively measures of the expected (i.e., user-specified) and actually achieved degrees of accuracy of the Laplace transform inversion. For details, the reader is referred to the NAG manual [NUM].

10.3.2 The two-unit semi-Markov model revisited

As a second application, we consider Model 7 from Section 1.2.2, some asymptotic aspects of which are already known from Section 9.4.2. The quantity under consideration is the probability mass function of the number of system 'up' periods during $[0, t]$, i.e., $\{ \Pr\{ M_{A_1}(t) = m \} : m = 0, 1, ... \}$. The assumptions of Section 9.4.2 apply. It will be assumed here that ξ, the time needed to identify the faulty unit, is uniformly distributed over some interval $[t_0, t_1] \subset [0, \infty)$, $t_0 < t_1$, and thus

$$
\Pr\{ \xi \le t \} =
\begin{cases}
1 & \text{for } t \ge t_1, \\[2mm]
\dfrac{t - t_0}{t_1 - t_0} & \text{for } t \in (t_0, t_1), \\[2mm]
0 & \text{for } t \le t_0.
\end{cases}
$$

The Laplace transform of ξ is thus given for $\tau \in \mathbb{C}^+$ by

$$
\xi^*(\tau) =
\begin{cases}
\dfrac{\exp\{ -t_0\, \tau \} - \exp\{ -t_1\, \tau \}}{(t_1 - t_0)\, \tau} & \text{for } \tau \ne 0, \\[3mm]
1 & \text{for } \tau = 0.
\end{cases}
\tag{10.38}
$$

(From a practical viewpoint, the above assumption about the distribution of ξ does not represent much of a restriction. This is because, as is well-known, any distribution can be approximated by a finite mixture of rectangular distributions. The Laplace transform of ξ will then be approximated by a corresponding convex-combination of expressions of the type (10.38).) We know $\mathbf{Q}^*(\tau)$ from Section 9.4.2. Thus,

$$
\mathbf{K}_{A_1 A_2}{}^*(\tau) =
\begin{array}{c}
\\ 1 \\ \\ 2
\end{array}
\begin{array}{cc}
\quad 3 \qquad\qquad 4 \quad \\
\left[
\begin{array}{cc}
\dfrac{2\lambda}{2\lambda + \tau} & 0 \\[4mm]
\dfrac{2\lambda\,\mu}{(\mu+\lambda+\tau)(2\lambda+\tau)} & \dfrac{\lambda}{(\mu+\lambda+\tau)}
\end{array}
\right]
\end{array},
$$

$$
\mathbf{K}_{A_2 A_1}{}^*(\tau) =
\begin{array}{c}
\\ 3 \\ 4
\end{array}
\begin{array}{c}
\quad 1 \qquad\quad 2 \quad \\
\left[
\begin{array}{cc}
0 & \xi^*(\tau) \\[2mm]
0 & 2\mu/(2\mu+\tau)
\end{array}
\right]
\end{array}.
$$

This model, too, has been implemented on the VAX 8650 with the NAG library routines C06LBF and C06LCF. The results are shown in Table 10.3 for t = 8760 hours (= 1 year), t_0 = 10 hours, t_1 = 15 hours, λ = 4.0 / year, μ = 0.05 / hour. (This is equivalent to saying that the component mean time to failure is $1 / \lambda$ = 3 months and that the component mean repair time is $1 / \mu$ = 20 hours.)

		NAG parameters					
m	Pr{ $M_{A_1}(t)$ = m | Y_0 = 1 }	σ_0	σ	b	ε_{tol}	ERRVEC(1)	ℓ
1	$3.461\ 10_4$	1.0_6	1.0_5	5.0_2	1.0_2	$4.110\ 2_3$	128
2	$2.766\ 86_3$	1.0_6	1.0_5	5.0_2	1.0_3	$6.706\ 9_5$	256
3	$1.122\ 28_2$	1.0_6	1.0_5	5.0_2	1.0_3	$7.142\ 8_5$	256
4	$3.018\ 98_2$	1.0_6	1.0_5	5.0_2	1.0_3	$5.266\ 8_4$	256
5	$6.059\ 22_2$	1.0_6	1.0_5	5.0_2	1.0_3	$2.140\ 9_{10}$	512
6	$9.678\ 33_2$	1.0_6	1.0_5	5.0_2	1.0_3	$1.815\ 0_9$	512
7	$1.281\ 55_1$	1.0_6	1.0_5	5.0_2	1.0_3	$1.279\ 9_8$	512
8	$1.446\ 97_1$	1.0_6	1.0_5	5.0_2	1.0_3	$7.744\ 6_8$	512
9	$1.422\ 05_1$	1.0_6	1.0_5	5.0_2	1.0_3	$4.106\ 9_7$	512
10	$1.235\ 80_1$	1.0_6	1.0_5	5.0_2	1.0_3	$1.939\ 5_6$	512
11	$9.614\ 93_2$	1.0_6	1.0_5	5.0_2	1.0_3	$8.261\ 5_6$	512
(11	$9.614\ 93_2$	1.0_6	1.0_5	5.0_2	2.0_{11}	$1.754\ 7_{13}$	1024)
12	$6.765\ 08_2$	1.0_6	1.0_5	5.0_2	1.0_3	$1.356\ 6_{13}$	1024
13	$4.340\ 41_2$	1.0_6	1.0_5	5.0_2	1.0_3	$6.976\ 2_{14}$	1024
14	$2.557\ 08_2$	1.0_6	1.0_5	5.0_2	1.0_3	$6.650\ 4_{14}$	1024
15	$1.391\ 51_2$	1.0_6	1.0_5	5.0_2	1.0_3	$1.384\ 8_{13}$	1024
16	$7.030\ 37_3$	1.0_6	1.0_5	5.0_2	1.0_3	$2.623\ 4_{13}$	1024
17	$3.312\ 47_3$	1.0_6	1.0_5	5.0_2	1.0_3	$4.134\ 7_{13}$	1024
18	$1.461\ 18_3$	1.0_6	1.0_5	5.0_2	1.0_3	$6.342\ 8_{13}$	1024
19	$6.055\ 35_4$	1.0_6	1.0_5	5.0_2	1.0_3	$9.908\ 0_{13}$	1024
20	$2.364\ 82_4$	1.0_6	1.0_5	5.0_2	1.0_3	$1.434\ 8_{12}$	1024
21	$8.727\ 32_5$	1.0_6	1.0_5	5.0_2	1.0_3	$2.347\ 2_{12}$	1024
22	$3.051\ 23_5$	1.0_6	1.0_5	5.0_2	1.0_3	$1.579\ 6_{11}$	1024
23	$1.012\ 89_5$	1.0_6	1.0_5	5.0_2	1.0_3	$6.655\ 3_{11}$	1024
24	$3.199\ 21_6$	1.0_6	1.0_5	5.0_2	1.0_3	$2.481\ 4_{10}$	1024
25	$9.632\ 34_7$	1.0_6	1.0_5	5.0_2	1.0_3	$8.673\ 5_{10}$	1024

Table 10.3 Pr{ $M_{A_1}(t)$ = m | Y_0 = 1 } by numerical Laplace transform inversion and the corresponding NAG parameter values with t = 8760 hours (= 1 year); semi-Markov example.
(Negative powers of 10 are indicated by a subscript, e.g., 9.432_1 stands for 9.432×10^{-1}.)

Y is assumed to start at t = 0 in state No. 1. We have used simulation to validate the results in Table 10.3; the results are shown in Table 10.4. Notice that once the correct NAG parameter combination had been established for *one* choice of m, (by comparing the result with those from the simulation in Table 10.4) it could be essentially reused for all other values of m. The case m = 11 is duplicated in Table 10.3 in order to illustrate the effect of change in the (user-supplied) parameter ε_{tol}. A higher required accuracy of the Laplace transform inversion has resulted in more terms being included in the truncated version of (10.35). The case m = 11 in

Table 10.3 also demonstrates, however, that the actually computed value may be unaffected even with a higher number of terms in the truncated version of (10.25). The simulation results in Table 10.4 have been obtained by a FORTRAN 77 program on the VAX 8650. The CPU seconds needed to produce each column are also shown in Table 10.4. As the computation of each of the entries in Table 10.3 was almost instantaneous, the Laplace transform inversion method is once again shown to be a viable method.

m	$n = 10$	$n = 10^2$	$n = 10^3$	$n = 10^4$	$n = 10^5$	$n = 10^6$
1	0.0	0.0	1.00_3	3.000_4	$4.200\ 0_4$	$3.190\ 00_4$
2	0.0	0.0	2.00_3	2.800_3	$2.630\ 0_3$	$2.742\ 00_3$
3	0.0	0.0	8.00_3	8.200_3	$1.145\ 0_2$	$1.142\ 10_2$
4	0.0	5.0_2	3.30_2	3.010_2	$2.999\ 0_2$	$3.043\ 30_2$
5	2.0_1	1.0_1	5.90_2	6.030_2	$6.058\ 0_2$	$6.002\ 70_2$
6	1.0_1	5.0_2	9.30_2	9.760_2	$9.687\ 0_2$	$9.679\ 20_2$
7	1.0_1	1.7_1	1.33_1	1.280_1	$1.282\ 2_1$	$1.285\ 22_1$
8	1.0_1	1.4_1	1.68_1	1.474_1	$1.434\ 4_1$	$1.448\ 49_1$
9	3.0_1	1.4_1	1.28_1	1.426_1	$1.419\ 9_1$	$1.424\ 24_1$
10	0.0	1.4_1	1.26_1	1.282_1	$1.249\ 9_1$	$1.231\ 32_1$
11	0.0	8.0_2	8.00_2	9.340_2	$9.559\ 0_2$	$9.595\ 60_2$
12	1.0_1	6.0_2	6.90_2	6.750_2	$6.754\ 0_2$	$6.779\ 60_2$
13	0.0	1.0_2	3.90_2	4.150_2	$4.385\ 0_2$	$4.344\ 20_2$
14	1.0_1	2.0_2	2.40_2	2.680_2	$2.645\ 0_2$	$2.576\ 90_2$
15	0.0	2.0_2	1.70_2	1.350_2	$1.315\ 0_2$	$1.368\ 00_2$
16	0.0	2.0_2	1.40_2	6.800_3	$7.140\ 0_3$	$6.998\ 00_3$
17	0.0	0.0	3.00_3	3.500_3	$3.380\ 0_3$	$3.250\ 00_3$
18	0.0	0.0	3.00_3	9.000_4	$1.460\ 0_3$	$1.430\ 00_3$
19	0.0	0.0	0.00	5.000_4	$4.900\ 0_4$	$6.370\ 00_4$
20	0.0	0.0	0.00	1.000_4	$2.500\ 0_4$	$2.460\ 00_4$
21	0.0	0.0	0.00	0.000	$7.000\ 0_5$	$8.800\ 00_5$
22	0.0	0.0	0.00	0.000	$3.000\ 0_5$	$3.200\ 00_5$
23	0.0	0.0	0.00	0.000	$2.000\ 0_5$	$1.000\ 00_5$
24	0.0	0.0	0.00	0.000	$0.000\ 0$	$4.000\ 00_6$
25	0.0	0.0	0.00	0.000	$0.000\ 0$	$1.000\ 00_6$
CPU-time [secs]	12.0	12.1	13.1	22.1	112.0	1015.9

Table 10.4 Pr{ $M_{A_1}(t) = m \mid Y_0 = 1$ } by simulation; n is the number of samples of Y in [0, t] with t = 8760 hours (= 1 year); semi-Markov example.
(Negative powers of 10 are indicated by a subscript, e.g., 9.432_1 stands for 9.432×10^{-1}.)

In our initial choice of the NAG parameters, we were guided by the NAG manual's recommendations; some trial and error was also involved. Later, the experience gained has contributed to a more informed selection of the parameters. For example, comparing the parameter values in Tables 10.1 and 10.3 will reveal that for the semi-Markov example, the NAG parameter 'b' had to be chosen much smaller than for Markov case. This is because the time there was measured (internally) by hours, whereas years have been used for the Markov model. This interdependence between 'b' and the time unit used in the course of the computations is apparent from (10.35): 'b' has the dimension [1/time]. If the routines C06LBF and C06LCF are used, it is essential to explore the behaviour of the Laplace transform (subroutine) prior to the actual computations.

10.4 Implementation issues

10.4.1 The NAG library

In this section, we make some additional observations about the NAG FORTRAN library in general and about the inversion scheme in particular. The intention is to put in perspective the potential benefit which may be derived from using the NAG program library.

First, the number of routines in the NAG library is enormous (several hundred). They are all structured, however, according to the same principle. The argument list of the subroutines comprises (see the source code in Section 10.4.2):

(1)   The user-supplied FUNCTION name (if any) which is invoked by the CALLed NAG subroutine. In our example, for the routine C06LBF, this is FUN, the FUNCTION supplying the Laplace transform.

(2)   The list of the user-supplied *input* parameters. For the routine C06LBF, these are SIGMA0, SIGMA, B, EPSTOL, and MMAX.

(3)   The list of the *output* parameters. For the routine C06LBF, these are MACT ($= \ell =$ the number of terms in the truncated version of (10.36)), ACOEF, and ERRVEC.

(4)   The last in the parameter list of every NAG SUBROUTINE is the INTEGER variable IFAIL. This serves to indicate on return to the user the relative success of the intended computation.

The routines C06LBF and C06LCF have to be invoked in succession since while the first one calculates the Taylor coefficients ACOEF ($= a_0, a_1, \ldots$) in (10.36), the second one accepts them as an input parameter and calculates from them the value of the inverse, FINV, by summation. The user cannot specify the number of terms to be included in the truncated version of (10.35). He is instead expected to specify the 'pseudo-accuracy' parameter EPSTOL which is related to the estimated accuracy of the computed inverse; for details, see the manual. Generally, there will be additional parameters which one will want to pass to a FUNCTION invoked by a NAG routine. In our case for example, C06LBF uses the user-supplied FUNCTION FUN which, in addition to its argument TAU, also depends on RATES, U12, and U21. The values of such variables are passed by a named COMMON block, by COMMON /MATRICES/ here.

The NAG routines are 'black box routines', i.e., the user cannot inspect and/or modify their source code. This compares NAG rather unfavourably with the MATLAB toolboxes. These are collections of MATLAB functions for specific application areas; they are visible to the user. The advantage of the NAG library over MATLAB, on the other hand, is the large number of algorithms available in it. Notice, in passing, that the other NAG library routine in the FORTRAN implementation of the Markov example in Section 10.3.2 is the matrix inversion

SUBROUTINE **F04ADF** (CALLed in SUBROUTINE U). NAG library routine CALLs are boxed in the code below. For an introduction to the NAG library, see Phillips' book [PHI].

10.4.2 Input data file and FORTRAN 77 code for the Markov model

Input data file

```
0.0000D0 0.4380D2 0.5000D0 0.0000D0 0.5000D0 0.0000D0 0.0000D0 0.0000D0
0.5840D4 0.0000D0 0.2500D2 0.0000D0 0.2500D2 0.0000D0 0.0000D0 0.0000D0
0.1168D4 0.0000D0 0.0000D0 0.4380D2 0.0000D0 0.0000D0 0.5000D0 0.0000D0
0.0000D0 0.1168D4 0.5840D4 0.0000D0 0.0000D0 0.0000D0 0.0000D0 0.2500D2
0.1168D4 0.0000D0 0.0000D0 0.0000D0 0.0000D0 0.4380D2 0.5000D0 0.0000D0
0.0000D0 0.1168D4 0.0000D0 0.0000D0 0.5840D4 0.0000D0 0.0000D0 0.2500D2
0.0000D0 0.0000D0 0.1168D4 0.0000D0 0.1168D4 0.0000D0 0.0000D0 0.4380D2
0.0000D0 0.0000D0 0.0000D0 0.1168D4 0.0000D0 0.1168D4 0.5840D4 0.0000D0
```

FORTRAN code

```fortran
      PROGRAM MARKOV
C
C Main routine. It is interactive and it creates a 'log' file of
C the session.
C Unit 5, 6: for keyboard input, VDU output.
C Unit   10: for output to 'log' file.
C Unit   11: for input from data file.
C
      CHARACTER*30 PARAM_FILE, !* For name of the input file *!
     &             SCR_FILE    !* For screen output file*!
C
      INTEGER NSTATES, !* Number of states of the Markov chain *!
     &        NA1,     !* Cardinality of A1 *!
     &        NA2,     !* Cardinality of A2 *!
     &        ROW,     !* Row index *!
     &        COL,     !* Column index *!
     &        M,       !* Number of visits to A1 during [0, t] *!
     &        S,       !* Initial state (in {1, ..., 8}) *!
     &        I,
C
C for C06LBF ...
C
     &        MMAX, !* >= Number of terms to be used in C06LBF *!
     &        MACT, !* Number of terms actually used by C06LBF *!
     &        IFAIL !* Error indicator for Nag routine C06LBF *!
C
      DOUBLE PRECISION RATES(8,8), !* Transition rate matrix *!
     & TEMP,    !* Working variable *!
     & RETAU,   !* Real part of tau *!
     & IMTAU,   !* Imaginary part of tau *!
C
C for C06LBF ...
C
     & SIGMA0, !* Abscissa of convergence of the LT *!
     & SIGMA,  !* Parameter in the Laguerre expansion *!
     & B,      !* Parameter in the Laguerre expansion *!
     & EPSTOL, !* Required pseudo-accuracy *!
     & ACOEF(1024), !* Will contain the Laguerre Coefs *!
     & ERRVEC(8),   !* Diagnostic information *!
C
```

194

```
C for C06LCF ...
C
     & T,    !* Argument for which inv. LT is wanted *!
     & FINV !* Calculated value of Prob *!
C
     COMPLEX*16 TAU,
     &          U12(6,2), U21(2,6), !* Matrices in (8.15)   *!
     &          HSTARA1(6), HSTARA2(2), !* Vectors of L.T.s *!
     &          LAP,       !* Value of Laplace transform   *!
C
C for C06LBF...
C
     &          FUN !* Function name of L.T. *!
C
     CHARACTER*1 ANSW
C
     COMMON /MATRICES/ RATES, U12, U21
     COMMON /CARD/ NSTATES, NA1, NA2
     COMMON /INOUT/ M,
     &              HSTARA1, HSTARA2,
     &              S
C
     EXTERNAL  C06LBF, C06LCF,!* NAG routines for inverse LT. *!
     &          FUN     !* L.T. for C06LBF to be called. It    *!
C                       !* uses HSTAR; initial state is in 'S' *!
     INTRINSIC DCMPLX,
     &          DREAL,
     &          DIMAG
C
     DATA NSTATES, NA1, NA2 /8,6,2/
C
   1 WRITE(6,*) 'Enter screen file name'
     READ(5,2,ERR=1,END=1) SCR_FILE
   2 FORMAT(A30)
     OPEN(UNIT=10,STATUS='NEW',ERR=1,FILE=SCR_FILE)
C
   3 WRITE(6,4)
     WRITE(10,4)
   4 FORMAT(' Enter input file name with parameters')
     READ(5,5,END=3,ERR=3) PARAM_FILE
   5 FORMAT(A30)
     WRITE(10,6) PARAM_FILE
   6 FORMAT(' ',A30)
     OPEN(UNIT=11,FILE=PARAM_FILE,STATUS='OLD',ERR=3)
     REWIND(11)
C
     DO ROW=1,NSTATES
        READ(11,*) (RATES(ROW,COL), COL=1,NSTATES)
     END DO
C
     WRITE(6,7)  NSTATES
     WRITE(10,7) NSTATES
   7 FORMAT(' The cardinality of the state space is ',I2)
     WRITE(6,10)  NA1
     WRITE(10,10) NA1
  10 FORMAT(' The cardinality of A1 is ',I2)
     WRITE(6,11)  NA2
     WRITE(10,11) NA2
  11 FORMAT(' The cardinality of A2 is ',I2)
C
C Calculate transition rate matrix...
C
     DO ROW=1,NSTATES
        TEMP=0.0D0
        DO COL=1,NSTATES
           TEMP=TEMP+RATES(ROW,COL)
```

```
            END DO
            RATES(ROW,ROW)=-TEMP
        END DO
C
C Display transition rate matrix...
C
        WRITE(6,*)   ' '
        WRITE(10,*)  ' '
        WRITE(6,12)
        WRITE(10,12)
   12 FORMAT(' The transition rate matrix is as follows:')
        DO ROW=1,NSTATES
            WRITE(6,13)  (RATES(ROW,COL), COL=1,NSTATES)
            WRITE(10,13) (RATES(ROW,COL), COL=1,NSTATES)
   13     FORMAT(' ',8(D12.5,X))
        END DO
C
   25 WRITE(6,*)  'Test Laplace transform ? (Y/N)'
        WRITE(10,*) 'Test Laplace transform ? (Y/N)'
        READ(5,28,END=25,ERR=25) ANSW
   28 FORMAT(A1)
        WRITE(10,26)  ANSW
   26 FORMAT(' ',A1)
        IF(ANSW.EQ.'N') GO TO 99
        IF(ANSW.NE.'Y') GO TO 25
C
C Enter parameters for inverse Laplace transform...
C
   14 WRITE(6,*)   'Enter the real part of tau'
        WRITE(10,*)  'Enter the real part of tau'
        READ(5,*,END=14,ERR=14) RETAU
   15 WRITE(6,*)   'Enter the imaginary part of tau'
        WRITE(10,*)  'Enter the imaginary part of tau'
        READ(5,*,END=14,ERR=15) IMTAU
        TAU=DCMPLX(RETAU,IMTAU)
        WRITE(6,16)   TAU
        WRITE(10,16)  TAU
   16 FORMAT(' tau = ',D12.5,' + ',D12.5,'i')
C
        WRITE(6,*)   ' '
        WRITE(10,*)  ' '
   21 WRITE(6,*)   'Enter m'
        WRITE(10,*)  'Enter m'
        READ(5,27,END=21,ERR=21) M
   27 FORMAT(I2)
        WRITE(6,22)   M
        WRITE(10,22)  M
   22 FORMAT(' m = ',I2)
C
        WRITE(6,*)   ' '
        WRITE(10,*)  ' '
   29 WRITE(6,*)   'Enter the initial state s (in {1, ..., 8})'
        WRITE(10,*)  'Enter the initial state s (in {1, ..., 8})'
        READ(5,30,END=29,ERR=29) S
   30 FORMAT(I2)
        WRITE(6,31)   S
        WRITE(10,31)  S
   31 FORMAT(' s = ',I2)
C
        LAP=FUN(DCMPLX(RETAU,IMTAU))
C
        WRITE(6,23)   M, S, LAP
        WRITE(10,23)  M, S, LAP
   23 FORMAT(' The Laplace transform of t --> Pr[MA1(t) =',
       &I2,' | Y0 = ',I2,' ] is ',D12.5,' + ',D12.5,'i')
```

```
C
      GO TO 25
   99 CONTINUE
C
  100 WRITE(6,*)  ' '
      WRITE(10,*) ' '
  101 WRITE(6,*)  'Inverse Laplace transform required ? (Y/N)'
      WRITE(10,*) 'Inverse Laplace transform required ? (Y/N)'
      READ(5,102,END=101,ERR=101) ANSW
  102 FORMAT(A1)
      WRITE(10,103)  ANSW
  103 FORMAT(' ',A1)
      IF(ANSW.EQ.'N') GO TO 999
      IF(ANSW.NE.'Y') GO TO 101
C
C Laplace transform inversion ...
C
      WRITE(6,*)  ' '
      WRITE(10,*) ' '
  104 WRITE(6,*)  'Enter the initial state s (in {1, ..., 8})'
      WRITE(10,*) 'Enter the initial state s (in {1, ..., 8})'
      READ(5,105,END=104,ERR=104) S
  105 FORMAT(I2)
      IF((S.LT.1).OR.(S.GT.8)) GO TO 104
      WRITE(6,106)  S
      WRITE(10,106) S
  106 FORMAT(' s = ',I2)
C
  118 WRITE(6,*)  'Enter m = 0, 1, ... '
      WRITE(10,*) 'Enter m = 0, 1, ... '
      READ(5,119,END=118,ERR=118) M
  119 FORMAT(I2)
      WRITE(6,120)  M
      WRITE(10,120) M
  120 FORMAT(' m = ',I2)
C
  107 WRITE(6,*)  'Enter time t in years'
      WRITE(10,*) 'Enter time t in years'
      READ(5,108,END=107,ERR=107) T
  108 FORMAT(D12.5)
      WRITE(6,109)  T
      WRITE(10,109) T
  109 FORMAT(' t = ',D12.5,' years')
C
  110 WRITE(6,*)  'Enter SIGMA0'
      WRITE(10,*) 'Enter SIGMA0'
      READ(5,111,END=110,ERR=110) SIGMA0
  111 FORMAT(D12.5)
      WRITE(6,112)  SIGMA0
      WRITE(10,112) SIGMA0
  112 FORMAT(' SIGMA0 = ',D12.5)
C
  113 WRITE(6,*)  'Enter SIGMA'
      WRITE(10,*) 'Enter SIGMA'
      READ(5,114,END=113,ERR=113) SIGMA
  114 FORMAT(D12.5)
      WRITE(6,115)  SIGMA
      WRITE(10,115) SIGMA
  115 FORMAT(' SIGMA = ',D12.5)
C
  130 WRITE(6,*)  'Enter the parameter B'
      WRITE(10,*) 'Enter the parameter B'
      READ(5,131,END=130,ERR=130) B
  131 FORMAT(D12.5)
      WRITE(6,132)  B
      WRITE(10,132) B
```

```
    132 FORMAT(' B = ',D12.5)
C
    133 WRITE(6,*)  'Enter EPSTOL'
        WRITE(10,*) 'Enter EPSTOL'
        READ(5,134,END=133,ERR=133) EPSTOL
    134 FORMAT(D12.5)
        WRITE(6,135)  EPSTOL
        WRITE(10,135) EPSTOL
    135 FORMAT(' EPSTOL = ',D12.5)
C
        MMAX=1024
C
        IFAIL=0
C
        CALL C06LBF(FUN,SIGMA0,SIGMA,B,EPSTOL,MMAX,MACT,
       &                              ACOEF,ERRVEC,IFAIL)
C
        WRITE(6,116)  IFAIL
        WRITE(10,116) IFAIL
    116 FORMAT(' IFAIL = ',I2,' on exit from C06LBF')
C
        WRITE(6,136)  SIGMA
        WRITE(10,136) SIGMA
    136 FORMAT(' SIGMA = ',D12.5,' on exit from C06LBF')
        WRITE(6,137)  B
        WRITE(10,137) B
    137 FORMAT(' B = ',D12.5,' on exit from C06LBF')
        WRITE(6,138)  MACT
        WRITE(10,138) MACT
    138 FORMAT(' MACT = ',I4,' on exit from C06LBF')
        DO I=1,8
           WRITE(6,139)  I, ERRVEC(I)
           WRITE(10,139) I, ERRVEC(I)
    139    FORMAT(' On exit from C06LBF ERRVEC(',I1,') = ',D12.5)
        END DO
C
        IFAIL = 0
C
        CALL C06LCF(T,SIGMA,B,MACT,ACOEF,ERRVEC,FINV,IFAIL)
C
        WRITE(6,140)  IFAIL
        WRITE(10,140) IFAIL
    140 FORMAT(' IFAIL = ',I2,' on exit from C06LCF')
C
        WRITE(6,117)  T, M, S, FINV
        WRITE(10,117) T, M, S, FINV
    117 FORMAT(' Pr[ MA1(',D12.5,') = ',I2,'| Y0 = ',
       &I2,' ] = ',D22.15)
C
        GO TO 100
    999 CONTINUE
C
        STOP
        END
```

```
        COMPLEX*16 FUNCTION FUN(TAU)
C
C This function calculates the L.T. of hs(t;m) =
C Pr{ MA1(t)= m | Y0 = s }, t >= 0. It is called by the NAG
C routine C06LBF.
C
```

```
      INTEGER NSTATES,  !* Number of states of the Markov chain *!
     &        NA1,       !* Cardinality of A1 (=6) *!
     &        NA2,       !* Cardinality of A2 (=2) *!
     &        ROW,       !* Row index *!
     &        M,         !* Number of visits to A1 during [0, t] *!
     &        S          !* Initial state (in {1, ..., 8}) *!
C
      DOUBLE PRECISION RATES(8,8) !* Transition rate matrix *!
C
      COMPLEX*16 TAU,          !* Argument in COMPLEX*16 form *!
     &           U12(6,2),     !* U matrix (see paper) *!
     &           U21(2,6),     !* U matrix (see paper) *!
     &           HSTARA1(6),   !* Vector of L.T.s       *!
     &           HSTARA2(2)    !* Vector of L.T.s       *!
C
      COMMON /MATRICES/ RATES, U12, U21
      COMMON /CARD/ NSTATES, NA1, NA2
      COMMON /INOUT/ M,
     &              HSTARA1,
     &              HSTARA2,
     &              S
C
      EXTERNAL U,    !* Calculates the matrices U12 and U21 *!
     &         HSTAR !* Delivers the  v e c t o r  of L.T.s *!
C
      CALL U(TAU)
      CALL HSTAR(TAU)
C
      IF(S.LE.6) THEN
         FUN=HSTARA1(S)
      ELSE
         FUN=HSTARA2(S-NA1)
      END IF
C
      RETURN
      END
```

```
      SUBROUTINE U(TAU)
C
C This subroutine calculates the matrices U12 = KA1A2* and U21 =
C KA2A1* for a given tau. COMMON /MATRICES/ is used to pass on
C the matrices. COMMON /CARD/ is used to pass the cardinalities.
C
      DOUBLE PRECISION RATES(8,8),
     &                 WKSPCE(36) !* Nag workspace *!
      COMPLEX*16 TAU,
     &           U12(6,2), U21(2,6),
     &           TEMP,                     !* Temporary scalar   *!
     &           TEMP1(6,6), TEMP2(2,2),   !* Temporary matrices *!
     &           ID1(6,6), ID2(2,2),       !* Identity matrices *!
     &           C1(6,6), C2(2,2) !* Inverses of TEMP1 & TEMP2 *!
C
      INTEGER NSTATES, NA1, NA2, !* Cardinalities (=8, 6, 2) *!
     &        ROW, COL, !* Indeces *!
     &        IFAIL     !* NAG failure indicator *!
C
      COMMON /MATRICES/ RATES,    !* 'Input' matrix *!
     &                  U12, U21 !* 'Result' matrices *!
      COMMON /CARD/ NSTATES, NA1, NA2
C
      EXTERNAL F04ADF !* Nag matrix inversion subroutine *!
C
      INTRINSIC DCMPLX
```

```
C
C Initialisation...
C
      DO ROW=1,NA1
         DO COL=1,NA1
            IF(ROW.EQ.COL) THEN
               TEMP1(ROW,COL)=TAU-DCMPLX(RATES(ROW,COL),0.0D0)
               ID1(ROW,COL)=(1.0D0,0.0D0)
            ELSE
               TEMP1(ROW,COL)=-DCMPLX(RATES(ROW,COL),0.0D0)
               ID1(ROW,COL)=(0.0D0,0.0D0)
            END IF
         END DO
      END DO
C
      DO ROW=1,NA2
         DO COL=1,NA2
            IF(ROW.EQ.COL) THEN
               TEMP2(ROW,COL)=
     &         TAU-DCMPLX(RATES(NA1+ROW,NA1+COL),0.0D0)
               ID2(ROW,COL)=(1.0D0,0.0D0)
            ELSE
               TEMP2(ROW,COL)=
     &         -DCMPLX(RATES(NA1+ROW,NA1+COL),0.0D0)
               ID2(ROW,COL)=(0.0D0,0.0D0)
            END IF
         END DO
      END DO
C
C For matrix inversion, invoke F04ADF from the Nag library...
C
      IFAIL=0
      CALL F04ADF(TEMP1,NA1,ID1,NA1,NA1,NA1,C1,NA1,WKSPCE,IFAIL)
C
      IF(IFAIL.NE.0) THEN
         WRITE(6,1)  IFAIL
         WRITE(10,1) IFAIL
    1    FORMAT(' IFAIL = ',I2,' on first exit from F04ADF')
      ELSE
         DO ROW=1,NA1
            DO COL=1,NA2
               TEMP=(0.0D0,0.0D0)
               DO I=1,NA1
               TEMP=TEMP+C1(ROW,I)*DCMPLX(RATES(I,COL+NA1),0.0D0)
               END DO
               U12(ROW,COL)=TEMP
            END DO
         END DO
      END IF
C
C Once again F04ADF for matrix inversion...
C
      IFAIL=0
      CALL F04ADF(TEMP2,NA2,ID2,NA2,NA2,NA2,C2,NA2,WKSPCE,IFAIL)
C
      IF(IFAIL.NE.0) THEN
         WRITE(6,2)  IFAIL
         WRITE(10,2) IFAIL
    2    FORMAT(' IFAIL = ',I2, ' on second exit from F04ADF')
      ELSE
         DO ROW=1,NA2
            DO COL=1,NA1
               TEMP=(0.0D0,0.0D0)
               DO I=1,NA2
```

```
                TEMP=TEMP+C2(ROW,I)*DCMPLX(RATES(I+NA1,COL),0.0D0)
             END DO
             U21(ROW,COL)=TEMP
          END DO
       END DO
    END IF
C
    RETURN
    END
```

```
             SUBROUTINE  HSTAR(TAU)
C
C This subroutine calculates hstar. COMMON /MATRICES/ is used
C to pass on the matrices. COMMON /CARD/ is used to pass the
C cardinalities. COMMON /INOUT/ is used to pass on M and the
C result vectors HSTARA1 and HSTARA2. It uses the subroutines
C U and MATPOW.
C
      COMPLEX*16 TAU,
     & U12(6,2),   U21(2,6),
     & ID1(6,6),   ID2(2,2),    !* Identity matrices *!
     & HSTARA1(6), HSTARA2(2),  !* Result vectors    *!
     & TEMP,
     & TEMP1(6,6), TEMP2(2,2),
     & TEMPA1(6),  TEMPA2(2),
     & RES1(6,6), RES2(2,2) !* Matrix powers from by MATPOW *!
C
      DOUBLE PRECISION RATES(8,8)
C
      INTEGER M,         !* 'm' in Pr[ MA1(t) = m | Y0=s ] *!
     &        S,         !* Initial state (in {1, ..., 8} *!
     &        ROW, COL,
     &        I          !* Working index *!
C
      COMMON /MATRICES/ RATES,
     &                  U12, U21
      COMMON /CARD/ NSTATES, NA1, NA2
      COMMON /INOUT/ M,
     &               HSTARA1, HSTARA2,
     &               S
C
      EXTERNAL U,      !* Subroutine for the matrices U12 & U21 *!
     &         MATPOW !* Power of a COMPLEX*16 matrix *!
C
C Initialisation of the identity matrices...
C
      DO ROW=1,NA1
         DO COL=1,NA1
            IF(ROW.NE.COL) THEN
               ID1(ROW,COL)=(0.0D0,0.0D0)
            ELSE
               ID1(ROW,COL)=(1.0D0,0.0D0)
            END IF
         END DO
      END DO
C
      DO ROW=1,NA2
         DO COL=1,NA2
            IF(ROW.NE.COL) THEN
               ID2(ROW,COL)=(0.0D0,0.0D0)
            ELSE
               ID2(ROW,COL)=(1.0D0,0.0D0)
            END IF
         END DO
      END DO
```

```
C
C Calculate result vectors HSTARA1 and HSTARA2...
C
      CALL U(TAU)
      IF(M.EQ.0) THEN
        DO ROW=1,NA1                        !* ... by (10.11) *!
          HSTARA1(ROW)=(0.0D0,0.0D0)
        END DO
        DO ROW=1,NA2                        !* ... by (10.12) *!
          TEMP=(0.0D0,0.0D0)
          DO COL=1,NA1
            TEMP=TEMP+U21(ROW,COL)
          END DO
          HSTARA2(ROW)=((1.0D0,0.0D0)-TEMP)/TAU
        END DO
      ELSE
C
C Implement (10.17)...
C
      DO ROW=1,NA1                  !* ... create TEMP1=U12*U21 *!
        DO COL=1,NA1
          TEMP=(0.0D0,0.0D0)
          DO I=1,NA2
            TEMP=TEMP+U12(ROW,I)*U21(I,COL)
          END DO
          TEMP1(ROW,COL)=TEMP
        END DO
      END DO
C
      DO ROW=1,NA1      !* ... create TEMPA1=(I - U12*U21)*1 *!
        TEMP=(0.0D0,0.0D0)
        DO COL=1,NA1
          TEMP=TEMP+(ID1(ROW,COL)-TEMP1(ROW,COL))
        END DO
        TEMPA1(ROW)=TEMP
      END DO
C
      CALL MATPOW(TEMP1,RES1,NA1,M-1)
C
      DO ROW=1,NA1             !* ... HSTARA1 from TEMPA1 & RES1 *!
        TEMP=(0.0D0,0.0D0)
        DO I=1,NA1
          TEMP=TEMP+RES1(ROW,I)*TEMPA1(I)
        END DO
        HSTARA1(ROW)=TEMP/TAU
      END DO
C
C Implement (10.18)...
C
      DO ROW=1,NA2                  !* ... create TEMP2=U21*U12 *!
        DO COL=1,NA2
          TEMP=(0.0D0,0.0D0)
          DO I=1,NA1
            TEMP=TEMP+U21(ROW,I)*U12(I,COL)
          END DO
          TEMP2(ROW,COL)=TEMP
        END DO
      END DO
C
      DO ROW=1,NA2          !* ... TEMPA2=U21*(I - U12*U21)*1 *!
        TEMP=(0.0D0,0.0D0)  !*              =U21*TEMPA1          *!
        DO I=1,NA1
          TEMP=TEMP+U21(ROW,I)*TEMPA1(I)
        END DO
        TEMPA2(ROW)=TEMP
```

```fortran
          END DO
C
          CALL MATPOW(TEMP2,RES2,NA2,M-1)
C
          DO ROW=1,NA2
             TEMP=(0.0D0,0.0D0)
             DO I=1,NA2
                TEMP=TEMP+RES2(ROW,I)*TEMPA2(I)
             END DO
             HSTARA2(ROW)=TEMP/TAU
          END DO
       END IF
C
       RETURN
       END
```

```fortran
          SUBROUTINE  MATPOW(MATIN,MATOUT,MATORD,M)
C
C The next subroutine calculates the Mth power of a square
C COMPLEX*16 matrix. Input: M, MATIN, MATORD. Output: MATOUT.
C M should not be > 20.
C
       INTEGER MATORD,
     &         ROW, COL,  !* Index *!
     &         I, J,      !* Index *!
     &         M          !* Matrix exponent *!
C
       COMPLEX*16 MATIN(MATORD,MATORD),
     &            MATOUT(MATORD,MATORD),
     &            TEMP,                   !* Temporary variable *!
     &            TEMPMAT(20,20)          !* Temporary matrix *!
C
       DO ROW=1,MATORD        !* Put MATOUT = identity matrix *!
          DO COL=1,MATORD
             IF(ROW.EQ.COL) THEN
                MATOUT(ROW,COL)=(1.0D0,0.0D0)
             ELSE
                MATOUT(ROW,COL)=(0.0D0,0.0D0)
             END IF
          END DO
       END DO
C
       DO I=1,M
C
          DO ROW=1,MATORD                !* MATOUT = MATOUT*MATIN *!
             DO COL=1,MATORD
                TEMP=(0.0D0,0.0D0)
                DO J=1,MATORD
                  TEMP=TEMP+MATOUT(ROW,J)*MATIN(J,COL)
                END DO
                TEMPMAT(ROW,COL)=TEMP
             END DO
          END DO
C
          DO ROW=1,MATORD
             DO COL=1,MATORD
                MATOUT(ROW,COL)=TEMPMAT(ROW,COL)
             END DO
          END DO
C
       END DO
C
       RETURN
       END
```

10.4.3 MATLAB implementation of the Laplace transform inversion algorithm

The MATLAB implementation of the numerical Laplace transform inversion technique from Section 10.2.1 will now be illustrated by the example of the exponential density with unit rate parameter, i.e., for the function exp{ - t }, t ≥ 0. The MATLAB code itself is shown in Section 10.4.4.

The new MATLAB language elements are as follows. First, the first argument of the function `InverseLaplace` is a function name, namely `glap`; therefore, `InverseLaplace` is said to be a 'function function' (see [MAT2]). `glap` is a local variable of `InverseLaplace` and access to values of what stands for `glap` is by using the MATLAB function evaluator `feval`; the line concerned is boxed in the code. When `InverseLaplace` is invoked, the actual parameter is the function name `Lapexp`, the Laplace transform of the exponential density with unit rate. Notice that `Lapexp` has to be included in the list of parameters in inverted commas when `InverseLaplace` is invoked. The other new features of MATLAB are the inverse fast Fourier transform `ifft`, and the complex conjugate `conj`; for details on these, see [MAT2].

Table 10.5 shows some results obtained by this MATLAB implementation. It is seen that the relative error is not uniform for different values of t, not even for the same choice of N. The computing times are noted in the last column of Table 10.5. They relate to an implementation using the *full* version of MATLAB.

N	t = 1.0	t = 2.0	t = 3.0	t = 4.0	CPU-secs
4	4.018×10^{-1} (9.23 %)	2.575×10^{-1} (90.28 %)	7.531×10^{-2} (51.26 %)	-2.707×10^{-2} (- 247.78 %)	1
8	4.859×10^{-1} (32.07 %)	7.929×10^{-2} (- 41.41 %)	2.288×10^{-2} (- 54.04 %)	5.226×10^{-2} (185.34 %)	1
16	3.098×10^{-1} (- 15.78 %)	1.726×10^{-1} (27.55 %)	3.692×10^{-2} (- 25.83 %)	3.214×10^{-3} (- 82.45 %)	1
32	4.058×10^{-1} (10.30 %)	1.231×10^{-1} (- 9.06 %)	5.157×10^{-2} (3.59 %)	1.460×10^{-2} (- 20.26 %)	2
64	3.563×10^{-1} (- 3.14 %)	1.329×10^{-1} (- 1.77 %)	3.900×10^{-2} (- 21.67 %)	2.070×10^{-2} (13.04 %)	3
128	3.658×10^{-1} (- 0.56 %)	1.372×10^{-1} (1.36 %)	4.570×10^{-2} (- 8.21 %)	1.610×10^{-2} (- 12.12 %)	6
256	3.696×10^{-1} (0.46 %)	1.329×10^{-1} (- 1.79 %)	5.341×10^{-2} (7.28 %)	1.725×10^{-2} (- 5.79 %)	12
512	3.654×10^{-1} (- 0.67 %)	1.342×10^{-1} (- 0.85 %)	4.791×10^{-2} (- 3.77 %)	1.649×10^{-2} (- 9.99 %)	32
∞	3.679×10^{-1} (0.00 %)	1.353×10^{-1} (0.00 %)	4.979×10^{-2} (0.00 %)	1.832×10^{-2} (0.00 %)	

Table 10.5 Some values of the exponential density exp{ - t }, t ≥ 0, by numerical Laplace transform inversion. (The percentages are the relative error.)

10.4.4 MATLAB code

MATLAB *function* Lapexp (in file: 'Lapexp.m')

```
function G = Lapexp(tau)
% Laplace transform of the exponential distribution
% with rate 1

G = 1/(1 + tau);
```

MATLAB *function function* InverseLaplace (in file: 'InverseLaplace.m')

```
function y = InverseLaplace(glap,t,T,N,c)

disp('Started building up the vector X at ... ');
fix(clock)
X = [0];
for j=1:(N-1)
  theta = j*2*pi/N;
  omega = T/(2*tan(theta/2));
  X = [X (i*omega + T/2)*feval(glap,c+i*omega)];
end

disp('Finished building up the vector X at ... ');
fix(clock)
C = conj(ifft(conj(X)));
disp('Finished FFT at ... ');
fix(clock)

% Lagpol is the vector of the Laguerre polynomial
% defined by the Rodrigues formula (10.19). Recursive
%`evaluation is based on (10.32) - (10.34).
% Lagpol(1,n) is the (n-1)st Laguerre polynomial
% (n = 1, 2, ...,N).

x = t*T;
Lagpol = [1.0 (1.0 - x)];
for m=2:(N-1)
  Lagpol(1,m+1) = ...
  ((2*(m-1) + 1 - x)*Lagpol(1,m) - ...
  (m-1)*Lagpol(1,m-1))/m;
end
disp('Obtained the Laguerre polynomials at ... ');
fix(clock)
disp('Result is ... ');
y = (C*Lagpol')*exp((c - T/2)*t);
```

MATLAB *script* (in file: 'invlap')

```
exp(-t)
N=2;
for j=2:9
    N=N*2
    x=InverseLaplace('Lapexp',t,T,N,c)
    100*(x-exp(-t))/exp(-t)
end
```

CHAPTER 11

THE NUMBER OF SPECIFIC SERVICE LEVELS OF A REPAIRABLE SEMI-MARKOV SYSTEM DURING A FINITE TIME INTERVAL: JOINT DISTRIBUTIONS

In Chapter 10, the finite state space S of the irreducible semi-Markov process $Y = \{ Y_t : t \geq 0 \}$ was partitioned into two disjoint non-empty sets A_1 and A_2, and the random variable under consideration was $M_{A_1}(t)$, the number of visits of Y to A_1 during the time interval [0, t]. In this chapter, we are concerned with the corresponding results for the joint distribution of occupation frequencies in the case that the state space S is partitioned into *three* subsets: $S = A_1 \cup A_2 \cup A_3$. The variable under consideration is thus the tuple $(M_{A_1}(t), M_{A_2}(t))$.

To illustrate the practical relevance of this variable in the reliability context, let us consider the Markov Model 1 from Section 1.2.1. There, the state space is such that the set $A_1 = \{1, 2\}$ corresponds to both units being 'up'. The set $A_2 = \{3, 4, 5, 6\}$ comprises all system states in which exactly one of the units is 'up'. $A_3 = \{7, 8\}$ is the set of system 'down' states. Clearly, the state space S is partitioned here such that each of its subsets corresponds to a certain 'service level' of the system. $(M_{A_1}(t), M_{A_2}(t))$ is an account of the frequency with which the system has resided in the two working service levels during the first t time units.

In Section 11.1, a recursive scheme will be established for the Laplace transforms of

$$\mathbf{h}(t; m_1, m_2) = \{ \Pr\{ M_{A_1}(t) = m_1, M_{A_2}(t) = m_2 \mid Y_0 = s \}, s \in S \}, \qquad (11.1)$$

for $m_1, m_2 = 0, 1, \ldots$. Closed form formulae for the Laplace transforms do not appear achievable. Instead, a method based on the dynamic programming approach is described in Section 11.2 for their computation.

It should be noted that the present chapter is restricted to theory, i.e., no computational implementation will be presented here. The intention here is to explore the feasibility of the methodology of Chapter 10 in the multivariate case. The material in this chapter is new.

11.1 A recurrence relation for $\mathbf{h}(t; m_1, m_2)$ in the Laplace transform domain

The notation and terminology from the Chapters 8 - 10 will be retained here. The Laplace transforms of the fuctions in (11.1) form the three double arrays

$$\{ \mathbf{h}^*_{A_i}(\tau; m_1, m_2) : m_1, m_2 = 0, 1, \ldots \}, \quad i = 1, 2, 3,$$

with their entries given by

$$h^*_s(\tau; m_1, m_2) = \int_0^{+\infty} \exp\{-t\,\tau\}\, h_s(t; m_1, m_2)\, dt, \ \tau \in \mathbb{C}^+.$$

The finite measures

$$\{ \kappa_{a_j,a_i} : i \in \{1, 2, 3\}, \ j \in \{1, 2, 3\} \setminus \{i\}, \ a_i \in A_i, \ a_j \in A_j \}$$

from (9.23) will be used here again. In a similar fashion to the reasoning in Section 10.1.1, the renewal argument now yields the following six equations (11.2) - (11.7).

$m_1 = m_2 = 0$:

$$h_s(t; 0, 0) = \begin{cases} 0 & \text{if } s \in A_1 \cup A_2, \\ \Pr\{ T_{A_3,1} > t \mid Y_0 = s \} & \text{if } s \in A_3, \end{cases} \tag{11.2}$$

$m_1 = 1, m_2 = 0$:

$$h_s(t; 1, 0) = \begin{cases} \Pr\{ T_{A_1,1} > t \mid Y_0 = s \} + \\ \quad \displaystyle\sum_{a_3 \in A_3} \int_{[0, t]} h_{a_3}(t - t_1; 0, 0)\, \kappa_{s,a_3}\{dt_1\} & \text{if } s \in A_1, \\ \\ 0 & \text{if } s \in A_2, \\ \\ \displaystyle\sum_{a_1 \in A_1} \int_{[0, t]} h_{a_1}(t - t_3; 1, 0)\, \kappa_{s,a_1}\{dt_3\} & \text{if } s \in A_3, \end{cases} \tag{11.3}$$

$m_1 = 0, m_2 = 1$:

$$h_s(t; 0, 1) = \begin{cases} \Pr\{ T_{A_2,1} > t \mid Y_0 = s \} + \\ \quad \displaystyle\sum_{a_3 \in A_3} \int_{[0, t]} h_{a_3}(t - t_2; 0, 0)\, \kappa_{s,a_3}\{dt_2\} & \text{if } s \in A_2, \\ \\ 0 & \text{if } s \in A_1, \\ \\ \displaystyle\sum_{a_2 \in A_2} \int_{[0, t]} h_{a_2}(t - t_3; 0, 1)\, \kappa_{s,a_2}\{dt_3\} & \text{if } s \in A_3, \end{cases} \tag{11.4}$$

$m_1 \geq 2, m_2 = 0$:

$$h_s(t\,;m_1,0) = \begin{cases} \displaystyle\sum_{a_3\in A_3}\int_{[0,\,t]} h_{a_3}(t-t_1;m_1-1,0)\,\mathsf{K}_{s,a_3}\{dt_1\} & \text{if } s \in A_1, \\[2ex] 0 & \text{if } s \in A_2, \\[2ex] \displaystyle\sum_{a_1\in A_1}\int_{[0,\,t]} h_{a_1}(t-t_3;m_1,0)\,\mathsf{K}_{s,a_1}\{dt_3\} & \text{if } s \in A_3, \end{cases}$$

(11.5)

$m_1 = 0, m_2 \geq 2$:

$$h_s(t\,;0,m_2) = \begin{cases} 0 & \text{if } s \in A_1, \\[2ex] \displaystyle\sum_{a_3\in A_3}\int_{[0,\,t]} h_{a_3}(t-t_2;0,m_2-1)\,\mathsf{K}_{s,a_3}\{dt_2\} & \text{if } s \in A_2, \\[2ex] \displaystyle\sum_{a_2\in A_2}\int_{[0,\,t]} h_{a_2}(t-t_3;0,m_2)\,\mathsf{K}_{s,a_2}\{dt_3\} & \text{if } s \in A_3, \end{cases}$$

(11.6)

$m_1, m_2 \geq 1$:

$$h_s(t;m_1,m_2) = \begin{cases} \displaystyle\sum_{a\in A_2\cup A_3}\int_{[0,\,t]} h_a(t-t_1;m_1-1,m_2)\,\mathsf{K}_{s,a}\{dt_1\} & \text{if } s \in A_1, \\[2ex] \displaystyle\sum_{a\in A_1\cup A_3}\int_{[0,\,t]} h_a(t-t_2;m_1,m_2-1)\,\mathsf{K}_{s,a}\{dt_2\} & \text{if } s \in A_2, \\[2ex] \displaystyle\sum_{a\in A_1\cup A_2}\int_{[0,\,t]} h_a(t-t_3;m_1,m_2)\,\mathsf{K}_{s,a}\{dt_3\} & \text{if } s \in A_3. \end{cases}$$

(11.7)

To accomplish the transition to the Laplace transform versions of (11.2) - (11.7), recall that by (10.7), the column-vector of Laplace transforms of the first sojourn times in A_i, $\{\,T_{A_i,1}{}^*(\tau\,;s) : s \in A_i\,\}$, now reads as follows

$$\{\,T_{A_i,1}{}^*(\tau\,;s) : s \in A_i\,\} =$$

$$K^*_{A_i(S \setminus A_i)}(\tau)\, \mathbf{1} = (\mathbf{I} - Q^*_{A_iA_i}(\tau))^{-1} \sum_{\substack{j=1 \\ j \neq i}}^{3} Q^*_{A_iA_j}(\tau)\, \mathbf{1} = \sum_{\substack{j=1 \\ j \neq i}}^{3} K^*_{A_iA_j}(\tau)\, \mathbf{1}. \qquad (11.8)$$

By (9.25) and (11.8), the equations (11.2) - (11.7) in the Laplace transform domain now read as follows:

$m_1 = m_2 = 0$:

$$\tau\, h^*_{A_i}(\tau; 0, 0) = \begin{cases} \mathbf{0} & \text{for } i \in \{1, 2\}, \\[2ex] \mathbf{1} - (K^*_{A_3A_1}(\tau)\, \mathbf{1} + K^*_{A_3A_2}(\tau)\, \mathbf{1}) & \text{for } i = 3, \end{cases} \qquad (11.9)$$

$m_1 = 1, m_2 = 0$:

$$\tau\, h^*_{A_i}(\tau; 1, 0) = \begin{cases} \begin{aligned} &\mathbf{1} - \\ &(K^*_{A_1A_2}(\tau)\, \mathbf{1} + K^*_{A_1A_3}(\tau)\, \mathbf{1} - \\ &\quad - K^*_{A_1A_3}(\tau)\, \tau\, h^*_{A_3}(\tau; 0, 0)) \end{aligned} & \text{for } i = 1, \\[2ex] \mathbf{0} & \text{for } i = 2, \\[2ex] K^*_{A_3A_1}(\tau)\, \tau\, h^*_{A_1}(\tau; 1, 0) & \text{for } i = 3, \end{cases} \qquad (11.10)$$

$m_1 = 0, m_2 = 1$:

$$\tau\, h^*_{A_i}(\tau; 0, 1) = \begin{cases} \begin{aligned} &\mathbf{1} - \\ &(K^*_{A_2A_1}(\tau)\, \mathbf{1} + K^*_{A_2A_3}(\tau)\, \mathbf{1} - \\ &\quad - K^*_{A_2A_3}(\tau)\, \tau\, h^*_{A_3}(\tau; 0, 0)) \end{aligned} & \text{for } i = 2, \\[2ex] \mathbf{0} & \text{for } i = 1, \\[2ex] K^*_{A_3A_2}(\tau)\, \tau\, h^*_{A_2}(\tau; 0, 1) & \text{for } i = 3, \end{cases} \qquad (11.11)$$

$m_1 \geq 2, m_2 = 0$:

$$\tau\, h^*_{A_i}(\tau; m_1, 0) = \begin{cases} K^*_{A_1A_3}(\tau)\, \tau\, h^*_{A_3}(\tau; m_1-1, 0) & \text{for } i = 1, \\[2ex] \mathbf{0} & \text{for } i = 2, \\[2ex] K^*_{A_3A_1}(\tau)\, \tau\, h^*_{A_1}(\tau; m_1, 0) & \text{for } i = 3, \end{cases} \qquad (11.12)$$

$m_1 = 0, m_2 \geq 2$:

$$\tau\, h^*_{A_i}(\tau; 0, m_2) = \begin{cases} 0 & \text{for } i = 1, \\ K^*_{A_2A_3}(\tau)\, \tau\, h^*_{A_3}(\tau; 0, m_2\text{-}1) & \text{for } i = 2, \\ K^*_{A_3A_2}(\tau)\, \tau\, h^*_{A_2}(\tau; 0, m_2) & \text{for } i = 3, \end{cases} \tag{11.13}$$

$m_1, m_2 \geq 1$:

$$\tau\, h^*_{A_1}(\tau; m_1, m_2) = K^*_{A_1A_2}(\tau)\, \tau\, h^*_{A_2}(\tau; m_1 - 1, m_2) +$$
$$K^*_{A_1A_3}(\tau)\, \tau\, h^*_{A_3}(\tau; m_1 - 1, m_2), \tag{11.14}$$

$$\tau\, h^*_{A_2}(\tau; m_1, m_2) = K^*_{A_2A_1}(\tau)\, \tau\, h^*_{A_1}(\tau; m_1, m_2 - 1) +$$
$$K^*_{A_2A_3}(\tau)\, \tau\, h^*_{A_3}(\tau; m_1, m_2 - 1), \tag{11.15}$$

$$\tau\, h^*_{A_3}(\tau; m_1, m_2) = K^*_{A_3A_1}(\tau)\, \tau\, h^*_{A_1}(\tau; m_1, m_2) +$$
$$K^*_{A_3A_2}(\tau)\, \tau\, h^*_{A_2}(\tau; m_1, m_2). \tag{11.16}$$

11.2 A computation scheme for the Laplace Transforms

By (11.9) - (11.16), a recurrence relation is given which uniquely defines the vectors $\{\tau\, h^*_{A_i}(\tau; m_1, m_2) : i = 1, 2, 3; m_1, m_2 \geq 0 \}$. Even though most computer languages do allow recursive procedure definitions (FORTRAN 77 being one of the few notable exceptions), to facilitate an *efficient* computational implementation, it will still be advisable to restructure the above equations such that recursion in that form is avoided. In this section a dynamic programming scheme (see, for example, [AHO]) will be established for the evaluation of the Laplace transforms in (11.9) - (11.16). More precisely, a substitute for (11.14) - (11.16) will be derived which will allow the quantities $\tau\, h^*_{A_i}(\tau; m_1, m_2)$ for $m_1, m_2 \geq 1$ to be evaluated *directly* once the boundary elements of the scheme (i.e. those with $m_1 = 1$ or $m_2 = 1$) are known. We eliminate $\tau\, h^*_{A_3}(\tau; ., .)$ by (11.16) from (11.14) to get for $m_1 \geq 2, m_2 \geq 1$,

$$\tau\, h^*_{A_1}(\tau; m_1, m_2) = K^*_{A_1A_3}(\tau)\, K^*_{A_3A_1}(\tau)\, \tau\, h^*_{A_1}(\tau; m_1 - 1, m_2) +$$
$$(K^*_{A_1A_2}(\tau) + K^*_{A_1A_3}(\tau)\, K^*_{A_3A_2}(\tau))\, \tau\, h^*_{A_2}(\tau; m_1 - 1, m_2). \tag{11.17}$$

Likewise, (11.15) and (11.16) imply for $m_1 \geq 1, m_2 \geq 2$,

$$\tau\, h^*_{A_2}(\tau; m_1, m_2) = K^*_{A_2A_3}(\tau)\, K^*_{A_3A_2}(\tau)\, \tau\, h^*_{A_2}(\tau; m_1, m_2 - 1) +$$
$$(K^*_{A_2A_1}(\tau) + K^*_{A_2A_3}(\tau)\, K^*_{A_3A_1}(\tau))\, \tau\, h^*_{A_1}(\tau; m_1, m_2 - 1). \tag{11.18}$$

We get now by induction from (11.17) and (11.18) respectively

$$\tau\, h^*_{A_1}(\tau; m_1, m_2) = (K^*_{A_1A_3}(\tau)\, K^*_{A_3A_1}(\tau))^{m_1-1}\, \tau\, h^*_{A_1}(\tau; 1, m_2) +$$

$$\sum_{k=1}^{m_1-1} (K^*_{A_1A_3}(\tau)\, K^*_{A_3A_1}(\tau))^{m_1-k-1} \times$$

$$(K^*_{A_1A_2}(\tau) + K^*_{A_1A_3}(\tau)\, K^*_{A_3A_2}(\tau))\, \tau\, h^*_{A_2}(\tau; k, m_2), \qquad (11.19)$$

$$\tau\, h^*_{A_2}(\tau; m_1, m_2) = (K^*_{A_2A_3}(\tau)\, K^*_{A_3A_2}(\tau))^{m_2-1}\, \tau\, h^*_{A_2}(\tau; m_1, 1) +$$

$$\sum_{k=1}^{m_2-1} (K^*_{A_2A_3}(\tau)\, K^*_{A_3A_2}(\tau))^{m_2-k-1} \times$$

$$(K^*_{A_2A_1}(\tau) + K^*_{A_2A_3}(\tau)\, K^*_{A_3A_1}(\tau))\, \tau\, h^*_{A_1}(\tau; m_1, k). \qquad (11.20)$$

(Notice that both (11.19) and (11.20) hold even for $m_1, m_2 \geq 1$.) Substitute now (11.20) into (11.19) and vice versa to get for $m_1, m_2 \geq 1$ respectively

$$\tau\, h^*_{A_1}(\tau; m_1, m_2) = (K^*_{A_1A_3}(\tau)\, K^*_{A_3A_1}(\tau))^{m_1-1}\, \tau\, h^*_{A_1}(\tau; 1, m_2) +$$

$$\sum_{k=1}^{m_1-1} (K^*_{A_1A_3}(\tau)\, K^*_{A_3A_1}(\tau))^{m_1-k-1} (K^*_{A_1A_2}(\tau) + K^*_{A_1A_3}(\tau)\, K^*_{A_3A_2}(\tau)) \times$$

$$(K^*_{A_2A_3}(\tau)\, K^*_{A_3A_2}(\tau))^{m_2-1}\, \tau\, h^*_{A_2}(\tau; k, 1) +$$

$$\sum_{k_1=1}^{m_1-1} \sum_{k_2=1}^{m_2-1} (K^*_{A_1A_3}(\tau)\, K^*_{A_3A_1}(\tau))^{m_1-k_1-1} \times$$

$$(K^*_{A_1A_2}(\tau) + K^*_{A_1A_3}(\tau)\, K^*_{A_3A_2}(\tau)) (K^*_{A_2A_3}(\tau)\, K^*_{A_3A_2}(\tau))^{m_2-k_2-1} \times$$

$$(K^*_{A_2A_1}(\tau) + K^*_{A_2A_3}(\tau)\, K^*_{A_3A_1}(\tau))\, \tau\, h^*_{A_1}(\tau; k_1, k_2), \qquad (11.21)$$

$$\tau\, h^*_{A_2}(\tau; m_1, m_2) = (K^*_{A_2A_3}(\tau)\, K^*_{A_3A_2}(\tau))^{m_2-1}\, \tau\, h^*_{A_2}(\tau; m_1, 1) +$$

$$\sum_{k=1}^{m_2-1} (K^*_{A_2A_3}(\tau)\, K^*_{A_3A_2}(\tau))^{m_2-k-1} (K^*_{A_2A_1}(\tau) + K^*_{A_2A_3}(\tau)\, K^*_{A_3A_1}(\tau)) \times$$

$$(K^*_{A_1A_3}(\tau)\, K^*_{A_3A_1}(\tau))^{m_1-1}\, \tau\, h^*_{A_1}(\tau; 1, k) +$$

$$\sum_{k_1=1}^{m_1-1} \sum_{k_2=1}^{m_2-1} (K^*_{A_2A_3}(\tau)\, K^*_{A_3A_2}(\tau))^{m_2-k_2-1} \times$$

$$(K^*_{A_2A_1}(\tau) + K^*_{A_2A_3}(\tau)\, K^*_{A_3A_1}(\tau))\, (K^*_{A_1A_3}(\tau)\, K^*_{A_3A_1}(\tau))^{m_1-k_1-1} \times$$

$$(K^*_{A_1A_2}(\tau) + K^*_{A_1A_3}(\tau)\, K^*_{A_3A_2}(\tau))\, \tau\, h^*_{A_2}(\tau; k_1, k_2). \tag{11.22}$$

The boundary elements for the scheme in (11.21) and (11.22) are $\tau\, h^*_{A_1}(\tau; 1, m_2)$ and $\tau\, h^*_{A_2}(\tau; m_1, 1)$ with $m_1, m_2 \geq 1$. From (11.14) and (11.15), they are seen to be respectively expressible in terms of the tuples of vectors $(\tau\, h^*_{A_2}(\tau; 0, m_2),\ \tau\, h^*_{A_3}(\tau; 0, m_2))$ and $(\tau\, h^*_{A_1}(\tau; m_1, 0),\ \tau\, h^*_{A_3}(\tau; m_1, 0))$. These latter quantities in turn are obtainable by (11.9) - (11.13). No recursion is involved in this procedure once the boundary elements are known. The scheme afforded by (11.21) and (11.22) is a simple instance of what is termed the 'dynamic programming' approach (see, e.g., [AHO], Chapter 10).

CHAPTER 12

FINITE TIME-HORIZON SOJOURN TIMES FOR FINITE SEMI-MARKOV PROCESSES

In the Chapters 5, 6 and 9 - 11, the underlying process was observed for a fixed and finite length of time, $t > 0$. However, the various system characteristics obtained in there did not include the finite-horizon version of the sojourn time variables. These variables will be examined in the present chapter.

We shall be concerned here with a semi-Markov process $Y = \{ Y_t : t \geq 0 \}$ whose state space S is partitioned as $S = A \cup B \cup C$ with C (possibly empty) consisting of all absorbing states of Y. If C is non-empty then Y is assumed to be absorbing (i.e., $A \cup B$ is then transient). Alternatively, Y may be irreducible in which case C is empty. Suppose now that the initial state of Y is some $s \in A \cup B$. The variable under consideration in this chapter is $T_{A,m}(t)$, the m*th* sojourn of Y in A *during* [0, t]; it is formally defined as follows

    (i) If Y is absorbed into C before its m*th* visit to A, or, if the m*th* visit of Y to A commences after t, then we put $T_{A,m}(t) = 0$;

    (ii) If Y's m*th* visit to A commences before t, at, say, $t' \leq t$, and finishes at, say, $t'' (> t')$, then we put $T_{A,m}(t) = \min\{t - t', t'' - t'\}$.

$T_{A,m}(t)$ can be viewed informally as the length of the 'm*th* sojourn of Y in A for an observer whose time-horizon is t'. The importance of allowing the time-horizon to be finite in the analysis of semi-Markov reliability models has also been pointed out recently by Jack [JAC1], Jack and Dagpunar [JAC2] and Christer and Jack [CHR]. Even though none of the results from the previous chapters will be used here to describe $T_{A,m}(t)$, from a methodological point of view this chapter is strongly related to what we have already seen: the 'renewal argument' will be used to obtain a recurrence relation for the vectors of double Laplace transforms

$$\mathbf{g}_m^{**}(\tau_1, \tau_2) = \{ g_m^{**}(\tau_1, \tau_2; s) : s \in S \setminus C \}, \quad m = 1, 2, \dots ,$$

the components of which are defined for $\tau_1, \tau_2 \in \mathbb{C}^+$ by

$$g_m^{**}(\tau_1, \tau_2; s) = \int_0^{+\infty} \int_0^{+\infty} \exp\{ - t_1 \tau_1 - t_2 \tau_2 \} \Pr\{ T_{A,m}(t_1) \leq t_2 \mid Y_0 = s \} \, dt_1 \, dt_2.$$

$$(12.1)$$

An alternative form of (12.1) is, of course,

$$h_m(\tau_1, \tau_2; s) = \int_0^{+\infty} \exp\{ - t_1 \tau_1 \} E[\exp\{ - \tau_2 T_{A,m}(t_1) \} \mid Y_0 = s] \, dt_1, \qquad (12.2)$$

where h_m is defined by

$$h_m(\tau_1, \tau_2; s) = \tau_2 \, g_m^{**}(\tau_1, \tau_2; s). \tag{12.3}$$

h_m is interrelated to the kth moment of $T_{A,m}(t_1)$ by

$$\left. \frac{\partial^k h_m(\tau_1, \tau_2; s)}{\partial \tau_2^k} \right|_{\tau_2 = 0} = (-1)^k \int_0^{+\infty} \exp\{-t_1 \, \tau_1\} \, E[\, (T_{A,m}(t_1))^k \mid Y_0 = s \,] \, dt_1.$$
$$\tag{12.4}$$

By virtue of (12.3) and (12.4), any recurrence relation for g_m^{**} also provides a tool for evaluating the Laplace transforms of the moments of $T_{A,m}(t_1)$ as a function of t_1. Laplace transform inversion with respect to the variable τ_1 in (12.4) yields values for the moments of $T_{A,m}(t_1)$ for fixed t_1.

The double Laplace transform technique is commonly used in the literature to analyse cumulative measures (such as the accumulated reward) arising in the theory of continuous-time processes; see, for example, [PUR1], [PUR2], [KUL1], [SMI], [REI2], and [ROSS]. Typically, though, these papers are devoted to the study of cumulative measures *totalled over the period* [0, t]. Using Puri's notation ([PUR1], [PUR2]), the quantities under investigation in those papers can be written in the form

$$\int_0^t f(Y_r, r) \, dr, \tag{12.5}$$

with some suitably defined measurable $f: S \times [0, \infty) \to [0, \infty)$. The sojourn time $T_{A,m}(t)$, to be addressed in this chapter, is, however, obviously not of the form (12.5).

In Section 12.1, the main result is presented and proved. It is a recurrence relation for the sequence of vectors defined in (12.1). In Section 12.2, the special case of an alternating renewal process is examined in more detail. Then, $S = \{1, 2\}$. The recurrence relation from the main theorem is used in Section 12.2 to arrive at an expression for the partial derivatives of the functions h_m at $\tau_2 = 0$ which in turn is used in conjunction with (12.4) to study the moments of $T_{A,m}(t)$. Assuming in the states '1' and '2' an L-stage Erlang, and an exponentially distributed holding time respectively, the first two moments of $T_{A,m}(t)$ are obtained by *symbolic* Laplace transform inversion with the computer algebra system MAPLE (see, e.g., [CHA1] and [CHA2]). Numerical results are obtained and discussed for L, m = 1, 2, 3.

This chapter is an enhanced version of the work reported in [CSE10]: in addition to what is in [CSE10], we show here the MAPLE code and discuss the relevant MAPLE features.

12.1 The double Laplace transform of finite-horizon sojourn times

With the notation of the Chapters 8 - 11, the main result is as follows.

THEOREM 12.1. *For* $\tau \in \mathbb{C}^+$ *the matrices* $\mathbf{I} - \mathbf{Q}^*_{AA}(\tau)$ *and* $\mathbf{I} - \mathbf{Q}^*_{BB}(\tau)$ *are invertible and for* $\tau_1, \tau_2 \in \mathbb{C}^+ \setminus \{0\}$ *we have the following recurrence relations*

$$g_1^{**}{}_A(\tau_1, \tau_2) = \frac{1}{\tau_1\tau_2(\tau_1+\tau_2)} (\mathbf{I} - \mathbf{Q}^*{}_{AA}(\tau_1+\tau_2))^{-1} \times$$

$$\left\{ \tau_1 (\mathbf{Q}^*{}_{AA}(0) - \mathbf{Q}^*{}_{AA}(\tau_1+\tau_2)) \mathbf{1} + (\tau_1 \mathbf{Q}^*{}_{AB}(0) - \tau_2 \mathbf{Q}^*{}_{AB}(\tau_1+\tau_2)) \mathbf{1} + \right.$$

$$\left. + (\tau_1 \mathbf{Q}^*{}_{AC}(0) - \tau_2 \mathbf{Q}^*{}_{AC}(\tau_1+\tau_2)) \mathbf{1} \right\} ; \tag{12.6}$$

and, for $m \in \{2, 3, ... \}$,

$$g_m^{**}{}_A(\tau_1, \tau_2) = (\mathbf{I} - \mathbf{Q}^*{}_{AA}(\tau_1))^{-1} \times$$

$$\left\{ \mathbf{Q}^*{}_{AB}(\tau_1) g_{m-1}^{**}{}_B(\tau_1, \tau_2) + \frac{1}{\tau_1\tau_2} \mathbf{Q}^*{}_{AC}(0) \mathbf{1} + \right.$$

$$\left. \frac{1}{\tau_1\tau_2} (\mathbf{Q}^*{}_{AA}(0) - \mathbf{Q}^*{}_{AA}(\tau_1)) \mathbf{1} + \frac{1}{\tau_1\tau_2} (\mathbf{Q}^*{}_{AB}(0) - \mathbf{Q}^*{}_{AB}(\tau_1)) \mathbf{1} \right\} ; \tag{12.7}$$

and, for $m \in \{1, 2, ... \}$,

$$g_m^{**}{}_B(\tau_1, \tau_2) = (\mathbf{I} - \mathbf{Q}^*{}_{BB}(\tau_1))^{-1} \times$$

$$\left\{ \mathbf{Q}^*{}_{BA}(\tau_1) g_m^{**}{}_A(\tau_1, \tau_2) + \frac{1}{\tau_1\tau_2} \mathbf{Q}^*{}_{BC}(0) \mathbf{1} + \right.$$

$$\left. \frac{1}{\tau_1\tau_2} (\mathbf{Q}^*{}_{BA}(0) - \mathbf{Q}^*{}_{BA}(\tau_1)) \mathbf{1} + \frac{1}{\tau_1\tau_2} (\mathbf{Q}^*{}_{BB}(0) - \mathbf{Q}^*{}_{BB}(\tau_1)) \mathbf{1} \right\} . \tag{12.8}$$

NOTE. (12.6) - (12.8) clearly provide a tool for a recursive evaluation of $g_m^{**}(\tau_1, \tau_2)$ for any $m \geq 1$. For irreducible Y, the above equations hold with $\mathbf{Q}^*{}_{AC} \equiv \mathbf{0}, \mathbf{Q}^*{}_{BC} \equiv \mathbf{0}$.

PROOF OF THEOREM 12.1. $\mathbf{I} - \mathbf{Q}^*{}_{BB}(\tau)$ and $\mathbf{I} - \mathbf{Q}^*{}_{AA}(\tau)$ are invertible by Theorem 8.1. To prove (12.6) - (12.8), we start by noting that for $t_1, t_2 \in [0, +\infty)$ and $s \in S \setminus C$ we have

$$\Pr\{ T_{A,m}(t_1) \leq t_2 \mid Y_0 = s \} = \sum_{s' \in S, s \neq s'} \Pr\{ T_{A,m}(t_1) \leq t_2, X_1 = s' \mid X_0 = s \}. \tag{12.9}$$

We also define the expressions $J_i(t_1, t_2)$, $i = 1, ..., 6$, for later use as follows

$$J_1(t_1, t_2) = r_{ss'} \int_{[0, t_2]} \Pr\{ T_{A,m}(t_1 - t) \leq t_2 - t \mid X_0 = s' \} F_{s,s'}\{dt\},$$

$$J_2(t_1, t_2) = r_{ss'} \int\limits_{[0,\, t_1]} \Pr\{\, T_{A,m}(t_1 - t) \le t_2 \mid X_0 = s'\} \, F_{s,s'}\{dt\},$$

$$J_3(t_1, t_2) = r_{ss'} \int\limits_{[0,\, t_1]} \Pr\{\, T_{A,m-1}(t_1 - t) \le t_2 \mid X_0 = s'\} \, F_{s,s'}\{dt\},$$

$$J_4(t_1, t_2) = r_{ss'} \, F_{s,s'}(t_2),$$

$$J_5(t_1, t_2) = r_{ss'},$$

$$J_6(t_1, t_2) = r_{ss'} \, (1 - F_{s,s'}(t_1)).$$

The domain of integration, $[0, \infty)^2$, of the double Laplace transform of a generic summand on the right hand side of (12.9) will be partitioned as $[0, +\infty)^2 = D_1 \cup D_2$, with D_1 and D_2 defined respectively by

$$D_1 = \{\, (t_1, t_2) : 0 \le t_2 \le t_1 < +\infty \,\}, \quad D_2 = \{\, (t_1, t_2) : 0 \le t_1 < t_2 < +\infty \,\}.$$

It is readily seen (by conditioning on the holding time spent by Y in the initial state $Y_0 = s$) that for $(t_1, t_2) \in D_1$ the probability $\Pr\{\, T_{A,m}(t_1) \le t_2, X_1 = s' \mid X_0 = s\}$ is equal to the values given in Table 12.1. Also notice that for $(t_1, t_2) \in D_2$ we have

$$\Pr\{\, T_{A,m}(t_1) \le t_2, X_1 = s' \mid X_0 = s\} = J_5(t_1, t_2).$$

The integrals which will be needed in the sequel are summarised in the following lemma.

	m = 1		m = 2	
	$s \in A$	$s \in B$	$s \in A$	$s \in B$
$s' \in A$	$J_1(t_1, t_2)$	$J_2(t_1, t_2) + J_6(t_1, t_2)$	$J_2(t_1, t_2) + J_6(t_1, t_2)$	$J_2(t_1, t_2) + J_6(t_1, t_2)$
$s' \in B$	$J_4(t_1, t_2)$	$J_2(t_1, t_2) + J_6(t_1, t_2)$	$J_3(t_1, t_2) + J_6(t_1, t_2)$	$J_2(t_1, t_2) + J_6(t_1, t_2)$
$s' \in C$	$J_4(t_1, t_2)$	$J_5(t_1, t_2)$	$J_5(t_1, t_2)$	$J_5(t_1, t_2)$

Table 12.1 $\Pr\{\, T_{A,m}(t_1) \le t_2, X_1 = s' \mid X_0 = s\}$ for $(t_1, t_2) \in D_1$

LEMMA 12.2 *It is for* $\tau_1, \tau_2 \in \mathbb{C}^+ \setminus \{\, 0 \,\}$,

$$\iint\limits_{D_1} \exp\{\, -t_1 \tau_1 - t_2 \tau_2 \,\} \, J_1(t_1, t_2) \, dt_1 \, dt_2 =$$

$$q_{ss'}{}^*(\tau_1+\tau_2) \left\{ g_m{}^{**}(\tau_1, \tau_2; s') - \frac{1}{\tau_2(\tau_1+\tau_2)} \right\}, \tag{12.10}$$

$$\iint\limits_{D_1} \exp\{ -t_1\,\tau_1 - t_2\,\tau_2 \}\, J_2(t_1, t_2)\, dt_1\, dt_2 = -q_{ss'}{}^*(\tau_1+\tau_2)\,\frac{1}{\tau_2(\tau_1+\tau_2)} +$$

$$q_{ss'}{}^*(\tau_1) \left\{ g_m{}^{**}(\tau_1, \tau_2; s') - \frac{1}{\tau_2(\tau_1+\tau_2)} + \frac{1}{\tau_1\tau_2} - \frac{1}{\tau_1(\tau_1+\tau_2)} \right\}, \tag{12.11}$$

$$\iint\limits_{D_1} \exp\{ -t_1\,\tau_1 - t_2\,\tau_2 \}\, J_3(t_1, t_2)\, dt_1\, dt_2 = -q_{ss'}{}^*(\tau_1+\tau_2)\,\frac{1}{\tau_2(\tau_1+\tau_2)} +$$

$$q_{ss'}{}^*(\tau_1) \left\{ g_{m-1}{}^{**}(\tau_1, \tau_2; s') - \frac{1}{\tau_2(\tau_1+\tau_2)} + \frac{1}{\tau_1\tau_2} - \frac{1}{\tau_1(\tau_1+\tau_2)} \right\}, \tag{12.12}$$

$$\iint\limits_{D_1} \exp\{ -t_1\,\tau_1 - t_2\,\tau_2 \}\, J_4(t_1, t_2)\, dt_1\, dt_2 = q_{ss'}{}^*(\tau_1+\tau_2)\,\frac{1}{\tau_1(\tau_1+\tau_2)}, \tag{12.13}$$

$$\iint\limits_{D_1} \exp\{ -t_1\,\tau_1 - t_2\,\tau_2 \}\, J_5(t_1, t_2)\, dt_1\, dt_2 = q_{ss'}{}^*(0) \left\{ \frac{1}{\tau_1\tau_2} - \frac{1}{\tau_2(\tau_1+\tau_2)} \right\}, \tag{12.14}$$

$$\iint\limits_{D_2} \exp\{ -t_1\,\tau_1 - t_2\,\tau_2 \}\, J_5(t_1, t_2)\, dt_1\, dt_2 = q_{ss'}{}^*(0)\,\frac{1}{\tau_2(\tau_1+\tau_2)}, \tag{12.15}$$

$$\iint\limits_{D_1} \exp\{ -t_1\,\tau_1 - t_2\,\tau_2 \}\, J_6(t_1, t_2)\, dt_1\, dt_2 =$$

$$q_{ss'}{}^*(0) \left\{ \frac{1}{\tau_1\tau_2} - \frac{1}{\tau_2(\tau_1+\tau_2)} \right\} - \frac{q_{ss'}{}^*(\tau_1)}{\tau_1\tau_2} + \frac{q_{ss'}{}^*(\tau_1+\tau_2)}{\tau_2(\tau_1+\tau_2)}. \tag{12.16}$$

Before proving Lemma 12.2, let us show that it allows us, rather straightforwardly, to conclude the proof of Theorem 12.1. Let us demonstrate this, for example, for (12.7). From the third column in Table 12.1, we have by (12.9), (12.11), (12.12), and (12.14) - (12.16) for $s \in A$ and $m \geq 2$ that

$$g_m{}^{**}(\tau_1, \tau_2; s) =$$

$$\sum_{\substack{s'\in A \\ s'\neq s}} \left\{ q_{ss'}{}^*(\tau_1) \left\{ g_m{}^{**}(\tau_1,\tau_2;s') - \frac{1}{\tau_2(\tau_1+\tau_2)} + \frac{1}{\tau_1\tau_2} - \frac{1}{\tau_1(\tau_1+\tau_2)} \right\} - \right.$$

$$- q_{ss'}{}^*(\tau_1+\tau_2)\frac{1}{\tau_2(\tau_1+\tau_2)} + q_{ss'}{}^*(0)\frac{1}{\tau_2(\tau_1+\tau_2)} +$$

$$\left. + q_{ss'}{}^*(0)\left\{ \frac{1}{\tau_1\tau_2} - \frac{1}{\tau_2(\tau_1+\tau_2)} \right\} - \frac{q_{ss'}{}^*(\tau_1)}{\tau_1\tau_2} + \frac{q_{ss'}{}^*(\tau_1+\tau_2)}{\tau_2(\tau_1+\tau_2)} \right\} +$$

$$\sum_{s'\in B} \left\{ q_{ss'}{}^*(\tau_1) \left\{ g_{m-1}{}^{**}(\tau_1,\tau_2;s') - \frac{1}{\tau_2(\tau_1+\tau_2)} + \frac{1}{\tau_1\tau_2} - \frac{1}{\tau_1(\tau_1+\tau_2)} \right\} - \right.$$

$$- q_{ss'}{}^*(\tau_1+\tau_2)\frac{1}{\tau_2(\tau_1+\tau_2)} + q_{ss'}{}^*(0)\frac{1}{\tau_2(\tau_1+\tau_2)} +$$

$$\left. + q_{ss'}{}^*(0)\left\{ \frac{1}{\tau_1\tau_2} - \frac{1}{\tau_2(\tau_1+\tau_2)} \right\} - \frac{q_{ss'}{}^*(\tau_1)}{\tau_1\tau_2} + \frac{q_{ss'}{}^*(\tau_1+\tau_2)}{\tau_2(\tau_1+\tau_2)} \right\} +$$

$$\sum_{s'\in C} \left\{ q_{ss'}{}^*(0)\left\{ \frac{1}{\tau_1\tau_2} - \frac{1}{\tau_2(\tau_1+\tau_2)} \right\} + \frac{q_{ss'}{}^*(0)}{\tau_2(\tau_1+\tau_2)} \right\}. \tag{12.17}$$

(12.7) now follows from the matrix form of (12.17) by solving (12.17) for $g_m{}^{**}{}_A(\tau_1,\tau_2)$. Similar considerations yield (12.6) and (12.8). ∎

PROOF OF LEMMA 12.2. (12.10) will be shown first. For $t_1 \geq 0$ we interchange the order of integration to get

$$\int_0^{t_1} \exp\{-t_2\,\tau_2\}\, J_1(t_1,t_2)\, dt_2 =$$

$$r_{ss'} \int_{[0,t_1]} \exp\{-t\,\tau_2\} \int_t^{t_1} \exp\{-\tau_2\,(t_2-t)\} \times$$

$$\Pr\{ T_{A,m}(t_1-t) \leq t_2 - t \mid X_0 = s'\}\, dt_2\, F_{s,s'}\{dt\}. \tag{12.18}$$

The inner integral in (12.18) can be rewritten as

$$\int_0^{+\infty} \exp\{-\tau_2\,v\}\, \Pr\{ T_{A,m}(t_1-t) \leq v \mid X_0 = s'\}\, dv - \int_{t_1}^{+\infty} \exp\{-\tau_2\,(t_2-t)\}\, dt_2$$

$$\tag{12.19}$$

since $u \leq v$ implies $T_{A,m}(u) \leq v$. Thus, denoting by $g_m^*(t_1, \tau_2; s')$ the Laplace transform of

$$g_m(t_1, t_2; s') = \Pr\{ T_{A,m}(t_1) \leq t_2 \mid X_0 = s' \}$$

with respect to the variable t_2, we get from (12.18) and (12.19)

$$\int_0^{t_1} \exp\{ - t_2 \, \tau_2 \} \, J_1(t_1, t_2) \, dt_2 =$$

$$r_{ss'} \int_{[0, t_1]} \exp\{ - t \, \tau_2 \} \{ g_m^*(t_1 - t, \tau_2; s') - \exp\{ - \tau_2 (t_1 - t) \} / \tau_2 \} \, F_{s,s'}\{dt\}.$$

$$(12.20)$$

From (12.20) it follows that

$$\iint_{D_1} \exp\{ - t_1 \, \tau_1 - t_2 \, \tau_2 \} \, J_1(t_1, t_2) \, dt_1 \, dt_2 =$$

$$r_{ss'} \int_0^{+\infty} \exp\{ - t_1 \tau_1 \} \int_{[0, t_1]} g_m^*(t_1 - t, \tau_2; s') \exp\{ - t \, \tau_2 \} \, F_{s,s'}\{dt\} \, dt_1 -$$

$$- r_{ss'} f_{s,s'}^*(\tau_1 + \tau_2) \frac{1}{\tau_2(\tau_1 + \tau_2)},$$ $$(12.21)$$

where $f_{s,s'}^*$ stands for the Laplace transform of the distribution with cumulative distribution function $F_{s,s'}$, i.e.,

$$f_{s,s'}^*(\tau) = \int_{[0, +\infty)} \exp\{ - \tau \, t \} \, F_{s,s'}\{dt\}.$$

The inner integral on the right hand side of (12.21) is a convolution integral which can be written as

$$\int_{[0, t_1]} g_m^*(t_1 - t, \tau_2; s') \, H_{s,s'}(dt; \tau_2),$$

where the family of complex-valued finite measures $\{ H_{s,s'}(\, . \, ; \tau) : \tau \in \mathbb{C}^+ \}$ is defined for measurable $D \subset [0, +\infty)$ by

$$H_{s,s'}(D; \tau) = \int_D \exp\{ - \tau \, t \} \, F_{s,s'}\{dt\}.$$

The Laplace transform of this measure is, of course,

$$\int_{[0, +\infty)} \exp\{ - \rho\, t \}\, H_{s,s'}(dt;\, \tau) = f_{s,s'}{}^*(\tau+\rho),\ \rho \in \mathbb{C}^+.$$

(12.21) can therefore be rewritten as

$$\iint_{D_1} \exp\{ - t_1\, \tau_1 - t_2\, \tau_2 \}\, J_1(t_1,\, t_2)\, dt_1\, dt_2 =$$

$$r_{ss'}\, f_{s,s'}{}^*(\tau_1+\tau_2) \left\{ g_m{}^{**}(\tau_1,\, \tau_2;\, s') - \frac{1}{\tau_2(\tau_1+\tau_2)} \right\},$$

from which (12.10) follows by the definition of $q_{s,s'}{}^*$.

We now turn to the proof of (12.11). It is $T_{A,m}(t_1 - t) \le t_2$ for $t > t_1 - t_2$ which implies that $J_2(t_1, t_2)$ can be rewritten as

$$J_2(t_1,\, t_2) =$$

$$r_{ss'} \int_{[0,\, t_1 - t_2]} \Pr\{\, T_{A,m}(t_1 - t) \le t_2 \mid X_0 = s'\}\, F_{s,s'}\{dt\} + r_{ss'}\, (F_{s,s'}(t_1) - F_{s,s'}(t_1 - t_2)).$$

(12.22)

From (12.22) it follows that

$$\int_0^{t_1} \exp\{ - t_2\, \tau_2 \}\, J_2(t_1,\, t_2)\, dt_2 =$$

$$r_{ss'} \int_{[0,\, t_1]} \int_0^{t_1 - t} \exp\{ - t_2\, \tau_2 \}\, \Pr\{\, T_{A,m}(t_1 - t) \le t_2 \mid X_0 = s'\}\, dt_2\, F_{s,s'}\{dt\} +$$

$$r_{ss'} \int_0^{t_1} \exp\{ - t_2\, \tau_2 \}\, dt_2\, F_{s,s'}(t_1) - r_{ss'} \int_0^{t_1} \exp\{ - t_2\, \tau_2 \}\, F_{s,s'}(t_1 - t_2)\, dt_2. \qquad (12.23)$$

The inner integral in the first term on the right hand side of (12.23) is as follows

$$\int_0^\infty \exp\{ - t_2\, \tau_2 \}\, \Pr\{\, T_{A,m}(t_1 - t) \le t_2 \mid X_0 = s'\}\, dt_2 - \int_{t_1-t}^\infty \exp\{ - t_2\, \tau_2 \}\, dt_2 =$$

$$g_m^*(t_1 - t, \tau_2; s') - \frac{1}{\tau_2} \exp\{ - \tau_2 (t_1 - t) \}.$$

Thus, (12.23) can be further evaluated as

$$r_{ss'} \int\limits_{[0,\, t_1]} \left\{ g_m^*(t_1 - t, \tau_2; s') - \exp\{ - \tau_2 (t_1 - t) \} / \tau_2 \right\} F_{s,s'}\{dt\} +$$

$$r_{ss'} \frac{1}{\tau_2} (1 - \exp\{ - t_1 \tau_2 \}) F_{s,s'}(t_1) - r_{ss'} \int_0^{t_1} \exp\{ - t_2 \tau_2 \} F_{s,s'}(t_1 - t_2) \, dt_2.$$

Multiplying the above by $\exp\{ - t_1 \tau_1 \}$ and integrating with respect to t_1 gives

$$\iint\limits_{D_1} \exp\{ - t_1 \tau_1 - t_2 \tau_2 \} J_2(t_1, t_2) \, dt_1 \, dt_2 =$$

$$r_{ss'} \, g_m^{**}(\tau_1, \tau_2; s') \, f_{s,s'}^*(\tau_1) -$$

$$- r_{ss'} \frac{1}{\tau_2} \int_0^\infty \int\limits_{[0,\, t_1]} \exp\{ - (\tau_1 + \tau_2)(t_1 - t) - t_1 \tau_1 \} F_{s,s'}\{dt\} \, dt_1 +$$

$$+ r_{ss'} \frac{1}{\tau_2} \int_0^\infty F_{s,s'}(t_1) \exp\{ - t_1 \tau_1 \} \, dt_1 - r_{ss'} \frac{1}{\tau_2} \int_0^\infty \exp\{ - (\tau_1 + \tau_2) t_1 \} F_{s,s'}(t_1) \, dt_1 -$$

$$- r_{ss'} \int_0^\infty \int_0^{t_1} \exp\{ - t_1 \tau_1 - t_2 \tau_2 \} F_{s,s'}(t_1 - t_2) \, dt_2 \, dt_1. \qquad (12.24)$$

(12.24) now implies (12.11) by observing that the two double integrals on the right hand side of (12.24) are given respectively by

$$\int_0^\infty \int\limits_{[0,\, t_1]} \exp\{ - (\tau_1 + \tau_2)(t_1 - t) - t_1 \tau_1 \} F_{s,s'}\{dt\} \, dt_1 =$$

$$\int\limits_{[0,\, +\infty)} \exp\{ - \tau_1 t \} \int_t^\infty \exp\{ - (\tau_1 + \tau_2)(t_1 - t) \} \, dt_1 \, F_{s,s'}\{dt\} = \frac{f_{s,s'}^*(\tau_1)}{\tau_1 + \tau_2},$$

and

$$\int_0^\infty \int_0^{t_1} \exp\{-t_1\,\tau_1 - t_2\,\tau_2\}\, F_{s,s'}(t_1 - t_2)\, dt_2\, dt_1 =$$

$$\int_0^\infty \exp\{-t_2\,\tau_2\} \int_{t_2}^\infty \exp\{-t_1\,\tau_1\}\, F_{s,s'}(t_1 - t_2)\, dt_1\, dt_2 = \frac{f_{s,s'}^*(\tau_1)}{\tau_1(\tau_1+\tau_2)}.$$

(12.12) is obtained from (12.11) by replacing m by m - 1. (From Table 12.1 it is seen that whenever $J_3(t_1, t_2)$ is evaluated, it is m - 1 ≥ 1.)

(12.13) follows from

$$\iint_{D_1} \exp\{-t_1\,\tau_1 - t_2\,\tau_2\}\, J_4(t_1, t_2)\, dt_1\, dt_2 =$$

$$r_{ss'} \int_0^\infty \exp\{-t_2\,\tau_2\}\, F_{s,s'}(t_2) \int_{t_2}^\infty \exp\{-t_1\,\tau_1\}\, dt_1\, dt_2 = \frac{r_{ss'}\, f_{s,s'}^*(\tau_1+\tau_2)}{\tau_1(\tau_1+\tau_2)}.$$

(12.14) and (12.15) are obtained by straightforward integration and by noting that $r_{ss'} = q_{ss'}^*(0)$.

Finally, (12.16) follows from (12.14) and

$$\iint_{D_1} \exp\{-t_1\,\tau_1 - t_2\,\tau_2\}\, J_6(t_1, t_2)\, dt_1\, dt_2 =$$

$$\iint_{D_1} \exp\{-t_1\,\tau_1 - t_2\,\tau_2\}\, (J_5(t_1, t_2) - r_{ss'}\, F_{s,s'}(t_1))\, dt_1\, dt_2 =$$

$$q_{ss'}^*(0) \left\{ \frac{1}{\tau_1\tau_2} - \frac{1}{\tau_2(\tau_1+\tau_2)} \right\} -$$

$$- r_{ss'} \int_0^\infty \exp\{-t_1\,\tau_1\} \int_0^{t_1} \exp\{-t_2\,\tau_2\}\, dt_2\, F_{s,s'}(t_1)\, dt_1. \qquad \blacksquare$$

12.2 An application: the alternating renewal process

12.2.1 Laplace transforms

Model 6 from Section 1.2.2 will now be examined. Thus, $S = \{1, 2\}$ and $C = \varnothing$. Assume that the state in $A = G = \{1\}$ stands for system 'up' while that in $B = \{2\}$ stands for system 'down'. We are interested in the m*th* system 'up' time assuming a finite time-horizon $t_1 > 0$, and assuming that the system has started in the 'up' state at $t = 0$. The Laplace transforms of the holding time distributions in A and B will be denoted by ϕ and $\mathcal{Y}$ respectively. We know $\mathbf{R}$ and $\mathbf{Q}^*(\tau)$ from Sections 1.2.2 and 9.4.1 respectively. (12.6) - (12.8) show that the (now scalar) functions $g_m^{**}{}_A(\tau_1, \tau_2)$ and $g_m^{**}{}_B(\tau_1, \tau_2)$ satisfy the following equations

$$g_1^{**}{}_A(\tau_1, \tau_2) = \frac{\tau_1 + \tau_2 \phi(\tau_1+\tau_2)}{\tau_1 \tau_2 (\tau_1+\tau_2)}, \tag{12.25}$$

$$g_m^{**}{}_A(\tau_1, \tau_2) = \phi(\tau_1) \, g_{m-1}^{**}{}_B(\tau_1, \tau_2) + \frac{1 - \phi(\tau_1)}{\tau_1 \tau_2}, \; m \geq 2, \tag{12.26}$$

$$g_m^{**}{}_B(\tau_1, \tau_2) = \mathcal{Y}(\tau_1) \, g_m^{**}{}_A(\tau_1, \tau_2) + \frac{1 - \mathcal{Y}(\tau_1)}{\tau_1 \tau_2}, \; m \geq 1. \tag{12.27}$$

(12.26) and (12.27) imply that

$$g_m^{**}{}_A(\tau_1, \tau_2) = \phi(\tau_1) \, \mathcal{Y}(\tau_1) \, g_{m-1}^{**}{}_A(\tau_1, \tau_2) + \frac{1 - \phi(\tau_1)\mathcal{Y}(\tau_1)}{\tau_1 \tau_2}, \; m \geq 2. \tag{12.28}$$

(12.25) and (12.28) can be rewritten in terms of h_m, which is defined by (12.3), to give

$$h_1(\tau_1, \tau_2; 1) = \frac{\tau_1 + \tau_2 \phi(\tau_1+\tau_2)}{\tau_1(\tau_1+\tau_2)}, \tag{12.29}$$

$$h_m(\tau_1, \tau_2; 1) = \phi(\tau_1) \, \mathcal{Y}(\tau_1) \, h_{m-1}(\tau_1, \tau_2; 1) + \frac{1 - \phi(\tau_1)\mathcal{Y}(\tau_1)}{\tau_1}, \; m \geq 2. \tag{12.30}$$

We get h_m in terms of h_1 from (12.30) by induction as follows

$$h_m(\tau_1, \tau_2; 1) = (\phi(\tau_1) \, \mathcal{Y}(\tau_1))^{m-1} \, h_1(\tau_1, \tau_2; 1) +$$

$$\frac{1 - \phi(\tau_1)\mathcal{Y}(\tau_1)}{\tau_1} \frac{1 - (\phi(\tau_1) \, \mathcal{Y}(\tau_1))^{m-1}}{1 - \phi(\tau_1)\mathcal{Y}(\tau_1)}, \; m \geq 2. \tag{12.31}$$

(It is of course assumed in (12.31) that $\phi(\tau_1)\mathcal{Y}(\tau_1) \neq 1$.)

By (12.29), (12.31), and (12.3), we have a closed form expression for the double Laplace transforms $g_m^{**}(\tau_1, \tau_2; 1)$, $m = 1, 2, \ldots$. Smith, Trivedi, and Ramesh [SMI] have discussed a numerical inversion scheme for the *double* Laplace transform of the cumulative reward arising in the theory of continuous-time Markov reward processes. There, the vector of Laplace transforms under investigation could be represented as the solution of a simple matrix equation involving the reward rate, and transition rate, matrices. This gave rise in [SMI] to a method built around the eigenvalue analysis of the matrices involved. No such scheme appears to be possible for the transforms h_m considered here; in the remaider of this section, we therefore consider some *analytical* implications of Theorem 12.1 for the alternating renewal process.

For fixed m, (12.4) establishes an interrelationship between the partial derivatives of h_m and the moments of $T_{A,m}(t)$. To use this, the next corollary gives the partial derivatives of h_m at $\tau_2 = 0$.

COROLLARY 12.3. *Assume that* $\phi(\tau_1)\mathcal{Y}(\tau_1) \neq 1$ *and that* ϕ *is sufficiently differentiable around the origin. Then, for* k = 1, 2, ... *and* m = 2, 3, ...

$$\left.\frac{\partial^k h_m(\tau_1, \tau_2; 1)}{\partial \tau_2^k}\right|_{\tau_2=0} = (\phi(\tau_1)\mathcal{Y}(\tau_1))^{m-1} \left.\frac{\partial^k h_1(\tau_1, \tau_2; 1)}{\partial \tau_2^k}\right|_{\tau_2=0}, \qquad (12.32)$$

where

$$\left.\frac{\partial^k h_1(\tau_1, \tau_2; 1)}{\partial \tau_2^k}\right|_{\tau_2=0} = (-1)^k \frac{k!}{\tau_1^{k+1}} - \sum_{\ell=1}^{k} \binom{k}{\ell} (-1)^\ell \frac{\ell!}{\tau_1^{\ell+1}} \phi^{(k-\ell)}(\tau_1). \qquad (12.33)$$

PROOF OF COROLLARY 12.3. (12.32) is straightforward from (12.31). (12.33) follows from (12.29) by Leibniz's Theorem for differentiation of a product. ∎

In what follows, we shall investigate the implications of Corollary 12.3 for the first two moments of $T_{A,m}(t)$. From (12.32) and (12.33) it is seen that

$$\tau_1 \left.\frac{\partial h_m(\tau_1, \tau_2; 1)}{\partial \tau_2}\right|_{\tau_2=0} = (\phi(\tau_1)\mathcal{Y}(\tau_1))^{m-1} \frac{\phi(\tau_1) - 1}{\tau_1}, \qquad (12.34)$$

$$\tau_1 \left.\frac{\partial^2 h_m(\tau_1, \tau_2; 1)}{\partial \tau_2^2}\right|_{\tau_2=0} = 2(\phi(\tau_1)\mathcal{Y}(\tau_1))^{m-1} \frac{1-\phi(\tau_1)+\phi'(\tau_1)\tau_1}{\tau_1^2}. \qquad (12.35)$$

Under the conditions of Corollary 12.3, the right hand side of (12.34) tends to $(\phi(\tau_1)\mathcal{Y}(\tau_1))^{m-1} \phi'(0)$ as $\tau_1 \to 0$. Similarly, it can be shown that the right hand side of (12.35) goes to $(\phi(\tau_1)\mathcal{Y}(\tau_1))^{m-1} \phi''(0)$ as $\tau_1 \to 0$. Hence, by applying the Final Value Theorem of Laplace Transforms (see, e.g., [GUE]), we see that for k = 1, 2,

$$\lim_{t \to +\infty} E[(T_{A,m}(t))^k \mid Y_0 = 1] = \int_{[0, +\infty)} t^k F_{1,2}\{dt\}, \qquad (12.36)$$

which is the k*th* moment of the holding time in state '1'. (12.36) is a result which is hardly surprising; it is simply an expression of the fact that the first two moments of the finite-horizon sojourn times approach the corresponding moments of the holding time in state '1' as the horizon goes to +∞. (12.36) is useful as a plausibility check on the correctness of our algebraic development. A similar reasoning will confirm (12.36) also for any k ≥ 3.

12.2.2 Symbolic inversion with MAPLE and computational experience

In certain special cases, (12.4) in conjunction with (12.34) and (12.35) allows the first two moments of $T_{A,m}(t)$ to be obtained in a closed form. Assume now that the holding time in '1' has an L-stage Erlang distribution with mean $1/\rho$ and the holding time in '2' is exponentially distributed with rate μ. It is therefore

$$\phi(\tau) = \left(\frac{L\rho}{L\rho + \tau}\right)^{L}, \quad \psi(\tau) = \frac{\mu}{\mu + \tau}.$$

The computer algebra package MAPLE (Version 4.2.1) has been used to obtain closed form expressions for the first two moments of $T_{A,m}(t)$ (with $Y_0 = 1$ and for L, m = 1, 2, 3) by *symbolic* Laplace transform inversion. (MAPLE is described in, for example, [CHA1] and [CHA2].) The MAPLE code is shown in Section 12.2.3. The CPU times on the Apple Macintosh SE/30 are shown in Table 12.2. It is seen from Table 12.2, as one will expect from intuition, that the CPU times tend to increase as L or m increases; the complexity of the expressions themselves increases accordingly. The first few of the expression are as follows

$$E_{L=1}[\,T_{A,1}(t)\mid Y_0 = 1\,] = \frac{1 - \exp\{-\rho\,t\,\}}{\rho},$$

$$E_{L=2}[\,T_{A,1}(t)\mid Y_0 = 1\,] = \frac{1 - \exp\{-2\rho\,t\,\}}{\rho} - t\exp\{-2\rho t\,\},$$

$$E_{L=1}[\,T_{A,2}(t)\mid Y_0 = 1\,] =$$

$$\frac{1}{\rho} + \frac{(2\,\mu\,\rho - \mu^2)\exp\{-\rho\,t\,\}}{\rho\,\mu^2 - 2\rho^2\,\mu + \rho^3} + \frac{t\exp\{-\rho\,t\,\}\mu}{\rho - \mu} - \frac{\rho\exp\{-\mu\,t\,\}}{\mu^2 - 2\rho\,\mu + \rho^2}.$$

	m = 1			m = 2			m = 3		
L = ...	1	2	3	1	2	3	1	2	3
For $E[T_{A,m}(t)]$	3.4	4.0	6.1	8.7	25.1	47.0	23.8	59.7	112.1
For $E[(T_{A,m}(t))^2]$	8.3	6.2	8.3	17.0	35.1	60.4	36.1	77.6	191.8

Table 12.2 Processing times (in seconds) on the Apple Macintosh SE/30 for
the symbolic Laplace transform inversion with MAPLE

MAPLE allows numerical values to be substituted for variables once a closed form symbolic expression is available. This device has been used to obtain the Figures 12.1 - 12.7 (they are shown at the end of this section). The numerical values for the rate parameters are $\rho = 1.0$ and $\mu = 5.0$. Figures 12.1 - 12.3 show that the mean of $T_{A,m}(t)$ decreases as m increases. The influence of L is less marked; it is shown in Figure 12.4 for m = 3. Figures 12.5 - 12.7 show the variance of $T_{A,m}(t)$. For t → +∞, both the mean and the variance of $T_{A,m}(t)$ are seen to be converging to their respective counterpart in the infinite-horizon case.

To indicate the limitations of the present approach, let us finally add that even though symbolic Laplace transform inversion seems to have been successfully completed by MAPLE for L = 4 and m = 3, the resulting numerical values for the moments were erroneous in the vicinity of t = 0. The difficulties are probably due to the extreme complexity of the closed form expressions produced by MAPLE; finding the exact cause of this is an outstanding problem that requires further work. To illustrate the degree of complexity of the formulae produced by MAPLE, Table 12.3 shows the CPU times needed to compute both the mean *and* the variance of $T_{A,m}(t)$ with MAPLE for 21 equidistant values of t in [0, 10] by numerically *evaluating* the closed form expressions.

	m = 1			m = 2			m = 3		
L = ...	1	2	3	1	2	3	1	2	3
CPU time [sec]	7.3	5.9	9.8	15.6	49.9	116.7	41.6	140.6	351.6

Table 12.3 Processing times (in seconds) on the Apple Macintosh SE/30 to compute one pair of curves by substitution in MAPLE

In conclusion, let us briefly examine the MAPLE features used in the code shown in the next section. MAPLE has all the language constructs of what one would expect from a programming language; we use the `for` loop, the redirection of output to a file by `writeto`, printing by `lprint`, etc. The most commonly used MAPLE procedures are automatically loaded into MAPLE when they are used for the first time. Partial differentiation by using `diff` is such a function. However, the Laplace transform inversion function `invlaplace` must be loaded by entering `readlib(laplace)`. Finally, `time()` is used to measure the elapsed time.

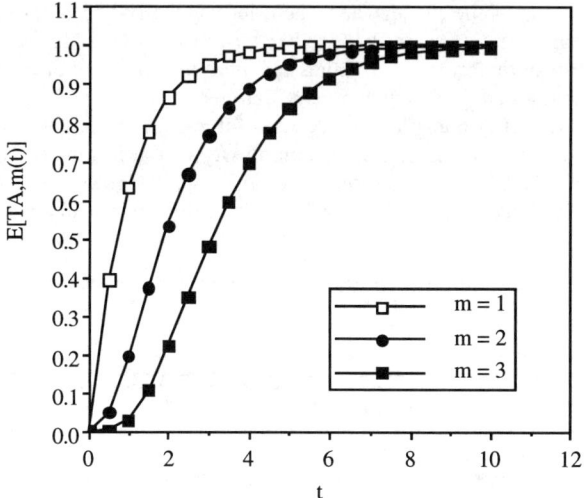

Figure 12.1 The mean of $T_{A,m}(t)$ for $L = 1$

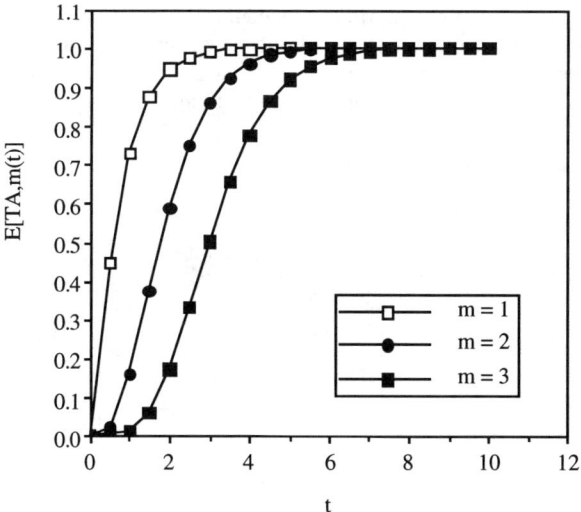

Figure 12.2 The mean of $T_{A,m}(t)$ for $L = 2$

227

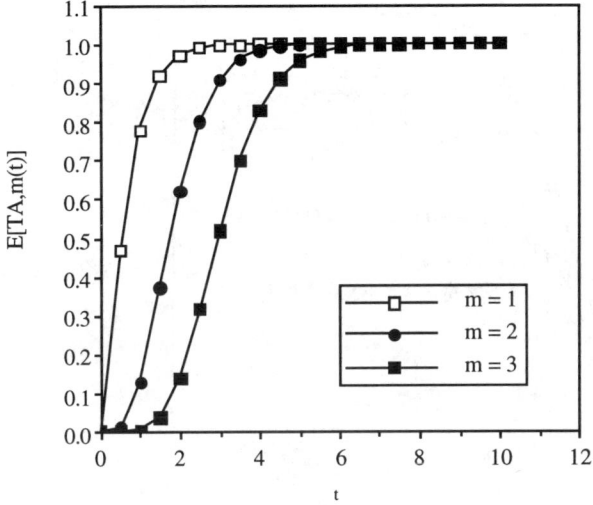

Figure 12.3 The mean of $T_{A,m}(t)$ for $L = 3$

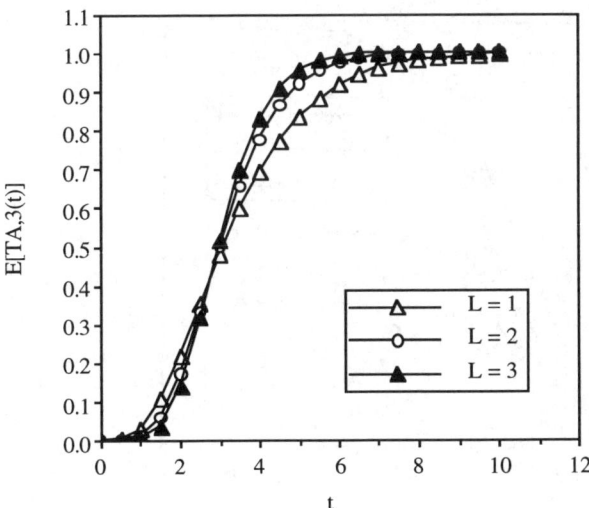

Figure 12.4 The mean of $T_{A,3}(t)$

228

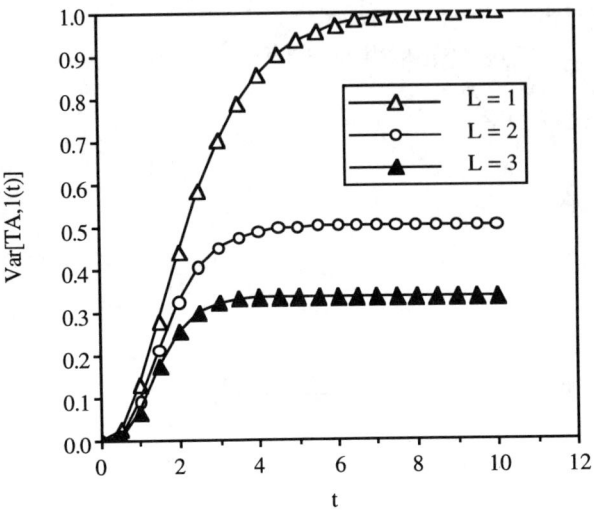

Figure 12.5 The variance of $T_{A,1}(t)$

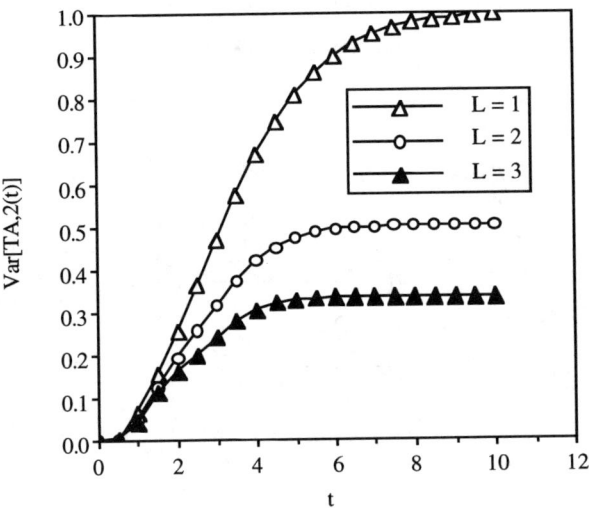

Figure 12.6 The variance of $T_{A,2}(t)$

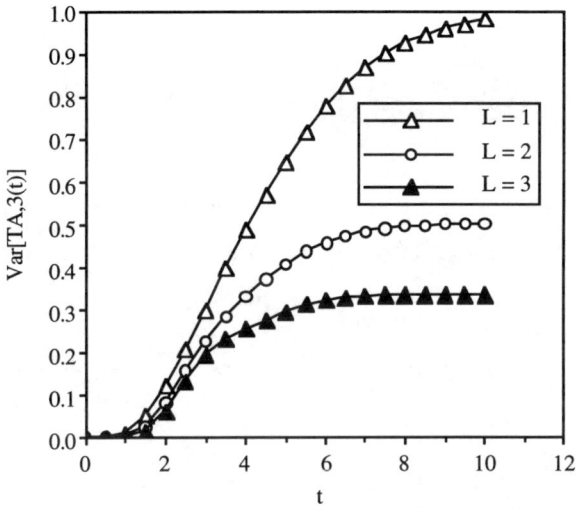

Figure 12.7 The variance of $T_{A,3}(t)$

12.2.3 MAPLE code

```
##############################################################################
#                                                                            #
#      This MAPLE program performs the Laplace transform inversion with      #
# respect to the variable TAU1 in (12.34) and (12.35). It is assumed         #
# that the repair time distribution is exponential with mean 1/MU and        #
# the failure time distribution is L-stage Erlang with mean 1/RHO, L =       #
# 1, 2, ... . Up state is '1', 'down' state is '2'. Thus, the holding        #
# time distribution in '1' is L-stage Erlang with mean 1/RHO and the         #
# holding time disribution in '2' is exponential with mean 1/MU. The         #
# process starts in state '1'.                                               #
#                                                                            #
##############################################################################

lprint(`Started executing ``moments```);
writeto(`moments.log`);
lprint(`The program started at`,time(),`CPU seconds`);

readlib(laplace):
PHI  := (L*RHO/(L*RHO+TAU))^L:
GAM  := MU/(MU+TAU):

for L from 1 to 4 do
  lprint(`The failure time distribution is L-stage Erlangian with L = `,L);
  for m from 1 to 4 do
    lprint(`m = `,m);
    lprint(`Started calculating ETSTAR and ET2STAR; CPU-time =`,
           time(),`sec`);
    ETSTAR := ((PHI*GAM)^(m-1))*(PHI-1)/(TAU^2):
    lprint(`Finished with ETSTAR and started ET2STAR; CPU-time =`,
```

```
               time(),`sec`);
      ET2STAR:= ((PHI*GAM)^(m-1))*2*(1-PHI+TAU*diff(PHI,TAU$1))/(TAU^3);
      lprint(`Finished with ET2STAR and started ET; CPU-time =`,
               time(),`sec`);
      ET      := invlaplace(ETSTAR,TAU,T):
      lprint(`Finished with ET and started ET2; CPU-time =`,time(),`sec`);
      ET2     := invlaplace(ET2STAR,TAU,T):
      lprint(`Finished with ET2 at CPU-time =`,time(),`sec`);

      EXPEC[L,m]   := - ET:
      EXPEC2[L,m]  := ET2:
   od;
od;
lprint(`CPU-time at finish of symbolic calculations is `,time(),`sec`);

RHO := 1.0:
lprint(`Rho = `,RHO);
MU := 5.0:
lprint(`Mu = `,MU);

for L from 1 to 4 do
   lprint(`The failure time distribution is L-stage Erlangian with L = `,L);
   for m from 1 to 4 do
      lprint(`m = `,m);
      lprint(`The CPU-time is `,time(),`sec`);
      for k from 1 to 20 do
        T:= 0.5*k:
        lprint(`ET(`,T,`yr) =`,EXPEC[L,m],\
                `VART(`,T,`yr) =`,EXPEC2[L,m]-EXPEC[L,m]*EXPEC[L,m],`yr^2`);
      od;
   od;
od;

lprint(`CPU-time at finish of numerical calculations is `,time(),`sec`);

RHO    := 'RHO':
MU     :='MU':
T      := 'T':

writeto(terminal);
close(`moments.log`);
lprint(`Finished now`);
```

POSTSCRIPT

In this concluding part of the work, some topics for future research will be discussed.

Chapter 11 is an obvious candidate for future work. The method described there should be implemented and combined with a numerical Laplace transform inversion scheme to study the method's practical viability.

Consider the following situation. A system modelled by the (Markov or semi-Markov) process $Y = \{ Y_t : t \geq 0 \}$ is in one of the following three groups of states

A_1 = 'up' states,
A_2 = 'down' states corresponding to *minor* repairs,
A_3 = 'down' states corresponding to *major* overhauls.

This is well within our present framework. Assume, however, that we are not merely interested in the sequence of 'up' times $T_{A_1,1}, T_{A_1,2}, \ldots$ but would also wish to include knowledge of the commencements of the operation of the system when 'overhauled'; thus, we would be interested in the sequence of sojourns of Y in A_1 *as punctuated by visits to* A_3. This will give rise to a *double array* of A_1-sojourn times $\{ T^{(n)}_{A_1,i} : i \geq 1, n \geq 0 \}$, where $T^{(n)}_{A_1,i}$ stands for the length of Y's *ith* visit to A_1 as it has visited A_3 exactly n times. Here, counting of the number of visits to A_1 is started after each visit to A_3 anew. The distribution theory of this double array will require the renewal argument to be used three times in succession. Also notice that for fixed n, the sequence $\{ T^{(n)}_{A_1,i} : i \geq 1 \}$ is non-zero for a finite (random) number of i only. Thus, the Laplace transform

$$E[\exp\{ - \sum_{i=1}^{\infty} \tau_i T^{(n)}_{A_1,i} \}]$$

exists. The results could be fashioned along the lines in this monograph.

Two system characteristics have been dealt with here in both the Markov and semi-Markov contexts; they are: the sojourn time vector and the number of visits to a subset of the state space during a finite time interval. Not much is known about the semi-Markov versions of all the other variables which we have examined within the Markov framework. Let us have a closer look at the compound dependability measure from Chapter 6. It was defined for $t \in (0, t_0)$ and $m \in \{ 0, 1, \ldots \}$, by $\Pr\{ T_G(t_0) > t, M_B(t_0) \leq m \}$. In Chapter 6, we have used randomization to arrive at a closed form expression for this probability. Randomization, however, is not applicable to semi-Markov processes. In this case, it might be expedient to express $\Pr\{ T_G(t_0) > t, M_B(t_0) \leq m \}$ in terms of the sojourn time vector since the Laplace transform of the latter is known from Theorem 8.2. This is accomplished in the following proposition.

PROPOSITION P.1. *For an irreducible (semi-)Markov processs* $Y = \{ Y_t : t \geq 0 \}$ *whose state space is partitioned as* $S = G \cup B$, *we have for* $t' \in (0, t_0)$,

$$\Pr\{ T_G(t_0) \geq t', M_B(t_0) = k \} = \Pr\{ Y_0 \in G, TS_{G,1} \geq t_0 \}, \tag{P.1}$$

if $k = 0$, *and if* $k \geq 1$, *then*

$\Pr\{ T_G(t_0) \geq t', M_B(t_0) = k \} =$

$\Pr\{ Y_0 \in G, TS_{G,k} + TS_{B,k-1} < t_0, TS_{G,k} \geq t' \} -$

$- \Pr\{ Y_0 \in G, TS_{G,k} + TS_{B,k} < t_0, TS_{G,k} \geq t' \} +$

$+ \Pr\{ Y_0 \in G, TS_{G,k} + TS_{B,k} < t_0, TS_{B,k} \leq t_0 - t' \} -$

$- \Pr\{ Y_0 \in G, TS_{G,k+1} + TS_{B,k} < t_0, TS_{B,k} \leq t_0 - t' \} +$

$+ \Pr\{ Y_0 \in B, TS_{G,k-1} + TS_{B,k-1} < t_0, TS_{G,k-1} \geq t' \} -$

$- \Pr\{ Y_0 \in B, TS_{G,k-1} + TS_{B,k} < t_0, TS_{G,k-1} \geq t' \} +$

$+ \Pr\{ Y_0 \in B, TS_{G,k-1} + TS_{B,k} < t_0, TS_{B,k} \leq t_0 - t' \} -$

$- \Pr\{ Y_0 \in B, TS_{G,k} + TS_{B,k} < t_0, TS_{B,k} \leq t_0 - t' \}.$ (P.2)

NOTE. (P.1) and (P.2) are easily seen to imply an expression for the desired probability by summing over $k = 0, ..., m$ and by taking $t' \downarrow t$.

PROOF OF PROPOSITION P.1. (P.1) follows from

$\{ T_G(t_0) \geq t', M_B(t_0) = 0 \} = \{ Y \text{ does not visit } B \text{ during } [0, t_0] \} = \{ Y_0 \in G, TS_{G,1} \geq t_0 \}.$

To see (P.2), we first note that for $k \geq 1$,

$\Pr\{ T_G(t_0) \geq t', M_B(t_0) = k \} =$

$\Pr\{ T_G(t_0) \geq t', M_B(t_0) \geq k \} - \Pr\{ T_G(t_0) \geq t', M_B(t_0) \geq k + 1 \}.$ (P.3)

The event $\{ M_B(t_0) \geq k \}$ can be partitioned as

$\{ Y_0 \in G, TS_{G,k} + TS_{B,k-1} < t_0 \} \cup \{ Y_0 \in B, TS_{G,k-1} + TS_{B,k} < t_0 \},$

from which it is seen by (P.3) that

$\Pr\{ T_G(t_0) \geq t', M_B(t_0) = k \} =$

$\Pr\{ Y_0 \in G, T_G(t_0) \geq t', TS_{G,k} + TS_{B,k-1} < t_0 \} +$

$+ \Pr\{ Y_0 \in B, T_G(t_0) \geq t', TS_{G,k-1} + TS_{B,k-1} < t_0 \} -$

$- \Pr\{ Y_0 \in G, T_G(t_0) \geq t', TS_{G,k+1} + TS_{B,k} < t_0 \} -$

- $Pr\{ Y_0 \in B, T_G(t_0) \geq t', TS_{G,k} + TS_{B,k} < t_0 \} =$

$Pr\{ Y_0 \in G, T_G(t_0) \geq t', TS_{G,k} + TS_{B,k-1} < t_0 \leq TS_{G,k+1} + TS_{B,k} \} +$

$Pr\{ Y_0 \in B, T_G(t_0) \geq t', TS_{G,k-1} + TS_{B,k-1} < t_0 \leq TS_{G,k} + TS_{B,k} \}.$ (P.4)

To evaluate the right hand side of (P.4), $T_G(t_0)$ will now be expressed in terms of the sojourn times. The first event to be considered is

$\{ Y_0 \in G, TS_{G,k} + TS_{B,k-1} < t_0 \leq TS_{G,k+1} + TS_{B,k} \} =$

$\{ Y_0 \in G, TS_{G,k} + TS_{B,k-1} < t_0 \leq TS_{G,k} + TS_{B,k} \} \cup$

$\{ Y_0 \in G, TS_{G,k} + TS_{B,k} < t_0 \leq TS_{G,k+1} + TS_{B,k} \}.$ (P.5)

It is easily seen that $T_G(t_0) = TS_{G,k}$ on the first set on the right hand side of (P.5), while $T_G(t_0) = t_0 - TS_{B,k}$ on the second set on the right hand side of (P.5). Thus, the first probability on the right hand side of (P.4) is

$Pr\{ Y_0 \in G, TS_{G,k} + TS_{B,k-1} < t_0 \leq TS_{G,k} + TS_{B,k}, TS_{G,k} \geq t' \} +$

$Pr\{ Y_0 \in G, TS_{G,k} + TS_{B,k} < t_0 \leq TS_{G,k+1} + TS_{B,k}, t_0 - TS_{B,k} \geq t' \},$

which gives first four terms on the right hand side of (P.2). The remaining other four probabilities on the right hand side of (P.2) are obtained in a similar fashion by writing

$\{ Y_0 \in B, TS_{G,k-1} + TS_{B,k-1} < t_0 \leq TS_{G,k} + TS_{B,k} \} =$

$\{ Y_0 \in B, TS_{G,k-1} + TS_{B,k-1} < t_0 \leq TS_{G,k-1} + TS_{B,k} \} \cup$

$\{ Y_0 \in B, TS_{G,k-1} + TS_{B,k} < t_0 \leq TS_{G,k} + TS_{B,k} \},$ (P.6)

and by noting that $T_G(t_0) = TS_{G,k-1}$ and $T_G(t_0) = t_0 - TS_{B,k}$ on the first and second set respectively on the right hand side of (P.6). ∎

The Laplace transforms of the variables $(TS_{G,i} + TS_{B,j}, TS_{G,k})$ $(i, j, k = 1, ..., m)$ in (P.1) and (P.2) are available by Theorem 8.2. Thus it should be possible, in principle, to furnish a software implementation of $Pr\{ T_G(t_0) > t, M_B(t_0) \leq m \}$ based on a two-dimensional numerical Laplace transform inversion scheme.

The last point illustrates once more the importance of Laplace transform inversion algorithms. This problem has been covered in the literature by the following references: [AHM], [BEL1], [BEL2, Chapter 6], [CHOU], [CRU], [DUB], [DUR], [FEL, Chapter VII.6], [GAV], [GAR1], [GAR2], [JAG1], [JAG2], [LON], [MAR1] - [MAR3], [SING], [STEH], [STEP1], [STEP2], [STEP3, Chapters 8 and 9], [TAL], and [WEE]. (Use of the methods from [SING] and [STEH] has been reported in [LAP] and [CHOI] respectively.) It appears, however, that no systematic comparison of these methods has ever been attempted. It is known that no single method can be devised which will perform numerical Laplace transform inversion to a given accuracy [BEL2, Section 1.10]. This is one more reason for exploring the existing algorithms, in particular with a view to applying them to reliability problems. The criteria for evaluating the methods should be

- Suitability;
- Ease of implementation, possibly aided by the (local) availability of software components;
- Speed of execution in the given computing environment;
- Ease of operation (determined, for example, by the method's auxiliary parameters if there are any).

Our work has three main themes: discrete-time Markov chains, continuous-time Markov processes, and semi-Markov processes. This is the traditional set of topics and order of presentation adhered to in the literature. We have chosen not to develop a separate theory for *discrete-time* semi-Markov processes since they can be subsumed within the continuous-parameter formulation. However, discrete-time semi-Markov processes are worth considering separately if one is interested in exploring computational implementations which do not make use of Laplace transforms; see, Mode and Pickens [MOD]. The discrete-parameter processes are then used for solving the continuous-time problem approximately. It would be interesting to see the theory of sojourn times developed for discrete-parameter semi-Markov processes; the theory would be formulated in the 'time domain', i.e., we would be working directly with the probability mass functions of the holding time distributions. First results on these issues have been reported in [CSE12] and [CSE13]; some further work is in preparation.

Asymptotic results did not concern us much in this monograph. The following is intended merely to provide a first source of reference for those wanting to pursue this question. The cumulative sojourn time sequence $\{ TS_{A_1,m} : m \geq 1 \}$ can be modelled by what is known as an 'additive process on a finite Markov chain'; see, [KEI1] - [KEI3]. For these variables (under suitable conditions) the Central Limit Theorem holds; this follows from [KEI1] - [KEI3]. A direct proof of this (via characteristic functions) might be possible by using results from [PUR3]. No asymptotic theory seems to be available for the other variables considered here. In particular, a closed form (asymptotic) expression for the dependability measure from Chapter 6 would be useful since the present method has computational difficulties associated with it.

Most systems whose component failure, and repair-time distributions are arbitrary, cannot be modelled by semi-Markov processes. (The alternating renewal process which modells a one-unit system is a notable exception to this.) For such systems the theory of (semi-)regenerative and non-regenerative processes will have to be used ([BIR1], [BIR2]). These processes may be converted by the introduction of what is known as 'supplementary variables' into a Markov process whose state space is, however, not discrete (see [COX], [DIC], [KOS], [SIN1]). It would be desirable to obtain a theory of sojourn times in this context too. (Another method which is commonly referred to as the 'device of stages' [SIN1], [SIN2], is also used (perhaps more routinely by engineers) for *approximately* solving reliability models of non-Markovian systems. The method replaces the original model by a Markovian one which has an enlarged state space. If a model is approached by this technique, no new theory is called for.)

In the course of our work, we encountered many relevant references on *reliability in power engineering*; of those, we would like to list the following selection for future work on the applied side: [BIL2], [BIL3], [CAO], [CHE], [DHI1], [DHI2], [FON], [GON], and [SHI]. [BIL4] is a comprehensive collection of original research papers in this area covering the period 1947 - 1991.

REFERENCES

[AHM]   A. Ahmad, P. Johannet, and Ph. Auriol, Efficient inverse Laplace transform algorithm for transient overvoltage calculation. *IEE Proceedings-C: Generation, Transmission and Distribution* **139** (1992), 117-121.

[AHO]   A. V. Aho, J. E. Hopcroft, and J. D. Ullman, *Data Structures and Algorithms*. Addison-Wesley, Reading, Massachusetts, 1983.

[BEA]   M. Beasley, *Reliability for Engineers*. Macmillan, Houndmills, Basingstoke, London, 1991.

[BEL1]   R. E. Bellman, R. E. Kalaba, and J. A. Lockett, *Numerical Inversion of the Laplace Transform*. Elsevier, New York, 1966.

[BEL2]   R. E. Bellman and R. S. Roth, *The Laplace Transform*. World Scientific, Singapore, 1984.

[BHA1]   B. R. Bhat, Some properties of regular Markov chains. *Annals of Mathematical Statistics* **32** (1961), 59-71.

[BHA2]   U. N. Bhat, *Elements of Applied Stochastic Processes*. John Wiley & Sons, New York, 1984.

[BIL1]   R. Billinton, *Power System Reliability Evaluation*. Gordon and Breach, New York, 1970.

[BIL2]   R. Billinton and Y. Kumar, Transmission line reliability models including common mode and adverse weather effects. *IEEE Transactions on Power Apparatus and Systems* **PAS-100** (1981), 3899-3910.

[BIL3]   R. Billinton, T. K. P. Medicherla, and M. S. Sachdev, Application of common-cause outage models in composite system reliability evaluation. *IEEE Transactions on Power Apparatus and Systems* **PAS-100** (1981), 3648-3657.

[BIL4]   R. Billinton, R. N. Allan, and L. Salvaderi (Eds), *Applied Reliability Assessment in Electric Power Systems*. IEEE Press, New York, 1991.

[BIR1]   A. Birolini, On the use of stochastic processes in modeling reliability problems. *Lecture Notes in Economics and Mathematical Systems* **252**. Springer-Verlag, Berlin, Heidelberg, New York, 1985.

[BIR2]   A. Birolini, *Qualität und Zuverlässigkeit technischer Systeme: Theorie, Praxis, Management. 3rd Edition*. Springer-Verlag, Berlin, Heidelberg, New York, 1991.

[CAO]   J. Cao, Reliability analysis of a repairable system in a changing environment subject to a general alternating renewal process. *Microelectronics and Reliability* **28** (1988), 889-892.

[CHA1]   B. W. Char, K. O. Geddes, G. H. Gonnet, M. B. Monagan, and S. M. Watt, *MAPLE Reference Manual. 5th Edition*. Symbolic Computation Group, Department of Computer Science, University of Waterloo, Waterloo, 1988.

[CHA2]   B. W. Char, K. O. Geddes, G. H. Gonnet, B. L. Leong, M. B. Monagan, and S. M. Watt, *First Leaves: A Tutorial Introduction to MAPLE V*. Springer-Verlag, New York, 1992.

[CHE]   K. Cheng, On the first failure of a system in a randomly varying environment. In: S. Osaki and J. Cao (eds.), *Proceedings of the China-Japan Reliability Symposium, September 13-25, 1987*, 34-43. World Scientific, Singapore, 1987.

[CHOI]   B. D. Choi, K. H. Rhee, and K. K. Park, The M/G/1 retrial queue with retrial rate control policy. *Probability in the Engineering and Informational Sciences* **7** (1993), 29-46.

[CHOU]   J.-H. Chou and I.-R. Horng, On a functional approximation for inversion of Laplace transforms via shifted Chebyshev series. *International Journal of Systems Science* **17** (1986), 735-739.

[CHR]   A. H. Christer and N. Jack, An integral equation approach for replacement modelling over finite time horizons. *IMA Journal of Mathematics Applied in Business and Industry* **3** (1991), 31-44.

[COO]   J. W. Cooley, P. A. Lewis, and P. D. Welch, The Fast Fourier Transform algorithm: programming considerations in the calculation of sine, cosine and Laplace transforms. *Journal of Sound and Vibration* **12** (1970), 315-337.

[COX]   D. R. Cox, The analysis of non-Markovian stochastic processes by the inclusion of supplementary variables. *Proceedings of the Cambridge Philosophical Society* **51** (1955), 433-441.

[CRU]   K. S. Crump, Numerical inversion of Laplace transforms using a Fourier series approximation. *Journal of the Association of Computing Machinery* **23** (1976), 89-96.

[CSE1]   A. Csenki, The joint distribution of sojourn times in finite semi-Markov processes. *Stochastic Processes and their Applications* **39** (1991), 287-299.

[CSE2]   A. Csenki, Some renewal-theoretic investigations in the theory of sojourn times in finite semi-Markov processes. *Journal of Applied Probability* **28** (1991), 822-832.

[CSE3]   A. Csenki, The number of working periods of a repairable Markov system during a finite time interval. *IEEE Transactions on Reliability* **43** (1994) *in press*.

[CSE4]   A. Csenki, The joint distribution of sojourn times in finite Markov processes. *Advances in Applied Probability* **24** (1992), 141-160.

[CSE5]   A. Csenki, Some new aspects of the transient analysis of discrete-parameter Markov models with an application to the evaluation of repair events in power transmission. *IMA Journal of Mathematics Applied in Business and Industry* **3** (1992), 193-206.

[CSE6]   A. Csenki, Sojourn times in Markov processes for power transmission dependability assessment with MatLab. *Microelectronics and Reliability* **32** (1992), 945-960.

[CSE7]   A. Csenki, Occupation frequencies for irreducible finite semi-Markov processes with reliability

236

applications. *Computers & Operations Research* **20** (1993), 249-259.

[CSE8]   A. Csenki, On a counting variable in the theory of discrete-parameter Markov chains. *Statistics & Probability Letters* **18** (1993), 105-112.

[CSE9]   A. Csenki, A compound measure of dependability for systems modelled by continuous-time absorbing Markov processes. Technical Report, Department of Computer Science and Applied Mathematics, Aston University, Birmingham, 1993. (Submitted for publication.)

[CSE10]   A. Csenki, Sojourn times with finite time-horizon in finite semi-Markov processes. *Applied Stochastic Models and Data Analysis* **9** (1993), 251-265.

[CSE11]   A. Csenki, A dependability measure for Markov models of repairable systems: solution by randomization and computational experience. *Computers & Mathematics with Applications* (1994) *in press.*

[CSE12]   A. Csenki, Cumulative operational time analysis of finite semi-Markov reliability models. *Reliability Engineering and System Safety* **44** (1994), 17-25.

[CSE13]   A. Csenki, The number of visits to a subset of the state space by a discrete-parameter semi-Markov process. *Statistics & Probability Letters* **23** (1994) *in press.*

[DAR]   J. N. Darroch and K. W. Morris, Some passage-time generating functions for discrete-time and continuous-time finite Markov chains. *Journal of Applied Probability* **4** (1967), 496-507.

[DES1]   E. De Souza e Silva and H. R. Gail, Calculating cumulative operational time distributions of repairable computer systems. *IEEE Transactions on Computers* **C-35** (1986), 322-332.

[DES2]   E. De Souza e Silva and H. R. Gail, Calculating availability and performability measures of repairable computer systems using randomization. *Journal of the Association for Computing Machinery* **36** (1989), 171-193.

[DES3]   E. De Souza e Silva and H. R. Gail, Performability analysis of computer systems. *Performance Evaluation* **14** (1992), 157-196.

[DHI1]   B. S. Dhillon and J. Natesan, Stochastic analysis of outdoor power systems in fluctuating environment. *Microelectronics and Reliability* **23** (1983), 867-881.

[DHI2]   B. S. Dhillon and H. C. Viswanath, Reliability analysis of a non-identical unit parallel system. *Microelectronics and Reliability* **31** (1990), 429-441.

[DIC]   C. Dichirio and C. Singh, Reliability analysis of transmission lines with common mode failures when repair times are arbitrarily distributed. *IEEE Transactions on Power Systems* **3** (1988), 1012-1019.

[DUB]   H. Dubner and J. Abate, Numerical inversion of Laplace transforms by relating them to the Finite Fourier Cosine transform. *Journal of the Association for Computing Machinery* **15** (1968), 115-123.

[DUR]   F. Durbin, Numerical Inversion of Laplace transforms: an efficient improvement to Dubner and Abate's method. *The Computer Journal* **17** (1974), 371-376.

[DYE]   D. Dyer, Unification of reliability/availability/repairability models for Markov Systems. *IEEE Transactions on Reliability* **R-38** (1989), 246-252.

[FEL]   W. Feller, *An Introduction to Probability Theory and Its Applications, Volume II* (Second Edn). John Wiley & Sons, New York, 1971.

[FON]   C. C. Fong, Reliability evaluation of transmission and distribution configurations with duration-dependent effects. *IEE Proceedings-C: Generation, Transmission and Distribution* **136** (1989), 64-67.

[FRA]   E. G. Frankel, *Systems Reliability and Risk Analysis.* Kluwer Academic Publishers, Dordrecht, Boston, London, 1988.

[GAR1]   B. S. Garbow, G. Giunta, and A. Murli, Software for an implementation of Weeks' method for the inverse Laplace transform problem. *ACM Transactions on Mathematical Software* **14** (1988), 163-170.

[GAR2]   B. S. Garbow, G. Giunta, J. N. Lyness, and A. Murli, Algorithm 662: A Fortran software package for the numerical inversion of the Laplace transform based on Weeks' method. *ACM Transactions on Mathematical Software* **14** (1988), 171-176.

[GAV]   D. P. Gaver, Observing stochastic processes, and approximate transform inversion. *Operations Research* **14** (1966), 444-459.

[GON]   T. Gönen, *Electric Power Transmission System Engineering: Analysis and Design.* John Wiley & Sons, New York, 1988.

[GRA]   F. A. Graybill, *Introduction to Matrices with Applications in Statistics.* Wadsworth & Brooks/Cole, Pacific Grove, California, 1983.

[GRO]   D. L. Grosh, *A Primer of Reliability Theory.* John Wiley & Sons, New York, 1989.

[GROS1]   D. Gross and D. R. Miller, Multiechelon repairable-item provisioning in a time-varying environment using the randomization technique. *Naval Research Logistics Quarterly* **31** (1984), 347-361.

[GROS2]   D. Gross and D. R. Miller, The randomization technique as a modeling tool and solution procedure for transient Markov processes. *Operations Research* **32** (1984), 343-361.

[GUE]   P. B. Guest, *Laplace Transforms and an Introduction to Distributions.* Ellis Horwood, New York, 1991.

[HIL]   D. R. Hill, *Experiments in Computational Matrix Algebra.* The Random House/Birkhauser

	Mathematics Series, Random House, New York, 1988.

[IOS] M. Iosifescu, *Finite Markov Processes and their Applications*. John Wiley & Sons, New York, 1980.

[JAC1] N. Jack, Repair replacement modelling over finite time horizons. *Journal of the Operational Research Society* **42** (1991), 759-766.

[JAC2] N. Jack and J. S. Dagpunar, Costing minimal repair-replacement policies over finite time horizons. *IMA Journal of Mathematics Applied in Business and Industry* **3** (1992), 207-217.

[JAG1] D. L. Jagerman, An inversion technique for the Laplace transform with application to approximation. *The Bell System Technical Journal* **57** (1978), 669-710.

[JAG2] D. L. Jagerman, An inversion technique for the Laplace transform. *The Bell System Technical Journal* **61** (1982), 1995-2002.

[KAR] S. Karlin and H. M. Taylor, *A First Course in Stochastic Processes*. Academic Press, New York, 1975.

[KEI1] J. Keilson and D. M. G. Wishart, A central limit theorem for processes defined on a finite Markov chain. *Proceedings of the Cambridge Philosophical Society* **60** (1964), 547-567.

[KEI2] J. Keilson and D. M. G. Wishart, Addenda to processes defined on a finite Markov chain. *Proceedings of the Cambridge Philosophical Society* **63** (1967), 187-193.

[KEI3] J. Keilson and S. S. Rao, A process with chain dependent growth rate. *Journal of Applied Probability* **7** (1970), 699-711.

[KEL] A. Z. Keller and I. S. Qamber, System availability synthesis. In: G. P. Libberton (ed.), *Tenth advances in reliability technology symposium*, 173-188. Elsevier Applied Science, London, New York, 1988.

[KEM] J. G. Kemeny and J. L. Snell *Finite Markov Chains*. Springer-Verlag, New York, Heidelberg, Berlin, 1976.

[KLA] K. B. Klaassen and J. C. L. Peppen, *System Reliability Concepts and Applications*. Edward Arnold, London, 1989.

[KOH] J. Kohlas, *Stochastic Methods of Operations Research*. Cambridge University Press, Cambridge, 1982.

[KOS] L. Kosten, *Stochastic Theory of Service Systems*. Pergamon Press, Oxford, New York, 1973.

[KUB] P. Kubat, U. Sumita, and Y. Masuda, Dynamic performance evaluation of communication/computer systems with highly reliable components. *Probability in the Engineering and Informational Sciences* **2** (1988), 185-213.

[KUL1] V. G. Kulkarni, V. F. Nicola, and K. S. Trivedi, Numerical evaluation of performability and job completion time in repairable fault-tolerant systems. *Journal of Systems and Software* **6** (1986), 175-182.

[KUL2] V. G. Kulkarni, V. F. Nicola, R. M. Smith, and K. S. Trivedi, On modelling the performance and reliability of multimode computer systems. *Proceedings of the 16th Annual Fault-Tolerant Computing Symposium*, 252-257. IEEE Press, 1986.

[KUZ] B. O. Kuzmanovic and N. Willems, *Steel Design for Structural Engineers*. Prentice-Hall, Englewood Cliffs, New Jersey, 1983.

[LAP] J.-C. Laprie, A. Costes, and C. Landrault, Parametric analysis of 2-unit redundant computer systems with corrective and preventive maintenance. *IEEE Transactions on Reliability* **R-30** (1981), 139-144.

[LEW] E. E. Lewis, *Introduction to Reliability Engineering*. John Wiley & Sons, New York, 1987.

[LIN] P. Linz, *Analytical and Numerical Methods for Volterra Equations*. Society for Industrial and Applied Mathematics, Philadelphia, 1985.

[LON] I. M. Longman, On the numerical inversion of the Laplace transform of a discontinuous original. *Journal of the Institute of Mathematics and its Applications* **4** (1968), 320-328.

[MAR1] W. Marszalek, The block-pulse functions method of the two-dimensional Laplace transform. *International Journal of Systems Science* **14** (1983), 1311-1317.

[MAR2] W. Marszalek, On the inverse Laplace transform of irrational and transcendental transfer functions via block-pulse functions method. *International Journal of Systems Science* **15** (1984), 869-876.

[MAR3] W. Marszalek, On the nature of block-pulse operational matrices. *International Journal of Systems Science* **15** (1984), 983-989.

[MAS1] Y. Masuda, J. G. Shantikumar, and U. Sumita, A general software availability/reliability model: numerical exploration via the matrix Laguerre transform. *Communications in Statistics - Stochastic Models* **2** (1986), 203-236.

[MAS2] Y. Masuda and U. Sumita, Analysis of a counting process associated with a semi-Markov process: number of entries into a subset of state space. *Advances in Applied Probability* **19** (1987), 767-783.

[MAS3] Y. Masuda and U. Sumita, Numerical analysis of gracefully degrading fault-tolerant computer systems: semi-Markov and Laguerre transform approach. *Computers and Operations Research* **18** (1991), 695-707.

[MAS4] Y. Masuda and U. Sumita, A multivariate reward process defined on a semi-Markov process and its first-passage-time distributions. *Journal of Applied Probability* **28** (1991), 360-373.

[MAT1] The MathWorks Inc., *MATLAB for Macintosh Computers User's Guide*. South Natick,

238

	Massachusetts, 1989.
[MAT2]	The MathWorks Inc., *The MATLAB Curriculum Series: The Student Edition of MATLAB - Student User Guide*. Prentice Hall, Englewood Cliffs, New Jersey, 1992.
[MOD]	C. J. Mode and G. T. Pickens, Computational methods for renewal theory and semi-Markov processes with illustrative examples. *The American Statistician* **42** (1988), 143-151.
[MOL]	C. B. Moler and C. F. Van Loan, Nineteen dubious ways to compute the exponential of a matrix. *SIAM Review* **20** (1979), 801-836.
[NAH]	J. Nahman and N. Mijuskovic, Reliability modeling of multiple overhead transmission lines. *IEEE Transactions on Reliability* **R-34** (1985), 281-285.
[NOL]	V. Nollau, *Semi-Markovsche Prozesse*. Verlag Harri Deutsch, Thun, Frankfurt a. M., 1981.
[NUM]	Numerical Algorithms Group, *Nag Fortran Library User Manual, Mark 14, Volume 1*. Oxford, 1990. (Routines: C06LAF, C06LBF, and C06LCF).
[PHI]	J. Phillips, *The NAG Library: A Beginner's Guide*. Oxford University Press, Oxford, 1986.
[PUR1]	P. S. Puri, A method for studying the integral functionals of stochastic processes with applications I. The Markov chain case. *Jornal of Applied Probability* **8** (1971), 331-343.
[PUR2]	P. S. Puri, A method for studying the integral functionals of stochastic processes with applications II. Sojourn time distributions for Markov chains. *Zeitschrift für Wahrscheinlichkeitstheorie und verwandte Gebiete* **23** (1972), 85-96.
[PUR3]	P. S. Puri, J. B. Robertson, and K. B. Sinha, A matrix limit theorem with applications to probability theory. *Sankhya* **52** (1990), 58-83.
[REI1]	A. Reibman and K. S. Trivedi, Numerical transient analysis of Markov models. *Computers and Operations Research* **15** (1988), 19-36.
[REI2]	A. Reibman, R. Smith, and K. S. Trivedi, Markov and Markov reward model transient analysis: An overview of numerical approaches, *European Journal of Operational Research* **40** (1989), 257-267.
[ROS]	S. M. Ross, *Stochastic Processes*. John Wiley & Sons, New York, 1983.
[ROSS]	M. H. Rossiter, The sojourn time distribution of an alternating renewal process, *Australian Journal of Statistics* **31** (1989), 143-152.
[RUB1]	G. Rubino and B. Sericola, Sojourn times in semi-Markov reward processes: application to fault-tolerant systems modeling. *Reliability Engineering and System Safety* **41** (1993), 1-4.
[RUB2]	G. Rubino and B. Sericola, *Accumulated Reward over the n First Operational Periods in Fault-Tolerant Computing Systems*. Tech. Rept. No. 1028, I.N.R.I.A., Campus de Beaulieu, 35042 Rennes Cedex, France, May 1989.
[RUB3]	G. Rubino and B. Sericola, Sojourn times in finite Markov processes. *Journal of Applied Probability* **26** (1989), 744-756.
[RUB4]	G. Rubino and B. Sericola, Successive operational periods as measures of dependability. In: A. Avizienis and J.-C. Laprie (eds.), *Dependable Computing and Fault-Tolerant Systems* **4** (1991), 239-254. Springer-Verlag, Wien and New York.
[RUB5]	G. Rubino and B. Sericola, Interval availability analysis using operational periods. *Performance Evaluation* **14** (1992), 257-272.
[SCH]	L. L. Scharf and R. T. Behrens, *A First Course in Electrical and Computer Engineering with MATLAB Programs and Experiments*. Addison-Wesley, Reading, Massachusetts, 1990.
[SER]	B. Sericola, Closed-form solution for the distribution of the total time spent in a subset of states of a homogeneous Markov process during a finite observation period. *Journal of Applied Probability* **27** (1990), 713-719.
[SHI]	D. Shi, Stationary distribution of the up (down) time of repairable systems. In: S. Osaki and J. Cao (eds.), *Proceedings of the China-Japan Reliability Symposium, September 13-25, 1987*, 328-337. World Scientific, Singapore, 1987.
[SIN1]	C. Singh and R. Billinton, *System Reliability Modelling and Evaluation*. Hutchinson, London, 1977.
[SIN2]	C. Singh and M. R. Ebrahimian, Non-Markovian models for common mode failures in transmission systems. *IEEE Transactions on Power Apparatus and Systems* **PAS-101** (1982), 1545-1550.
[SING]	K. Singhal and J. Vlash, Computation of time domain response by numerical inversion of the Laplace transform. *Journal of the Franklin Institute* **299** (1975), 109-126.
[SMI]	R. M. Smith, K. S. Trivedi, and A. V. Ramesh, Performability analysis: Measures, an algorithm, and a case study. *IEEE Transactions on Computers* **C-37** (1988), 406-417.
[STEH]	H. Stehfest, Algorithm: 368 Numerical inversion of Laplace transforms. *Communications of the ACM* **13** (1970), 47-49.
[STEP1]	E. Stepanek, Eine numerische Methode zur Umkehr der Laplacetransformation. *Wissenschaftliche Zeitschrift für Elektrotechnik* **17** (1971), 261-272.
[STEP2]	E. Stepanek, Erweiterung einer numerischen Methode zur Umkehr der Laplacetransformation. *Zeitschrift für Elektrische Informations und Energietechnik* **4** (1974), 41-44.
[STEP3]	E. Stepanek, *Praktische Analyse Linearer Systeme durch Faltungsoperationen*. Akademische Verlagsgesellschaft, Leipzig, 1976.
[SUM1]	U. Sumita and M. Kijima, Evaluation of minimum and maximum of a correlated pair of random

variables via the bivariate Laguerre transform. *Communications in Statistics - Stochastic Models* **2** (1986), 123-149.

[SUM2]   U. Sumita and Y. Masuda, Analysis of software availability/reliability under the influence of hardware failures. *IEEE Transactions on Software Engineering* **SE-12** (1986), 32-41.

[SUM3]   U. Sumita, J. G. Shantikumar, and Y. Masuda, Analysis of fault tolerant computer systems. *Microelectronics and Reliability* **27** (1987), 65-78.

[SUM4]   U. Sumita and Y. Masuda, An alternative approach to the analysis of finite semi-Markov and related processes. *Communications in Statistics - Stochastic Models* **3** (1987), 67-87.

[TAL]   A. Talbot, The accurate numerical inversion of Laplace transforms. *Journal of the Institute of Mathematics and its Applications* **23** (1979), 97-120.

[TIJ]   H. C. Tijms, *Stochastic Modelling and Analysis: A Computational Approach.* John Wiley & Sons, New York, 1990.

[TRI1]   K. S. Trivedi, A. Reibman, and R. Smith, Transient analysis of Markov and Markov reward models. In: G. Iazeolla, P. J. Courtois, and O. J. Boxma (eds.), *Computer Performance and Reliability,* 535-545. Elsevier Science Publishers, Amsterdam, 1988.

[TRI2]   K. S. Trivedi, J. K. Muppala, S. P. Woolet, and B. R. Haverkort, Composite performance and dependability analysis. *Performance Evaluation* **14** (1992), 197-215.

[WEB]   G. G. Weber, Life time distributions for coherent systems. In: J. K. Skwirzynski (ed.), *Proceedings of the NATO Advanced Study Institute on The Challenge of Advanced Computing Technology to System Design Methods,* Grey College, University of Durham, U.K., July 29 - August 10, 1985, 95-122. ASI Series, Vol. F22, Springer-Verlag, Berlin, Heidelberg, New York, 1986.

[WEE]   W. T. Weeks, Numerical inversion of Laplace transforms. *Journal of the Association of Computing Machinery* **13** (1966), 419-429.

240

SUBJECT INDEX

page

Lecture Notes in Statistics

For information about Volumes 1 to 6
please contact Springer-Verlag